Nuclear Choices for the Twenty-First Century

Nuclear Choices for the Twenty-First Century

A Citizen's Guide

Richard Wolfson and Ferenc Dalnoki-Veress

The MIT Press
Cambridge, Massachusetts
London, England

© 2021 Massachusetts Institute of Technology

All rights reserved. No part of this book may be reproduced in any form by any electronic or mechanical means (including photocopying, recording, or information storage and retrieval) without permission in writing from the publisher.

This book was set in Sabon by Westchester Publishing Services. Printed and bound in the United States of America.

Library of Congress Cataloging-in-Publication Data

Names: Wolfson, Richard, author. | Dalnoki-Veress, Ferenc, author.
Title: Nuclear choices for the twenty-first century : a citizen's guide / Richard Wolfson and Ferenc Dalnoki-Veress.
Description: Cambridge, Massachusetts : The MIT Press, 2021. | Includes bibliographical references and index.
Identifiers: LCCN 2020025433 | ISBN 9780262542036 (paperback)
Subjects: LCSH: Nuclear energy--Popular works.
Classification: LCC TK9145 .W592 2021 | DDC 621.48--dc23
LC record available at https://lccn.loc.gov/2020025433

10 9 8 7 6 5 4 3 2 1

Contents

Preface

Nuclear technology is an inescapable part of our lives. Nuclear reactors provide a significant share of our electrical energy. Techniques of nuclear medicine diagnose and treat our diseases. Nuclear processes help industry produce better and safer products, preserve our food and protect it from pests, and help archaeologists understand our past. And since the mid-twentieth century, nuclear weapons have purportedly kept the peace by threatening the annihilation of civilization.

With nuclear technology come dangers. Nuclear war is an obvious one. So are reactor accidents like those at Three Mile Island, Chernobyl, and Fukushima. Mining uranium, manufacturing nuclear weapons, and normal operation of nuclear power plants all release radioactivity to the environment. Nuclear medicine carries risks that must be weighed against its potential benefits. Even such non-nuclear technologies as aviation and house construction have nuclear dangers associated with them.

The news media regularly bring nuclear technology and its dangers to our attention. Nuclear technology provokes vigorous debates at the local, national, and global levels. Nuclear issues force us to make nuclear choices—individually in the voting booth, in citizens' forums, through our elected representatives, and through our leaders as they pursue international negotiations.

We've written this book on the premise that nuclear choices are best made by citizens who know something about the underlying issues, who understand the basics of nuclear technology, and who can judge for themselves statements advocating particular positions. In that spirit, the book demands no prior knowledge of nuclear matters. It does ask that readers be open to a range of opinions, be willing to grasp some basic science and technology, and be willing to bring informed judgment to their own nuclear choices.

We emphasize that *Nuclear Choices* does not expect of our readers any particular scientific or technological background. We'll supply the needed science and technology at levels appropriate for a general readership. Also, we don't expect our readers to toe a particular political line when it comes to nuclear issues, and we aren't going to push one on them. Although the book is written for individual citizens, it may also find use in college courses—although it's not designed as a textbook because it lacks problem sets, discussion questions, and other hallmarks of a textbook. We've written *Nuclear Choices* on the premise that citizens of today's industrialized societies cannot avoid nuclear technology, and whatever the context, the book should help citizens from all walks of life become familiar with nuclear technologies and the issues surrounding them and gain confidence in making nuclear choices.

Our specific goal is to introduce readers to ideas they'll need to understand nuclear issues as they're presented in contemporary news media and civic debate. By covering essentially all nuclear technologies in one book, we've been able to stress the connections among them—especially the multifaceted relation between nuclear power and nuclear weapons. Readers seeking a deeper understanding of individual nuclear technologies are referred to more thorough works listed at the end of each chapter.

Readers of a certain age may note a similarity between this book and RW's 1991 book *Nuclear Choices*, which was revised in 1993 to reflect the collapse of the Soviet Union but saw no further revisions. Yet the original *Nuclear Choices* remained in print, hopelessly dated, until we made the decision to create this new title in the spirit of the original book. So much has happened since that first *Nuclear Choices*! We've had new arms-control agreements that brought an order-of-magnitude drop in the world's nuclear arsenals; disturbing nuclear incidents including the Fukushima disaster; the September 11, 2001, attacks that heightened concern over nuclear terrorism; declassification of nuclear policy documents from the Cold War that give new insights into governmental thinking; a nuclear-armed Pakistan and the resulting India-Pakistan nuclear standoff; the astonishing rise of North Korea as a nuclear-armed state capable of threatening the United States with intercontinental ballistic missiles and thermonuclear weapons; the increasing recognition that Earth faces a global climate crisis with direct implications for our energy sources, including nuclear energy; a huge international effort to develop energy from nuclear fusion; a doubling of the background radiation dose in the United States; growth and then substantial decline in nuclear power's contribution to the world's electrical energy supply; and much, much more.

That vast scope of nuclear happenings challenges an author seeking comprehensive coverage of so broad and important a field. So RW invited FDV to join as coauthor on the new book and was delighted when the latter accepted. By bringing in FDV as coauthor, the book gains the expertise of a nuclear physicist who brings the unique perspective of a scientist working in what is largely a policy think tank (the Center for Nonproliferation Studies) devoted to issues around nuclear weapons proliferation as well as connections between nuclear power and nuclear weapons.

Many people and institutions contributed to the making of this book. We're particularly grateful to the Alfred P. Sloan Foundation's Books Program for a grant that helped support the project. (The original *Nuclear Choices* also received support from the Sloan Foundation, under its New Liberal Arts Program, which was intended to cultivate technological literacy in liberal arts environments.) We also received support from the One Middlebury Fund, whose purpose is to foster collaboration between Middlebury College in Vermont, where RW is based, and the college's graduate school, the Middlebury Institute of International Studies in Monterey, California, FDV's professional home. *Nuclear Choices* is probably the first book published jointly by scholars from Middlebury's two campuses on the East and West Coasts. In addition to these two generous grants, countless corporations, government agencies, national laboratories, universities, and individuals supplied factual information, photographs, and drawings. They're acknowledged individually where appropriate, and here we thank them all collectively. We're also grateful to our colleagues for their patience as this project drew us away from other obligations and, finally, to our families for their support.

Richard Wolfson
Middlebury, Vermont
March 2020

Ferenc Dalnoki-Veress
Monterey, California
March 2020

1

Nuclear Questions, Nuclear Choices

In 2017, 58 percent of Swiss voters supported a government-proposed ban on new nuclear power plants. A year later, Taiwanese citizens voted overwhelmingly to overturn a government policy that would have eliminated nuclear power on the island. So the Swiss voted against nuclear power and the Taiwanese voted for it. If you were a Swiss or Taiwanese citizen, how would you have voted? On what would you have based your vote? Concern for safety? For the environment? Economics?

You've hit your head in a fall and your doctor is worried about a possible concussion. She recommends a CT scan. This, you discover, will give you some 2 millisieverts of radiation—about 30 percent of your yearly background radiation dose. Should you have the scan, or is the radiation a greater risk than the possible concussion? And what's a millisievert, and what's background radiation, and why should you be subject to radiation at all?

In January 2018, amid rising tensions between the United States and North Korea over the latter's long-range ballistic missile tests, residents of Hawaii received texts reading "BALLISTIC MISSILE THREAT INBOUND TO HAWAII. SEEK IMMEDIATE SHELTER. THIS IS NOT A DRILL." Panic ensued until, more than half an hour later, a second announcement declared a false alarm. An employee of the Hawaii Emergency Management Agency had inadvertently sent the text, confusing what was to be a test message with the real thing. How would you, as a resident of Hawaii, have reacted to the message? Where could you have found shelter from a nuclear attack? And what of the rest of the United States' population, almost all now within range of North Korea's ballistic missiles? How serious is this threat, and how should the United States respond?

Since the late 2010s, the state of South Australia has seen controversy over whether to build nuclear waste facilities in the state's arid outback—including the possibility of importing waste from outside Australia. A

Royal Commission produced a report favoring nuclear waste storage in South Australia, citing economic benefits. Then a 300-member citizens jury voted two-to-one against the nuclear waste proposals. If you had been a member of that jury, how would you have voted? On what knowledge or instincts would you have based your vote? What is nuclear waste? How is it formed? How dangerous is it? How long will it remain hazardous? What if terrorists got their hands on it?

Your local natural-foods store is considering a ban on irradiated foods. Should you support the ban? Does irradiation make foods safer or less safe? Do the foods become radioactive? Is there a connection between food irradiation and the nuclear power industry or the nuclear weapons establishment?

In 2019, President Donald Trump withdrew the U.S. from the Intermediate-Range Nuclear Forces (INF) Treaty, a 1987 agreement between the United States and the countries of the former Soviet Union, including Russia. Under the agreement, the U.S. destroyed nearly 1,000 missiles while the other side destroyed nearly 2,000. The treaty is credited with reducing the risk of nuclear war, especially in late twentieth-century Europe. But had the INF Treaty become obsolete in today's world, where nuclear powers such as China aren't subject to the ban on intermediate-range missiles? Or has withdrawal from the INF sparked a new nuclear arms race? How would you, as a citizen or political leader, have decided whether or not to stick with the treaty?

Getting Informed

The questions in the preceding paragraphs are *nuclear* questions, and they call for *nuclear choices*—choices that need to be made by you, as a citizen, or by your elected leaders. As the multitude of questions suggests, nuclear issues are complex. They raise technical, political, moral, and practical questions. Those questions are far from academic; they demand answers and action from citizens, legislators, political activists, scientists, businesspeople, and national leaders. The answers we give and the choices we make have potentially major roles in shaping the future of civilization and of our planet itself.

We're called on to answer nuclear questions and to make nuclear choices, often without a clear sense of the relevant technical and political realities. How many voters really know what plutonium is, where it comes from, and why it's a crucial material in the nuclear age? How many people flipping on a light switch really understand what's going on at the (possibly nuclear) power plant that provides their electricity? How many people alive today because nuclear medicine techniques detected or even cured their cancers know that they owe their lives to radiation? We dread the reality of a nuclear-armed North Korea or the growing prospect of a nuclear-armed Iran, but how many of us understand how our country's own decisions might aid or hinder others' efforts to acquire nuclear weapons? Most of us harbor a deep fear of nuclear radiation, but do we know how its dangers compare with risks we willingly take, such as smoking, neglecting to use seat belts, or living with pollution from coal-fired power plants? Many of us yearn for a world free of nuclear weapons, but what about the political, strategic, and technical challenges on the path to that goal? And, ultimately, do we understand what it is that makes our nuclear technologies so fundamentally different from anything humanity has known before?

You might try to get answers to these and other questions from news media, from your peers, or from the Internet. But type "nuclear power" into a search engine and you'll get over 300 million hits. "Nuclear weapons" gets nearly 200 million. Even "plutonium" garners 10 million hits. Which of these are authoritative and which are propaganda for one side or the other of a particular nuclear issue?

This book is designed to provide citizens with a basic understanding of nuclear technology and of the controversies surrounding its use. The book is divided into three parts. Part I deals with the nature of the atom and its nucleus, with nuclear radiation, and with the fundamentals of nuclear energy. Part II examines nuclear power, including our use of energy, the operation of nuclear power plants, nuclear accidents, nuclear waste, and alternatives to nuclear power. Part III describes nuclear weapons, including their operation, their destructive effects, delivery systems for getting them to their targets, strategies for their use or nonuse, the feasibility of defense against them, the prospects for controlling the spread of these weapons to other countries and to terrorist groups, and ultimately how we might prevent nuclear war. But this division into three parts is in some respects only a convenience. Nuclear power and nuclear weapons share the same fundamental physics, and many of their technologies overlap. So do their histories: today's nuclear power plants are descendants of reactors originally developed for military purposes, and the 1950s "Atoms for Peace" slogan aimed to calm a public alarmed by the development of nuclear bombs—and perhaps to distract from the ongoing race to develop ever more destructive weapons. Some of the thorniest nuclear issues center on connections between nuclear power and nuclear weapons, and these issues will arise repeatedly throughout the book. So expect to find nuclear weapons in the section on nuclear power and vice versa.

This book is for citizens, not scientists or nuclear specialists. It assumes no particular background in science or in nuclear issues. It provides a simplified introduction to nuclear science and technology and the controversies that surround them. The book's goals are to instill a level of nuclear literacy that gives you an understanding of the nuclear issues you'll continually encounter and to help you make intelligent choices based on that understanding.

This is not an antinuclear book, nor is it a pronuclear book. It aims to provide you with an unbiased view of nuclear technology and the issues that surround it. That's easy where scientific and technological facts are concerned but harder when things get controversial. Engaging those controversies is as important as understanding the underlying technology, and therefore we'll make every attempt to present arguments on all sides. Your authors do have strong feelings on some of the issues and are quite open on others. Where a display of personal opinion is unavoidable, we'll clearly designate it as such; otherwise, any arguments presented or questions asked don't necessarily reflect the authors' own views.

Reading this book isn't an academic exercise. Nuclear technology is an unavoidable part of our world, with the potential to bring us substantial benefit or unimaginable disaster. You'll be called to make choices about nuclear issues, and this book should help you make them informed choices.

Will this book give you all the answers? Will it tell you what nuclear choices to make? No. You may, in fact, find it frustrating that you might come away from your reading less certain of your opinion on complex nuclear issues. That shouldn't be surprising—the questions presented at the beginning of this chapter suggest that even nuclear experts often disagree. So one thing you should take from this book is a healthy, critical skepticism about experts' or activists' nuclear opinions. If the experts agreed, resolving nuclear questions would be easy. But they don't agree. Yet the nuclear issues need resolution—and you're someone who has to help resolve them. This book is written on the premise that those issues are best resolved by citizens who understand the basis of nuclear technology.

I

The Nuclear Difference

2

Atoms and Nuclei

After the 1945 nuclear bombings of Japan, Albert Einstein remarked that "the unleashed power of the atom has changed everything save our modes of thinking."[1] What is it that makes nuclear technology—"the unleashed power of the atom"—extraordinary? Why is the nuclear age unlike any previous age? The answer, in a nutshell, is that nuclear processes release over a million times as much energy as the more familiar happenings of our everyday world. That's the **nuclear difference.** Whereas a coal-burning power plant consumes many 110-car trainloads of coal each week (figure 2.1a), a comparable nuclear plant requires only a few truckloads of uranium fuel a year (figure 2.1b). A single nuclear bomb can destroy a city, a job that thousands of conventional bombs can't accomplish. But why this nuclear difference? Where does the millionfold energy increase come from? And why, with nuclear processes, do we encounter the new phenomenon of radiation? The answers to these questions lie in the atom and its nucleus.

A World Made from Three Particles

The matter of our world exhibits tremendous variety, from the tenuousness of air to the solidity of steel, from the rugged density of rock to the delicacy of a snowflake, from the green slipperiness of a frog to the savory crunch of an apple. It's remarkable that all these things—and all other things on Earth and throughout the visible universe—are made from combinations of just three simple building blocks: the **neutron**, the **proton**, and the **electron**. In this chapter you'll see how neutrons, protons, and electrons join to form the atomic nuclei and then the atoms from which the matter of our world is made.

Physicists have identified scores of so-called subatomic particles, and experiments with ever larger, more energetic, and more expensive machines

Figure 2.1
(a) A truckload of uranium fuel arrives at the Vermont Yankee nuclear power plant. Four such truckloads supplied all the fuel needed for the now-closed plant's once-in-18-months refueling (Entergy Nuclear). (b) A 110-car trainload of coal arrives at a Kansas power plant. Fourteen such trainloads fuel the plant each week (Earl Richardson, *Topeka Capital-Journal*). The contrast between figures 2.1a and 2.1b is a manifestation of the nuclear difference.

continue in an effort to understand how these fundamental bits of matter are related. Yet most subatomic particles appear to be of little importance in the day-to-day interactions of matter. They do arise in the physicists' giant accelerators and in interactions of cosmic rays with Earth's atmosphere, and many played important roles in the early universe. But it's only a slight simplification to say that the composition and the behavior of ordinary matter—from a human heart to a nuclear bomb—involve only the interactions of neutrons, protons, and electrons.

Neutrons and protons are so tiny that it would take 13 trillion of them, lined up, to span an inch. Neutrons and protons have very nearly the same mass (for our purposes, the same thing as weight), and that mass is so small that a pound of either would contain 270 trillion trillion particles. You can envision these particles as small spheres, although a physicist might caution that concepts from your everyday world aren't entirely appropriate in the subatomic realm.

Neutrons and protons differ in an important respect: the neutron carries no **electric charge**, whereas the proton carries one unit of positive electric charge. Charge is a fundamental property of matter, and a subatomic particle either has it or doesn't. A particle with charge has either one unit of positive charge or one unit of negative charge; no other amount seems possible.[2] There's nothing missing or deficient about negative charge; positive and negative are just names we use to distinguish the two different kinds of charge.

So we have the neutron and the proton: particles of essentially the same size and mass, differing in that the proton carries one unit of positive electric charge, whereas the neutron, as its name implies, is electrically neutral. Together, neutrons and protons are called **nucleons.**

The third particle, the electron, is much less massive than the others—it would take about 2,000 electrons to equal the mass of a neutron or proton. The electron carries one unit of negative electric charge. Even though the electron is much less massive than the proton, its charge is exactly equal but opposite to that of the proton.

Combining Particles

To see how our complex world is made from neutrons, protons, and electrons, we need to understand how these particles join to make larger entities. Nature provides what appear as three fundamental **forces** by which particles can interact and stick together: the **gravitational force,** the **electric force,**[3] and the **nuclear force.** (One grand goal of physics is to learn whether these three "fundamental" forces are really aspects of a single interaction governing all that happens in the universe.)

The gravitational force is familiar: it keeps you rooted to Earth, makes you fall, and holds the Moon in its orbit around Earth and Earth in its orbit around the Sun. But gravity is the weakest of the forces, significant only for larger objects—things the size of people, missiles, mountains, planets, and stars. Gravity plays essentially no role in the subatomic world of nuclear interactions, and we'll neglect gravity as we explore basic nuclear phenomena. Gravity will become important again when we consider the trajectory of a missile or the meltdown of a reactor.

With gravity out of the picture, we have only the electric force and the nuclear force. For our purposes, the electric force manifests itself as an interaction between electrically charged particles. Particles with no charge—neutrons—don't experience the electric force. Two particles with the same charge, either positive or negative, experience a repulsive electric force. Two particles with opposite charges attract each other via the electric force. The strength of the attractive or repulsive force depends on the distance between the particles; move them farther apart and the force

What the world is made of . . .

○	neutrons (n)	No electric charge. Mass = 1. Feels nuclear force only.
⊕	protons (p)	Electric charge: +1. Mass = 1. Feels both electric and nuclear force
⊖	electrons (e)	Electric charge: –1. Mass = 1/2,000. Feels electric force only.

. . . and how it's stuck together

nuclear force	Acts between nucleons (n, p), (n, n), (p, p). Always attractive. Strong, but short range.
electric force	Acts between charged particles (e, p), (e, e), (p, p). Opposites attract; likes repel. Weak, but long range.

Figure 2.2
Three particles and the forces by which they interact.

weakens. However, it doesn't weaken all that rapidly with increasing distance, so the electric force is a **long-range force.**

The nuclear force acts only between nucleons—between protons and protons, between neutrons and neutrons, or between protons and neutrons. It's always attractive. When the particles are very close—roughly their own diameter apart—the nuclear force is extremely strong, but it falls off rapidly with distance, quickly becoming insignificant. The nuclear force is thus a strong but **short-range force.**

Figure 2.2 summarizes the three particles and the two forces by which they interact. The forces are characterized by their strength and range, and by the particles between which they act. Here we have everything we need to build the nuclei, atoms, and molecules from which all substances are made. We'll start with nuclei.

Building Nuclei

An **atomic nucleus** is a group of nucleons—neutrons and protons—bound together by the nuclear force. The simplest nucleus—that of hydrogen—is a single proton; every other nucleus contains a mixture of protons and neutrons. The next simplest nucleus is a combination of a proton

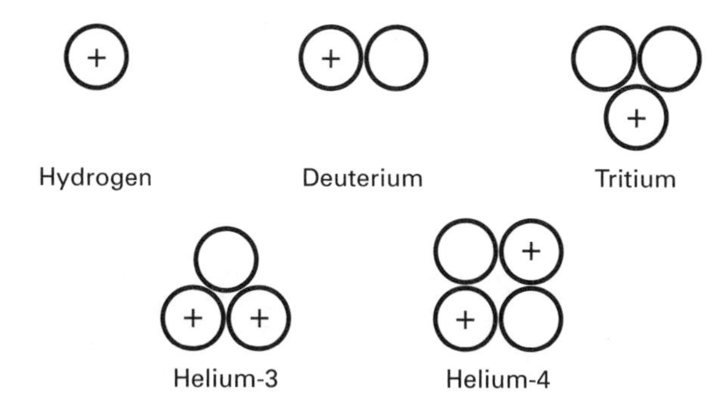

Figure 2.3
The hydrogen nucleus is a single proton. Other simple nuclei are deuterium (one proton and one neutron, also called a *deuteron*), tritium (one proton, two neutrons), helium-3 (two protons, one neutron), and helium-4 (two protons, two neutrons).

and a neutron, which constitutes a deuterium nucleus. A combination of one proton and two neutrons is tritium, a nucleus that can't last very long. Two other simple nuclei are helium-3, formed of two protons and one neutron, and helium-4, containing two protons and two neutrons. Figure 2.3 shows the five nuclei we've introduced so far, using the pictorial symbolism of figure 2.2 to represent protons and neutrons. We'll explore other nuclei shortly, but first we need to distinguish nuclei from atoms.

Building Atoms

A nucleus contains protons, which carry positive electric charge, and neutrons, which carry no electric charge. Therefore all nuclei are positively charged, and that means they attract negatively charged electrons. Normally, a nucleus surrounds itself with electrons equal in number to the protons in the nucleus. The resulting object is an **atom**. You can visualize an atom as a miniature solar system, with the nucleus surrounded by orbiting electrons, like the Sun by its planets. Gravity keeps the planets in orbit, while the electrical attraction of the nucleus plays the same role for the atomic electrons. Although a gross oversimplification, this picture contains the essence of what you'll need in order to understand the difference between nuclear and conventional energy sources. Figure 2.4 depicts atoms of hydrogen and helium, each consisting of a nucleus surrounded by the appropriate number of electrons.

Although figure 2.4 shows the essential configurations of atoms, it's misleading in its scale. In a real atom, the distance between the nucleus

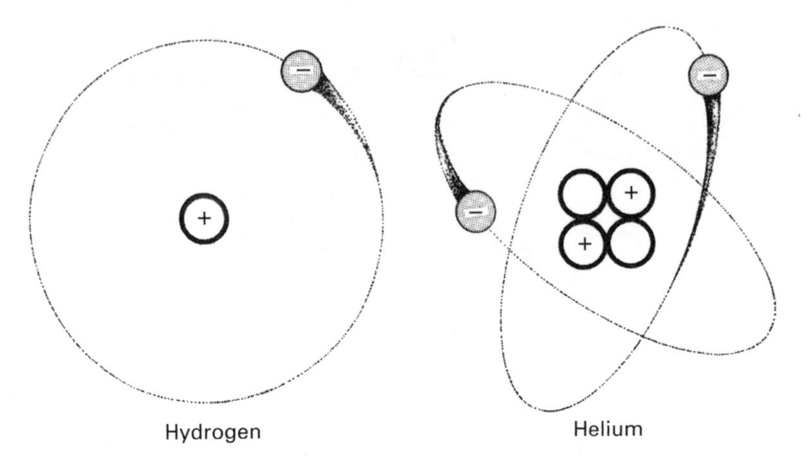

Figure 2.4
Atoms of hydrogen and helium. Note that in each case the number of orbiting electrons is the same as the number of protons in the nucleus.

and the surrounding electrons is far larger than our figure suggests—more than 10,000 times the diameter of the nucleus. If figure 2.4 were drawn to scale, the nucleus would be an invisible dot. If the nucleus of an atom were the size of a basketball, the electrons would be a mile away. Between the nucleus and its electrons would be a mile of emptiness. Atoms—and therefore everything that's made from atoms—are mostly empty space.

Building Molecules

Two or more atoms can join to form a **molecule**. The water molecule, for example, consists of two hydrogen atoms and an oxygen atom; its composition is reflected in its chemical formula, H_2O. This bonding of atoms to form molecules is what chemistry is all about, and the rearrangement of atoms into a new molecular configuration is a **chemical reaction**. Many conventional energy sources involve chemical reactions. Burning coal combines carbon atoms in the coal with oxygen from the atmosphere, producing carbon dioxide gas and in the process releasing energy (figure 2.5). Burning gasoline in your car's engine breaks up complicated molecules containing hydrogen and carbon, producing mostly carbon dioxide and water, along with energy. The high explosives used in conventional weapons undergo rapid chemical reactions, releasing their energy in a burst. The atoms in the food you eat are rearranged through chemical reactions in your body, supplying the energy that keeps you alive. Chemical reactions are important and commonplace in our lives.

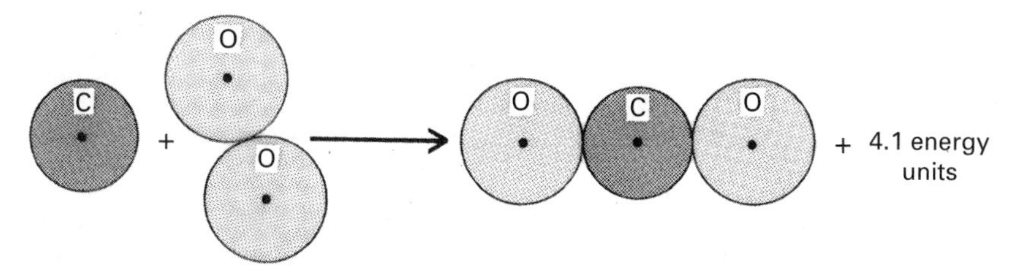

Figure 2.5
A chemical reaction. Here one atom of carbon joins two atoms of oxygen to form carbon dioxide. This is the basic reaction involved in burning coal. The energy released is 4.1 in units used by physicists working with atomic and nuclear processes. The black dots represent atomic nuclei, which aren't affected by the rearrangement of the atoms.

Chemical reactions involve electrons at the outer fringes of atoms. Those electrons interact through the relatively weak electric force, and therefore the energy involved in taking apart or putting together molecules is relatively small—so chemical reactions aren't highly energetic. Because atomic electrons are so far from the nucleus, the nuclei of interacting atoms remain widely separated. The nuclear force, which acts only at short distances, therefore plays no role in chemical reactions.

The Nuclear Difference

Nuclear reactions, in contrast, occur when the nucleus itself changes. This can happen if two nuclei join, if a nucleus ejects some of its nucleons, or if a nucleus is struck by another particle. Because the nuclear force is so strong at distances the size of a nucleus, nuclear reactions involve a lot of energy. Here, then, is the nuclear difference: it's the difference between the weaker but long-range electric force that governs the interactions of everyday chemical reactions, and the much stronger but short-range nuclear force that comes into play only in reactions involving the nucleus. Because of the relative strengths of the forces, that difference makes a typical nuclear reaction release several million times the energy of a chemical reaction (figure 2.6). That difference—based ultimately on the difference between the electric and nuclear forces—is, in turn, responsible for the dramatic differences between nuclear and conventional weapons, and between the fuel requirements of nuclear and fossil-fueled power plants (recall figure 2.1).

Whereas chemical reactions—burning coal or gasoline, metabolizing food, synthesizing plastics, and so on—are commonplace, nuclear reactions are rare under the conditions that prevail on Earth today. Our species

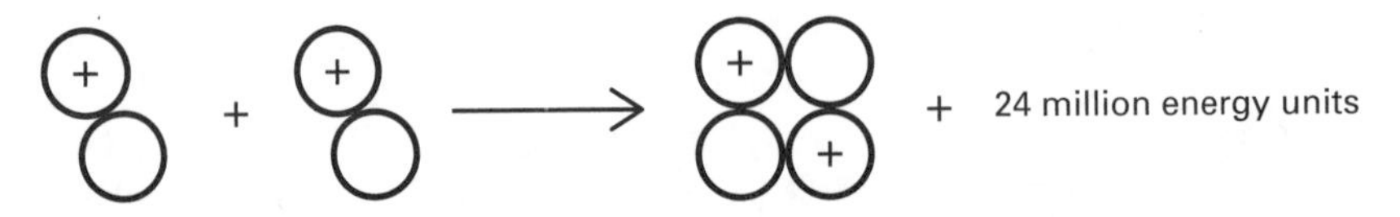

Figure 2.6
A nuclear reaction. Here two deuterium nuclei join to form a helium-4 nucleus. The nuclei are less than 1/10,000 the size of the atoms in figure 2.5, but the energy released is some 6 million times that of the chemical reaction shown in the preceding figure.

harnessed fire—a chemical reaction—in prehistoric times, but it wasn't until the mid-twentieth century that we learned to tend nuclear "fires" and their violent cousins, nuclear explosions. Nuclear reactions are common, though, elsewhere in the universe; in particular, the Sun and other stars shine because of nuclear reactions in their interiors.

You'll notice that this book never speaks of "atomic energy," "atomic bombs," "atomic power plants," or "atomic warfare," nor of "splitting the atom." The adjective *atomic* is ambiguous; since the interaction of atoms is involved in everyday chemical reactions, the energy they release might as well be called "atomic." Even Einstein's "unleashed power of the atom" suffers the same ambiguity. The reactions, the reactors, the bombs, the wars, the technologies that we're interested in here are distinctly *nuclear*, since their essence involves rearrangement of the atomic nucleus. And if we split anything, it will be a nucleus. We use the adjective *nuclear* to make all this absolutely clear.

Elements and Isotopes

An **element** is a substance that behaves chemically in a unique and identifiable way and whose most basic particle is a single atom. Oxygen is an element; so is hydrogen. Even a single oxygen atom exhibits the properties of elemental oxygen, but if you break that atom further it no longer behaves as oxygen. Water, H_2O, isn't an element; the smallest piece of water you can have is a single molecule, consisting of two hydrogen atoms and one oxygen atom. If you take the molecule apart, you have hydrogen and oxygen but no longer water.

What gives the atoms of a particular element their unique chemical behavior? "Chemical behavior" means how they interact with other atoms, forming the multitude of different substances that make up our world. You've seen that chemical reactions involve only the electrons that swarm in a distant cloud around the nucleus. So what determines the chemical behavior of an atom? Simply this: the number of electrons it contains. And what determines that? Since an atom forms when a nucleus attracts to itself as many electrons as it has protons, it's the number of protons in its nucleus that ultimately determines the chemical species to which an atom belongs.

The number of protons in a nucleus is the **atomic number**. Hydrogen, as figure 2.7 shows, has atomic number 1, helium has atomic number 2, and carbon has 6. Although figure 2.7 doesn't show all the individual particles, iron has 26 protons, gold has 79, and uranium has 92. An element's name and its atomic number are synonymous; to be oxygen is to have eight protons in your nucleus. In addition to its name and its atomic number, each element also has a unique one- or two-letter symbol; hydrogen is H, helium He, oxygen O, iron Fe, and uranium U. Table 2.1 gives the names, atomic numbers, and symbols of selected elements.

If we were chemists, we would be content with the atomic number of a nucleus—the number of protons—since that determines the species of chemical element. But here we're concerned with nuclear matters, so we need to characterize nuclei further. Figure 2.3 introduced the nuclei of hydrogen, deuterium, tritium, helium-3, and helium-4. Why the similar

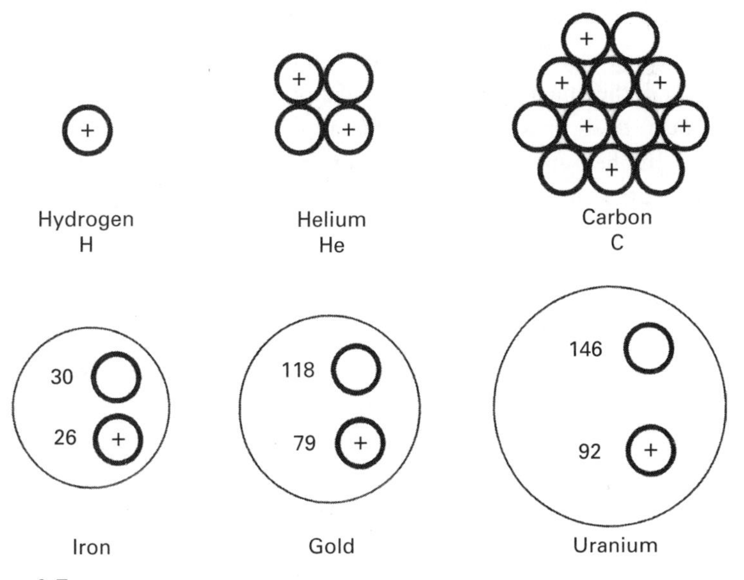

Figure 2.7
Some nuclei with element names and symbols. Figures are only suggestive; nucleons aren't locked into the fixed patterns shown here.

Table 2.1
Selected elements

Atomic number	Element name	Symbol
1	hydrogen	H
2	helium	He
6	carbon	C
7	nitrogen	N
8	oxygen	O
13	aluminum	Al
26	iron	Fe
38	strontium	Sr
79	gold	Au
86	radon	Rn
88	radium	Ra
92	uranium	U
94	plutonium	Pu

names for the last two? Because they're both nuclei of the same chemical element, helium. You can see in figure 2.3, and again in figure 2.8, why this is. Both contain two protons, and therefore both would form atoms with two electrons. Those atoms would exhibit similar chemical behavior, even though they have different numbers of neutrons. As far as the chemist is concerned, both are atoms of the same substance: helium. The names helium-3 and helium-4 reflect the total numbers of nucleons: two protons and one neutron in helium-3, two protons and two neutrons in helium-4. The total number of nucleons—protons and neutrons—is the **mass number** of a nucleus. Since protons and neutrons have nearly the same mass, the mass number gives approximately the total mass of a nucleus.

So helium-3 and helium-4 are both nuclei of helium, since the atoms they form have similar chemical behavior. But they *are* different, and that difference manifests itself in nuclear reactions. That's why we need to distinguish nuclei of the same element that have different numbers of neutrons and therefore different mass numbers. Such nuclei are called **isotopes**. Helium-3 and helium-4 are two isotopes of helium. And, as figure 2.8 shows, nuclei of hydrogen, deuterium, and tritium each have only one proton. So they're isotopes of the same element, namely hydrogen. Ordinary hydrogen could be called hydrogen-1; deuterium, hydrogen-2; and tritium, hydrogen-3. The use of separate isotope names is a confusion that, fortunately, is limited to hydrogen.

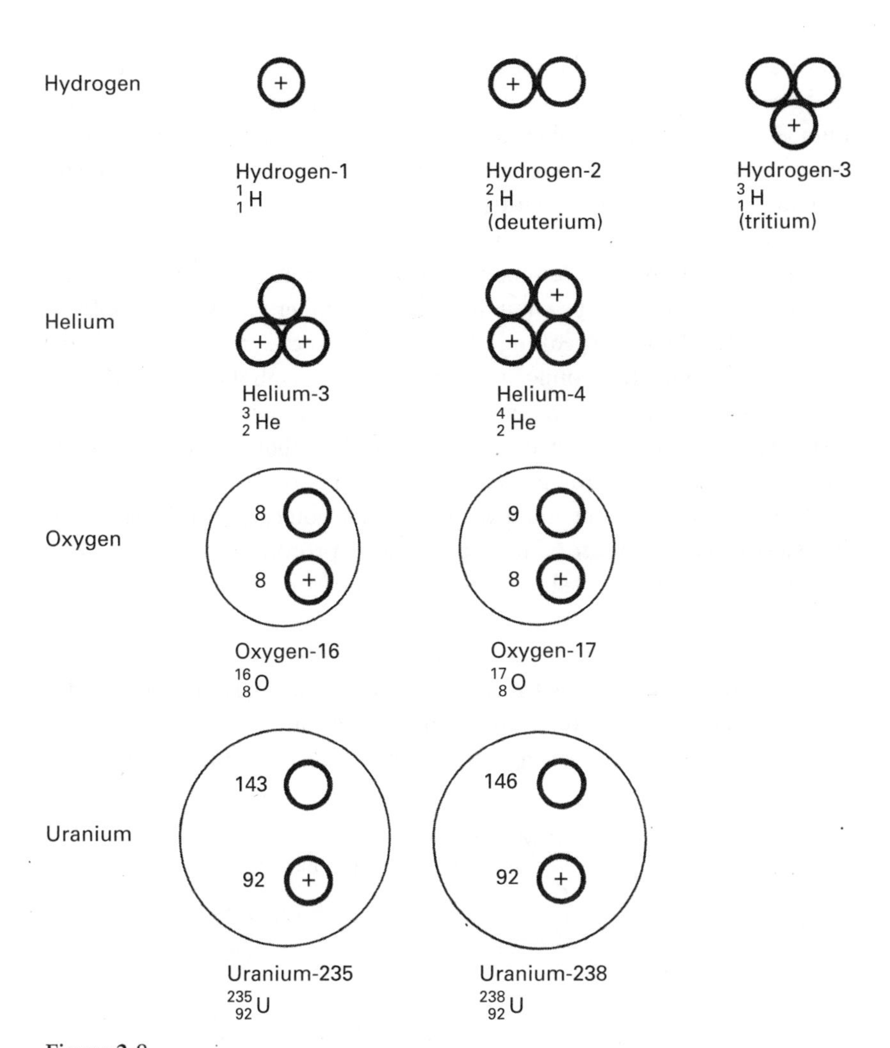

Figure 2.8
Isotopes of a given element have the same number of protons but different numbers of neutrons. Shown are three isotopes of hydrogen and two each of helium, oxygen, and uranium. Although isotopes have similar chemical behavior, their nuclear behavior can be very different. For example, only the rare isotope uranium-235 can serve directly as fuel for nuclear reactors and weapons.

To a chemist, He is the symbol for helium. But for nuclear purposes that doesn't tell us enough, so we elaborate by adding the atomic and mass numbers. The atomic number goes in front of the element symbol, at the bottom; the mass number goes in front at the top. Thus the helium isotopes helium-3 and helium-4 are written ${}^{3}_{2}He$ and ${}^{4}_{2}He$, respectively. Ordinary hydrogen is ${}^{1}_{1}H$, deuterium is ${}^{2}_{1}H$, and tritium is ${}^{3}_{1}H$. Soon we'll be very much concerned with two important isotopes of uranium, uranium-235 and uranium-238; since uranium has atomic number 92, their symbols are ${}^{235}_{92}U$ and ${}^{238}_{92}U$. Strictly speaking, the letter(s) and the atomic number in a symbol are redundant; atomic number 92 and the symbol U mean the same thing, namely uranium. Sometimes you'll see a nuclear symbol written with just the mass number, for example, ${}^{235}U$ or U-235; that's enough to tell the element (uranium) and the particular isotope (the one with a total of 235 nucleons). If you need the number of neutrons, you can get it by subtracting: since ${}^{235}_{92}U$ has 235 total nucleons and 92 protons, it has $235 - 92$ or 143 neutrons. Figure 2.9 shows the meaning of a nuclear symbol.

By now you've encountered a number of elements and some of their isotopes. Each isotope represents a unique combination of protons and neutrons, bound by the nuclear force to form a nucleus. How many different nuclei can we make? Is any combination of protons and neutrons a viable nucleus? No. The range of possible nuclei is distinctly limited.

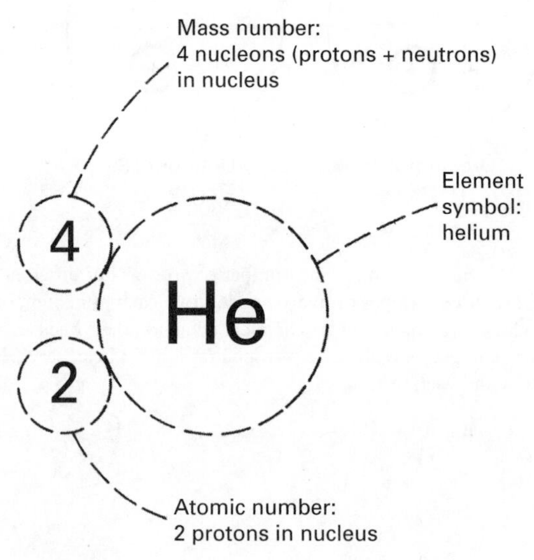

Figure 2.9
Anatomy of a nuclear symbol.

Nuclei consist of protons and neutrons held together by the nuclear force—a strong but short-range force. But the electric force is also present and acts to repel the protons in a nucleus. A combination of protons alone isn't possible; neutrons are needed to "dilute" the repulsive effect of the electric force. The lighter nuclei generally contain nearly equal numbers of protons and neutrons, and in the most common isotopes of helium ($^{4}_{2}He$), carbon ($^{12}_{6}C$), and oxygen ($^{16}_{8}O$), the numbers of protons and neutrons are exactly equal. Less common isotopes of these elements include helium-3 ($^{3}_{2}He$), carbon-13 ($^{13}_{6}C$), oxygen-17 ($^{17}_{8}O$), and oxygen-18 ($^{18}_{8}O$). In fact, these are the only nuclei of helium, carbon, and oxygen you can make and have stick together forever. (Nuclei that stick together indefinitely are **stable nuclei.**) If, for these and other light elements, you deviate too much from equal neutron and proton numbers, the resulting nucleus is **unstable**, meaning it won't stick together indefinitely. Try to make oxygen-19 ($^{19}_{8}O$), for example. It just won't stick. There are too many neutrons, and the nucleus will soon fly apart; you'll see just how in the next chapter. Try to make oxygen-14 ($^{14}_{8}O$), and again the nucleus comes apart, because now it has too few neutrons in relation to its eight protons.

You can see why a nucleus with too few neutrons might tend to come apart: there's less attractive nuclear force to counter the repulsive electric force between protons. With larger nuclei, this effect becomes more important. That's because protons at opposite sides of a large nucleus are so far apart that they don't feel the attractive but short-range nuclear force. But the long-range electric force still tends to repel them (figure 2.10). To counter this electric repulsion, the nucleus needs more nuclear "glue" in the form of neutrons that feel only the nuclear attraction. Therefore larger nuclei tend to have more neutrons than protons. Figure 2.11 plots number of neutrons versus number of protons for stable nuclei. Each little square represents a stable nucleus, specified by its neutron and proton numbers. For lighter nuclei (those with fewer nucleons, near the lower left of figure 2.11), the stable isotopes lie very close to the line representing equal numbers of protons and neutrons. But heavier nuclei deviate from this line as they require ever more neutrons to counter the electric repulsion of their widely separated protons. Thus the stable nuclei lie in a curved band that bends increasingly upward. You'll soon see how the shape of this band explains why the waste products of nuclear reactors and weapons are so dangerous. Above atomic number 83 (bismuth, with 83 protons) there are no stable isotopes. For these large nuclei, the repulsive electric force ultimately wins. Nuclei with more than 83 protons are all unstable, and, sooner or later, they come apart in one way or another.

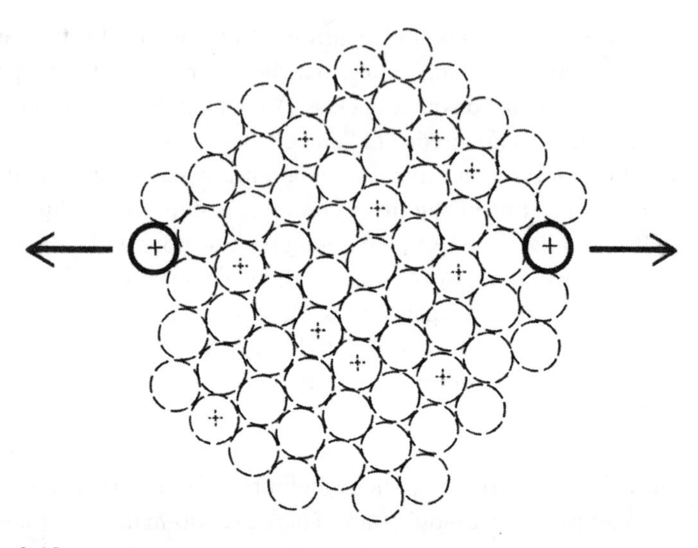

Figure 2.10
Two widely separated protons in a large nucleus experience mutual repulsion due to the long-range electric force. But because of its short range, the attractive nuclear force between them is insignificant. An excess of neutrons is therefore necessary to hold the nucleus together.

Summary

You've now met the few simple ingredients that make up our world: protons, neutrons, electrons, and the electric and nuclear forces that bind them into the nuclei, atoms, and molecules from which all else is made. You've seen how the relatively weak electric force is responsible for ordinary chemical reactions—interactions that involve only the electrons in distant orbits around their nuclei and that leave the nuclei unchanged. These chemical reactions are responsible for the energy released in burning coal or gasoline, in exploding TNT, and in metabolizing food.

Nuclear reactions, in contrast, involve rearrangement of the protons and neutrons that make up the atomic nucleus. Because these particles are so tightly bound by the strong nuclear force, the energy involved in nuclear reactions is millions of times that of chemical reactions. This single fact constitutes the nuclear difference, which explains why a coal-burning power plant consumes trainloads of coal each week whereas a few truckloads of uranium will fuel its nuclear counterpart for a year, and why a nuclear bomb can destroy a city whereas a conventional one takes

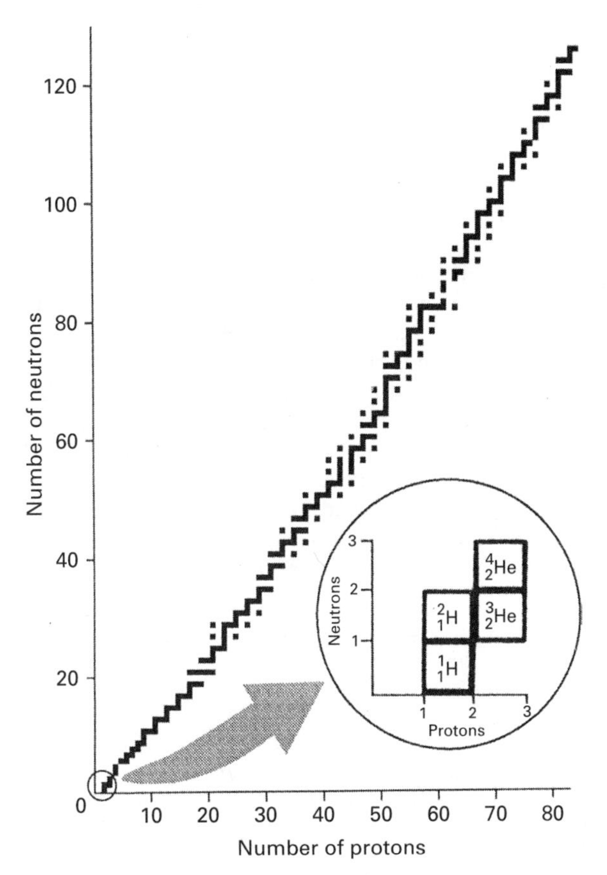

Figure 2.11
Neutron number versus proton number for stable nuclei, represented by small black squares. Most smaller nuclei have equal or nearly equal numbers of protons and neutrons, but larger nuclei have many more neutrons, for the reason explained in figure 2.10. Inset is a magnification showing the first four stable isotopes.

out only a few buildings. And it's this nuclear difference that makes warfare in the nuclear age a serious threat to human civilization.

Although we haven't yet explored nuclear reactions in detail, you've seen how protons and neutrons join together to form nuclei. Not all combinations result in stable nuclei that stick together indefinitely. So far we've considered stable nuclei, but you'll see in the next chapter that unstable nuclei also exist, although not forever. Those unstable nuclei are important factors—sometimes desirable, sometimes not—in most of today's nuclear technologies.

Superheavy Nuclei

The heaviest element naturally occurring in significant quantities is uranium, with atomic number 92. Starting in the 1930s, still heavier nuclei have been created artificially by bombarding heavy nuclei with other particles. The best known of these *transuranic elements* is plutonium (atomic number 94), which is produced in nuclear reactors and used in nuclear weapons. From the 1940s through early 1970s, elements 93 through 103 and element 106 were synthesized at the Lawrence Berkeley Laboratory in California; appropriately, elements 97 and 98 are, respectively, berkelium and californium. A group in the Soviet Union produced elements 104 and 105 in the 1960s. The 1980s and 1990s added elements 107 through 112, and the early twenty-first century extended the list through element 118. Laboratories in Russia, the United States, Germany, and Japan all contributed to the synthesis of these new elements. All the superheavy elements are unstable, with those numbered 115 and beyond having lifetimes well under one second. However, nuclear physicists anticipate an *island of stability*, with longer-lived isotopes appearing at atomic numbers around 120 and beyond.

Glossary

atom A nucleus surrounded by a number of electrons equal to the number of protons in the nucleus. An atom is the smallest particle of a chemical element.

atomic nucleus A cluster of protons and neutrons bound together by the nuclear force. Except for hydrogen (a single proton), all nuclei contain both protons and neutrons.

atomic number The number of protons in a nucleus. The atomic number determines the element; for example, hydrogen has atomic number 1, helium 2, oxygen 8, and uranium 92.

chemical reaction An event in which atoms are rearranged into a new molecular configuration, as in the joining of two hydrogen atoms and one oxygen to make a water molecule, H_2O. The nuclei of the interacting atoms are essentially unaffected in a chemical reaction, and an individual chemical reaction involves far less energy than a nuclear reaction.

electric charge A fundamental property of matter possessed by electrons and protons. Electric charge comes in two kinds, positive and negative.

electric force A force that acts between electrically charged particles. The electric force between oppositely charged particles is attractive; between particles of like charge, it's repulsive. At close range—roughly the size of a nucleon—the electric force is much weaker than the nuclear force, but with increasing distance it falls off less rapidly than the nuclear force. The electric force holds atoms together and is responsible for joining atoms to form molecules.

electron A subatomic particle whose mass is about 1/2,000 that of a proton or a neutron. The electron carries one unit of negative electric charge and feels only the electric force. Electrons surround nuclei to make complete atoms.

element A substance that behaves chemically in a unique and identifiable way, and whose most basic particle is a single atom. The chemical behavior of an element is determined by the number of electrons in its atoms, which in turn is determined by the number of protons in its nucleus. All atoms of a given element have the same number of protons in their nuclei, although they may differ in the number of neutrons. Hydrogen, helium, oxygen, and uranium are among the 92 naturally occurring elements.

force An interaction between particles of matter that manifests itself as an attraction or repulsion.

gravitational force An attractive force that exists between any two particles of matter. Gravity is the weakest of the fundamental forces and is insignificant at the atomic and nuclear scales.

isotope Isotopes of a given element are nuclei that differ in the number of neutrons they contain. Helium-3, for example, is an isotope of helium containing two protons and one neutron; helium-4 has two protons and two neutrons.

long-range force A force whose strength decreases relatively slowly with increasing distance between two particles. Both the electric force and the gravitational force are long-range forces.

mass number The total number of nucleons—protons and neutrons—in an atomic nucleus.

molecule A group of atoms joined through electrical interactions involving their outermost electrons. A molecule is the smallest particle of a chemical compound.

neutron A subatomic particle that's one constituent of the atomic nucleus. The neutron has nearly the same mass as the proton but carries no electric charge. It feels only the nuclear force.

nuclear difference Phrase we use to describe the roughly million-fold difference in energy released in nuclear reactions versus chemical reactions..

nuclear force A force that acts between nucleons. The nuclear force is always attractive. It's very strong at close range—roughly the diameter of a nucleon—but its strength drops rapidly with distance. The nuclear force binds nucleons together to form atomic nuclei.

nuclear reaction An event in which nucleons are rearranged, giving rise to one or more new nuclei. The joining of two deuterium nuclei to form a helium-4 nucleus is an example; so is the splitting of a uranium nucleus into smaller nuclei. Because the nuclear force is so strong, nuclear reactions involve a great deal of energy.

nucleon A proton or a neutron; either of the constituent particles of the atomic nucleus.

nucleus *See* atomic nucleus.

proton A subatomic particle that's one constituent of the atomic nucleus. The proton has nearly the same mass as the neutron. It carries one unit of positive electric charge and feels both the nuclear force and the electric force.

short-range force A force whose strength decreases rapidly with increasing distance between two particles. The nuclear force is a short-range force.

stable nucleus A nucleus that can exist indefinitely without spontaneously coming apart.

unstable nucleus A nucleus that cannot exist indefinitely but will eventually come apart.

Notes

1. Einstein quoted in *New York Times Magazine*, August 2, 1964, 54.

2. Quarks—more fundamental particles that make up protons and neutrons—have charges of one-third or two-thirds of the fundamental unit. Quarks aren't essential to an understanding of nuclear physics at the level needed here, so we won't consider them further.

3. We're using the term "electric force" to include not only the forces of electrical attraction and repulsion but also the magnetic force and the so-called weak nuclear force, all of which are now understood as aspects of the same fundamental force, called the *electroweak force*.

Further Reading

Challoner, John. *The Atom: A Visual Introduction*. Cambridge, MA: MIT Press, 2018. A prolific and respected science writer presents atomic and nuclear physics in laypersons' terms.

Chapman, Kit. *Superheavy: Making and Breaking the Periodic Table*. New York: Bloomsbury Sigma, 2019. An up-to-date account of the creation of elements that lie beyond uranium on the periodic table.

Close, Frank. *Nuclear Physics: A Very Short Introduction*. Oxford, UK: Oxford University Press, 2015. One of Oxford professor Frank Close's short introductions to physics topics, at a level considerably higher than *Nuclear Choices*.

Düllmann, Christoph E., and Michael Block. "The Quest for Superheavy Elements and the Island of Stability." *Scientific American* 318, no. 3 (March 2018): 46–53. An authoritative survey of the search for superheavy elements.

3

Radioactivity

The preceding chapter showed how atomic nuclei are made of protons and neutrons. But not all combinations of these nucleons stick together indefinitely. Those that do are the stable nuclei of figure 2.11; those that don't are unstable. Sooner or later, an unstable nucleus comes apart. The coming-apart process is **radioactive decay**, and unstable nuclei or materials containing them are **radioactive.** Some elements have both stable and unstable isotopes; the latter are **radioisotopes**. The physicist Marie Curie coined the term *radioactivity*. She won Nobel prizes in both physics and chemistry for her pioneering work on radioactive decay (figure 3.1).

Radioactive Decay

How does a radioactive nucleus come apart? Although there are many ways, we'll focus on the three most common: alpha decay, beta decay, and gamma decay. In **alpha decay**, a nucleus spits out two protons and two neutrons, bundled together as a helium-4 nucleus. This He-4 nucleus is called an **alpha particle**, a name that dates to the early 1900s, when it wasn't yet known that the particles were in fact helium nuclei. Alpha decay is common among the larger unstable nuclei, which need to rid themselves of excess protons (recall figure 2.10 and related discussion). The alpha particle carries off two protons, dropping the atomic number of the remaining nucleus by 2. Since the alpha particle contains a total of four nucleons, the mass number drops by 4. In a typical alpha decay, uranium-238 (${}^{238}_{92}U$) emits an alpha particle (${}^{4}_{2}He$), leaving a nucleus of thorium-234 (${}^{234}_{90}Th$). Figure 3.2 shows this decay, both pictorially and using nuclear symbols. Note how the sum of the atomic numbers on the left equals the atomic number on the right. The same is true for the mass numbers.

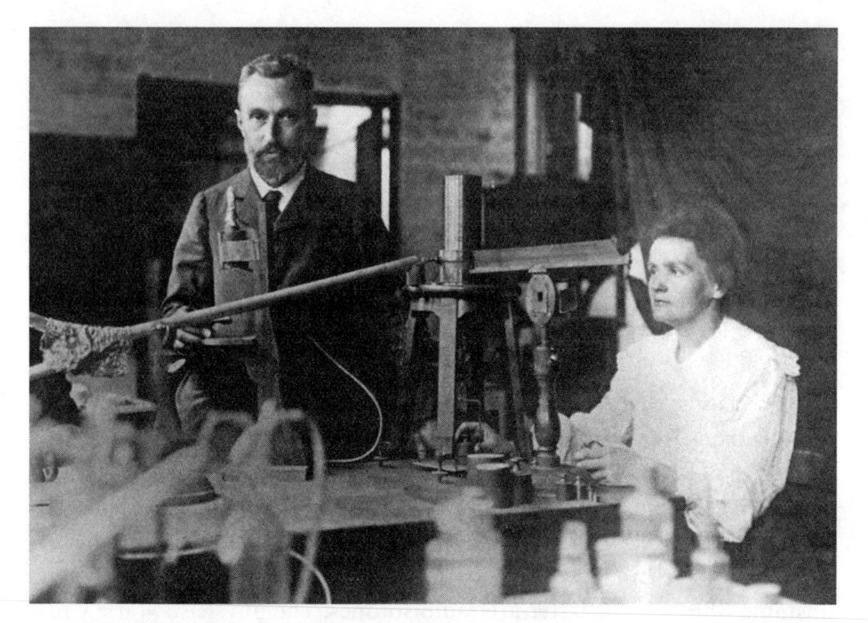

Figure 3.1
Marie and Pierre Curie in their Paris laboratory. (Science Source)

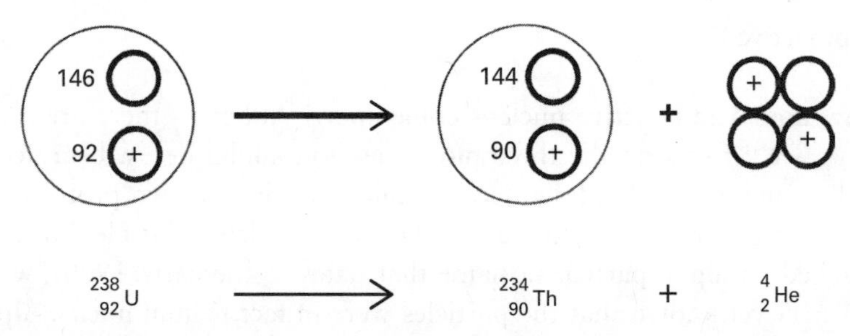

Figure 3.2
Alpha decay of uranium-238 yields an alpha particle (helium-4 nucleus) and thorium-234.

Unstable nuclei with too many neutrons would like either to rid themselves of neutrons or to gain protons. Remarkably, they do both at once. In the process of **beta decay,** a neutron turns into a proton and an electron. The electron, also called a **beta particle,** flies out of the nucleus, leaving the nucleus with one more proton and one fewer neutron than it previously had. Since there's one more proton, the atomic number increases by 1. But the total number of nucleons remains the same, so the mass number is unchanged. Figure 3.3 shows a typical beta decay,

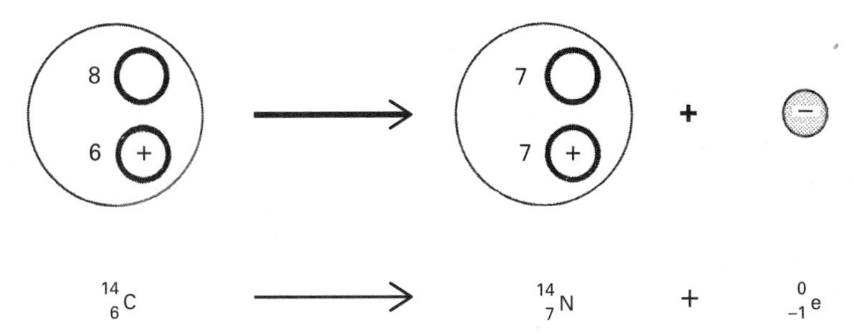

Figure 3.3
Beta decay of carbon-14. A neutron in C-14 turns into a proton and an electron; the electron is ejected, leaving nitrogen-14. Note that the mass number remains unchanged (6 + 8 for C-14 and 7 + 7 for N-14).

that of carbon-14. The end product, nitrogen-14, is the common stable isotope of nitrogen. In writing the decay symbolically, we've indicated the electron as ${}^{0}_{-1}e$; since it carries one unit of *negative* charge, the electron's atomic number is –1, and its mass number is 0 because its mass is far less than that of a nucleon. Using these numbers, the sum of the atomic numbers on the right is equal to that on the left, and similarly for the mass numbers. This equality must hold in any nuclear reaction.

You might wonder how the electron got mixed up with beta decay, given that nuclei contain only protons and neutrons. Was the electron somehow hiding in there? Or is a neutron really a combination of a proton and an electron? No, but a neutron can, through one manifestation of the forces we've subsumed under the term *electric*, spontaneously turn into a proton and an electron.[1] In fact, a free neutron—one that isn't part of a nucleus—will do so in less than an hour. And a neutron inside a nucleus with an excess of neutrons can also change into a proton and an electron—hence, beta decay. Only when they're constituents of stable nuclei can neutrons last indefinitely.

A variant of beta decay occurs when a nucleus emits a **positron**, the electron's positively charged antiparticle. This drops the atomic number by 1. Short-lived positron emitters are used in the medical imaging technique known as *positron emission tomography* (PET). Another process that lowers the atomic number is **electron capture**, in which a nucleus captures one of the innermost atomic electrons. The electron combines with a proton to make a neutron, in what's essentially the inverse of beta decay.

Sometimes a nucleus is struck by another particle that bounces off or goes right through without causing a nuclear reaction. Then the nucleus

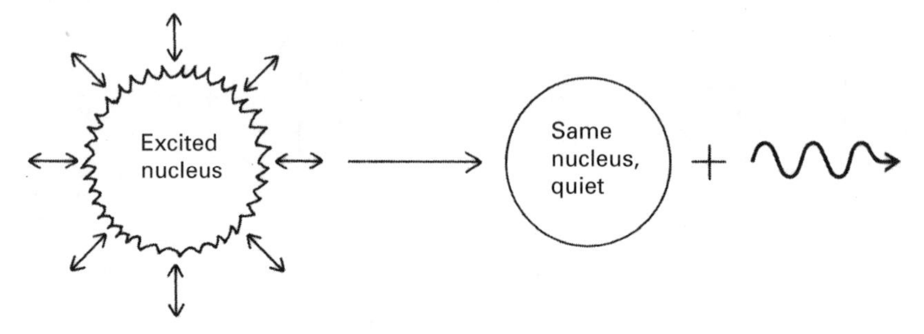

Figure 3.4
An excited nucleus (arrows suggest vibration) sheds excess energy by emitting a gamma ray (wavy line). No nucleons or electric charge leave the nucleus, so it retains its identity.

retains its identity, but it may be "shaken up" in the process, acquiring excess energy. The same thing can happen when a new nucleus forms as a result of a nuclear reaction such as electron capture, discussed above. The energetic nucleus is like a gong that's been struck by a hammer; originally it was quiet, but now it's vibrating. A nucleus with excess energy is said to be **excited**. Unlike the gong, which starts getting rid of its excess energy immediately in the form of sound, the nucleus can temporarily store the energy. It then emits the energy suddenly, in the form of a little energy bundle called a **gamma ray**. A gamma ray is a high-energy version of ordinary visible light and is yet another manifestation of the electric force. Once the nucleus has shed its excess energy by this process of **gamma decay**, it returns to its original quiet state. Figure 3.4 depicts a gamma decay.

Nuclear Radiation

Each radioactive decay results in a modified nucleus and a much smaller entity—either an alpha particle (He-4 nucleus), a beta particle (electron), or a gamma ray. Those entities are highly energetic: alpha particles from uranium-238 move at some 10,000 miles per second, beta particles from carbon-14 at some 150,000 miles per second, and gamma rays at the speed of light—186,000 miles per second. Each gamma ray "energy bundle" packs a million times the energy of a "bundle" of visible light. Again, the nuclear difference: the forces binding nuclei are so strong that large energies are involved any time a nucleus gets disrupted.

The energetic particles emitted in radioactive decay constitute **nuclear radiation**. Because of its high energy, nuclear radiation can damage atoms and molecules in its path. That's the reason for concern about radiation exposure to humans and other organisms. Even nonliving materials suffer radiation damage. We'll discuss radiation effects in the next chapter. Here, we explore further the physical aspects of radiation and radioactivity.

Although all nuclear radiation is highly energetic, the three forms of radiation differ in their ability to penetrate matter. Alpha particles, the slowest-moving, have relatively little penetrating power; typically, alpha particles can be stopped by a sheet of paper, a layer of clothing, or an inch of air. Thus it's easy to shield against alpha radiation—unless alpha-emitting material ends up on or inside the body. Radiation-induced lung cancers, for example, can result from alpha emitters lodging in the lungs.

Beta particles—electrons—are much lighter than alpha particles and move much faster. They can penetrate a fraction of an inch in solids and liquids (including the human body) and several feet in air.

Both alpha and beta particles are ultimately slowed because they're electrically charged particles that interact strongly with electrons in materials through which they pass. Gamma rays, in contrast, are electrically neutral and are therefore highly penetrating. The penetrating power of gamma rays depends on their energy: The highest-energy gamma rays encountered in nuclear technology may require several feet of dense shielding material.

Measuring Radiation

A technician refueling a nuclear reactor is accidentally exposed to highly radioactive spent fuel. A biologist working with radioactive material spills a carbon-14 solution in the lab. A Japanese citizen steps outdoors as a rainstorm brings down radioactive fallout from the 2011 Fukushima event. An airline pilot is regularly bombarded by cosmic rays. How serious are their radiation exposures? To answer that question, we need ways to describe amounts of radiation and radioactivity.

A simple way to characterize the level of radioactivity in a chunk of material is to give its **activity**—the number of radioactive decays that occur in a given time. The standard unit of activity is the **becquerel** (symbol Bq), defined as one decay per second. An older unit is the **curie** (Ci), named in honor of the Curies. One curie is 37 billion decays per second, approximately the activity of one gram of radium-226, an isotope that Marie Curie discovered. The 1979 Three Mile Island nuclear accident released some 15 curies, or 500 billion Bq, of iodine-131, while the 2011 release at Fukushima was a million times greater. In contrast, a typical banana has an activity of about 20 becquerels, or 500 picocuries (500 trillionths of a curie), because of its relativity high concentration of

Table 3.1
Units of radioactivity and radiation

Units	What they measure	How they're related
becquerel (Bq) curie (Ci)	Activity: rate of radioactive decay in a piece of radioactive material.	1 Bq = 1 decay/second 1 Ci = 37 billion Bq
gray (Gy) rad	Energy absorbed in material exposed to radiation	1 gray = 1 joule of energy per kilogram of material 1 rad = 0.01 Gy
sievert (Sv) rem	Like gray and rad but adjusted for biological effects	1 rem = 0.01 Sv

naturally occurring radioactive potassium-40. Thus some 20 decays of potassium-40 occur every second in that banana.

Activity, as measured in becquerels or curies, is a property of a source of radiation; it says nothing about objects exposed to the radiation. The effect of radiation on an exposed object depends on the energy the radiation deposits in the object. The **gray** (Gy) quantifies that energy. An older unit, the **rad**, is one one-hundredth of a gray—that is, 0.01 Gy. Radiation's biological effects depend not only on the energy absorbed, but also on the type of radiation and the organs being irradiated. Another unit, the **sievert** (Sv), measures the effective radiation dose adjusted for these factors. An older unit, the **rem**, is 0.01 Sv. Although the gray and the sievert (and the rad and the rem) are different, they're essentially interchangeable for most of the radiation we'll be discussing.

The next chapter describes the effects of various radiation doses; for now, note that the average citizen, worldwide, receives about 3 millisieverts (mSv; equivalently, 0.003 Sv or 0.3 rem) of radiation in the course of a year. A prompt dose—one received over a short time—of 4 Sv kills about half the people affected. Typical low-level radiation exposures are far less than that lethal dose, usually in the range of millisieverts or lower. Table 3.1 summarizes the units of radioactivity and radiation.

Detecting Radiation

Radiation is invisible, odorless, tasteless, and generally quite undetectable by human senses. That's one reason for the widespread public fear of radiation. In a few unfortunate accidents, people have received fatal radiation doses without knowing that anything was happening.

Nuclear History: Discovery of Radioactivity

In 1895, the German physicist Wilhelm Roentgen made the accidental discovery of mysterious, penetrating rays that emerged from a tube containing a beam of electrons. These were X-rays, and within months they were used to image bones within the human body. X-rays arise from accelerated electrons, not the atomic nucleus, so they don't involve radioactivity (although they have high enough energy that their biological effects are similar to those of nuclear radiation). But scientific excitement over X-rays soon led to the serendipitous discovery of radioactivity.

A mere four months after Roentgen's X-ray discovery, the French physicist Henri Becquerel wondered if uranium compounds might absorb sunlight energy and then emit X-rays. He performed experiments with uranium compounds that had been exposed to sunlight and then were placed on photographic plates wrapped in black paper to block light but pass X-rays. Becquerel's experiments, done on sunny days, seemed successful—until a cloudy stretch when he put his apparatus away in a desk drawer. To his amazement the plates nevertheless recorded a clear image of the uranium compounds. The uranium didn't require sunlight but somehow emitted penetrating radiation on its own! Becquerel had discovered radioactivity.

Other scientists took up the study of radioactivity, especially Marie and Pierre Curie—who quickly identified three new radioactive elements: polonium, thorium, and radium. The Curies shared the 1903 Nobel Prize for Physics with Becquerel, and later the radiation units curie and becquerel were named to honor these pioneers of radioactivity.

Technologically, though, radiation is easy to detect and to quantify. Radiation detectors range from simple, inexpensive devices that can tell if radiation is present to sophisticated instruments that measure the precise energies of the particles that make up the radiation. Here we'll examine just a few of these devices.

The **Geiger counter**, one of the simplest and oldest detectors, is still widely used. It consists of a gas-filled metal tube, with a window for radiation to enter (figure 3.5). The tube is connected to the negative terminal of a battery. A wire runs down the center of the tube and is connected to the positive terminal. A loudspeaker is also in the circuit. Normally the gas doesn't conduct electricity, so nothing happens. But when radiation enters the tube, it tears electrons off atoms of the gas. The electrons are attracted to the positive wire, causing a burst of electric current. With each burst the loudspeaker emits a click, so the click rate measures the rate at which radiation enters the tube. A meter gives a visual indication of the radiation level.

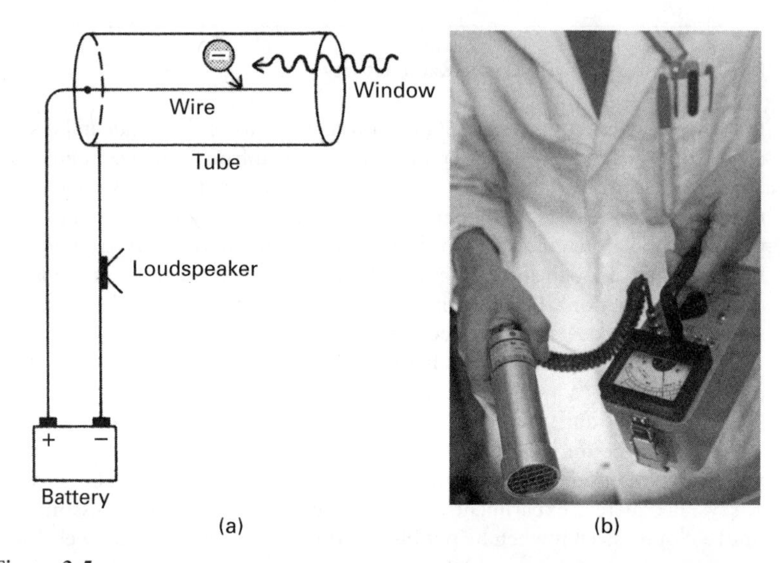

Figure 3.5
(a) Simplified diagram of a Geiger counter; (b) a Geiger counter in use. (Hank Morgan/Science Source)

Can't afford a Geiger counter? The light-sensing chip in your smartphone's camera can also detect gamma radiation. Apps are available that let you use your smartphone to measure gamma exposure. You have to cover the camera's lens to block visible light, and your phone won't be sensitive to alpha or beta radiation because they can't penetrate the lens.

Radiation often strips electrons from atoms. When electrons and atoms recombine, they emit a flash of light. Detectors using this phenomenon contain light sensors that record the flash and, by measuring its intensity, determine the radiation energy.

Radiation leaves permanent records in some materials. Radiation detectors made from such materials record long-term exposure. Workers subject to on-the-job radiation often wear these detectors. The presence of low-level radioactive contaminants such as radon gas in homes is detected by similar means.

Exposure and Contamination

Stand near radioactive material and you'll be exposed to radiation. The longer you stay, the greater your radiation dose. Move away and the exposure stops. If, on the other hand, radioactive material gets on you—or worse, inside you—then you're contaminated and continuously exposed to radiation. All unnecessary radiation exposure should be avoided. But **contamination** is especially dangerous, since the exposure will continue until the contaminant is removed. If radioactive material lands on your skin or clothing, washing may be enough for decontamination. But if you eat food containing biologically active radioisotopes—such as strontium-90, which is incorporated into bone, or iodine-131, which is absorbed by your thyroid—then decontamination can be particularly difficult.

Half-Life

For a single unstable nucleus, radioactive decay is a truly random event. But large numbers of nuclei show a pattern to their decays. Start with 1,000 radioactive nuclei and you'll find that after a certain time about 500 will have decayed. That time is the **half-life.** You can't predict which 500 will decay, and the number may vary slightly from one experiment to another, but on average the nuclei decay with remarkable regularity. If you wait another half-life, half the remaining nuclei will decay, leaving only 250 from the original sample. In another half-life, 125 of those will decay, leaving only 125. The process continues until all nuclei have decayed (figure 3.6).

Figure 3.6 shows that a radioactive sample has substantially decayed after a few half-lives have passed; after seven half-lives, only 1/128—less than 1 percent—of the original sample remains. Eight half-lives, and it's down to 1/256; nine, and it's only 1/512; 10 half-lives, and only 1/1,024 of the original nuclei remain undecayed. A good rule of thumb is that after 10 half-lives, only about one-thousandth of the original radioactive nuclei remain. Wait another 10 half-lives—for a total of 20—and only one-thousandth of those are left, or one-millionth of the original nuclei. After 30 half-lives you're down another factor of 1,000, with only a billionth of the original sample remaining. So it doesn't take very many half-lives for essentially all of a radioactive sample to decay.

Half-lives vary dramatically from one radioactive isotope to another. Excited states of some nuclei decay with half-lives around a thousandth of a trillionth of a second, whereas uranium-238's half-life is 4.5 billion years. Most nuclei lie between these extremes; half-lives of minutes to years are common. Table 3.2 lists the half-lives of some typical radioactive isotopes, many of which will concern us as we explore nuclear technologies and their consequences.

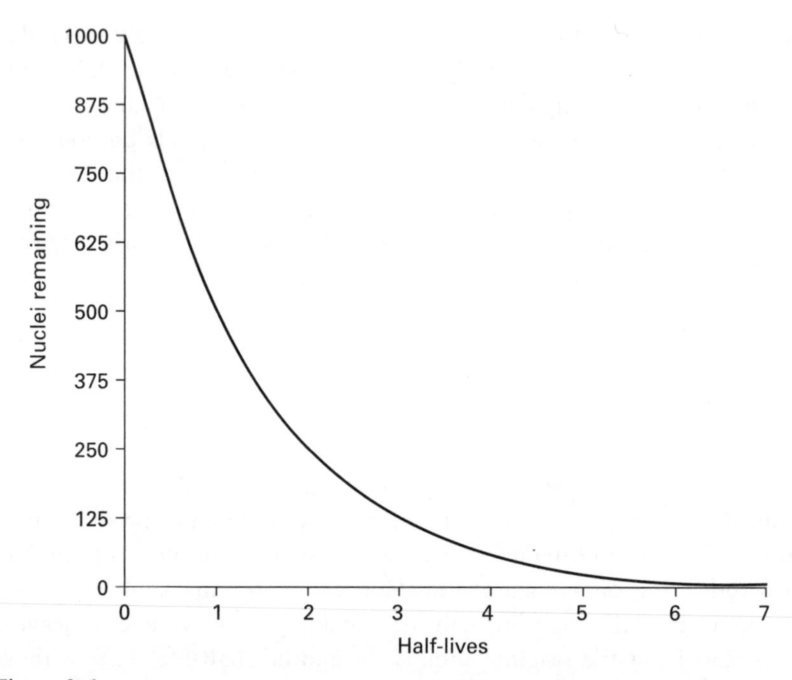

Figure 3.6
Decay of a radioactive material. In one half-life, half the nuclei still present will decay.

Table 3.2 shows that tritium decays with a half-life of about 12 years. You would have to wait 10 times this long, or about 120 years, for a given sample of tritium to decay to 1/1,000 of its original level. On the other hand, a few years is enough for a small but significant change in the amount of tritium present. Tritium used to boost the explosive yield of nuclear weapons must therefore be replenished regularly.

It would take 45 billion years (10 half-lives) for a sample of U-238 to drop to 1/1,000 of its original level. That's about three times the age of the universe! Plutonium-239 decays much more rapidly, with a 24,000-year half-life. But you would still have to wait 240,000 years for a plutonium sample to drop to 1/1,000 of its original level. Strontium-90, a significant component of nuclear waste, has a 29-year half-life, so you'd wait several human lifetimes for a factor-of-1,000 drop in activity. The 1986 Chernobyl accident contaminated much of Europe with iodine-131, which gets into cow's milk and then concentrates in the thyroid glands of humans who drink the milk. For some countries, contaminated milk had more than 10 times the allowed level of I-131. But I-131's eight-day half-life meant that a few weeks' wait were enough to bring I-131

Table 3.2
Half-lives of selected isotopes

Isotope	Half-life	Significance
carbon-14 $^{14}_{6}C$	5,730 years	Radioisotope formed in Earth's atmosphere by cosmic rays; used in radiocarbon dating.
iodine-131 $^{131}_{53}I$	8.04 days	Product of nuclear fission, released in weapons explosions and reactor accidents. Concentrates in milk and is absorbed in the thyroid.
oxygen-15 $^{15}_{8}O$	2.04 minutes	Short-lived oxygen isotope used in PET scans.
plutonium-239 $^{239}_{94}Pu$	24,110 years	Produced in nuclear reactors and used in most nuclear weapons. Sustains a vigorous chain reaction.
radium-226 $^{226}_{88}Ra$	1,600 years	Highly radioactive isotope discovered by Marie and Pierre Curie. Forms in the decay chain of U-238.
radon-222 $^{222}_{86}Rn$	3.82 days	Radioactive gas formed in the decay of Ra-226. Seeps into buildings, where it can give significant radiation exposure.
strontium-90 $^{90}_{38}Sr$	29 years	Fission product that mimics calcium, concentrating in bones. A particularly dangerous component of fallout from nuclear weapons.
tritium $^{3}_{1}H$	12.3 years	Used in biological studies and to enhance nuclear weapons yields.
uranium-235 $^{235}_{92}U$	704 million years	Scarce fissile isotope used as fuel in nuclear reactors and some nuclear weapons.
uranium-238 $^{238}_{92}U$	4.5 billion years	Predominant uranium isotope, making up 99.3 percent of natural uranium. Cannot sustain a chain reaction.
oganesson-294 $^{294}_{118}Og$	0.69 milliseconds	Isotope of the highest atomic number element (118) so far produced.

levels to within safety standards. Finally, oxygen-15, used in PET scans, has a two-minute half-life. Twenty minutes after injection with O-15, a patient's body contains only 1/1,000 of the initial radioactivity. After an hour—30 half-lives—only a billionth remains. For that reason, O-15 and other short-lived isotopes are particularly safe for medical studies.

Suppose we have a chunk of uranium-238 and chunk of strontium-90, with the same number of nuclei in each. How do their activities compare? It will take 4.5 billion years for half the U-238 nuclei to decay, but only 29 years for the Sr-90. So the strontium must decay at a much greater rate—greater by the ratio of 4.5 billion years to 29 years. Given equal quantities of different radioactive materials, those with the shorter half-lives will therefore be more highly radioactive. That's one reason why the relatively short-lived waste products from nuclear reactors are much more dangerously radioactive than the long-lived nuclear fuels that go into the reactors.

The Origin of Radioactive Materials

The lighter nuclei that make up our world—those up to about iron (atomic number 26)—were created through nuclear reactions in stars that existed long before Sun and Earth formed. (We'll explore this special status of iron in chapter 5.) The violently explosive deaths of those stars as *supernovas* spewed into space materials that would later become our solar system and ourselves. Most lighter elements formed while the stars shone steadily, through a process we'll explore in chapter 5. Until very recently physicists believed that heavier nuclei formed primarily during supernova explosions. But the 2017 detection of gravitational waves from a collision of two neutron stars—Sun-mass objects made almost entirely of neutrons—showed that some of the heavier elements arise in neutron-star collisions. Among the heavy elements produced in supernovae and neutron-star collisions are radioactive isotopes of uranium and other elements.

Earth and Sun formed about 5 billion years ago, and the nuclei that constitute them formed even earlier. So radioactive nuclei that were incorporated into our planet have had plenty of time to decay. Even uranium-238, with its 4.5-billion-year half-life, is only half as abundant as when Earth was new. For isotopes with half-lives substantially less than Earth's age, so many half-lives have passed that essentially all the nuclei originally present have long since decayed. For example, Earth's age is equal to nearly 200,000 half-lives of plutonium-239. Even if Earth had been pure Pu-239 (an impossibility for many reasons), dividing in half 200,000 times would have left none of the original Pu-239.

So why do we find radioisotopes with half-lives much less than the age of our planet? We do expect to find long-lived isotopes, such as uranium-238. When these decay they leave other nuclei, which themselves may be radioactive—and whose half-lives need not be long. Uranium-238, for example, decays by emitting an alpha particle, leaving thorium-234 (recall figure 3.2). Thorium-234 is also radioactive, decaying by beta emission with a half-life of about 24 days. So although its half-life is short, Th-234 is present wherever uranium is found. Beta decay turns a neutron into a proton and an electron, raising the atomic number by 1 while leaving the mass number unchanged (recall figure 3.3). Thus the thorium becomes protactinium-234 ($^{234}_{91}$Pa), which also undergoes beta decay. The resulting nucleus, uranium-234, marks the start of a chain of alpha decays that produces, among other things, radium-226 and **radon-222**. Figure 3.7 illustrates the **decay chain** that leads to radon. The chain continues, eventually reaching the stable isotope lead-206. A similar decay chain starts with thorium-232, a naturally occurring isotope whose 14-billion-year half-life is even longer than that of uranium-238.

Another way nature forms radioactive materials is with **cosmic rays**, high-energy particles emitted in astrophysical processes. They're energetic

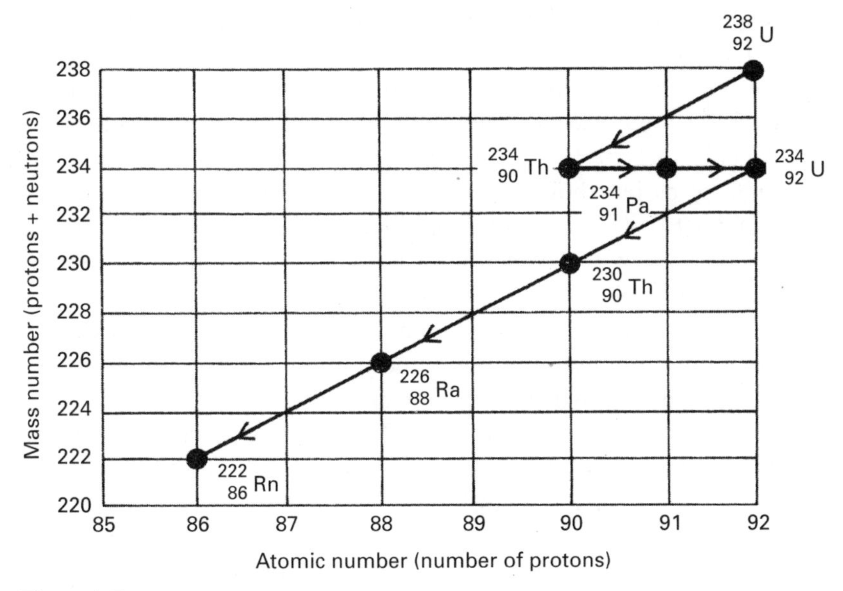

Figure 3.7
The decay of uranium-238 results in shorter-lived nuclei that are found wherever uranium is present. Here the decay chain is shown on a chart of mass number versus atomic number and is carried as far as radon-222. The chain continues until it reaches the stable isotope lead-206.

enough to cause nuclear reactions when they strike Earth's atmosphere, and some of these reactions result in radioactive nuclei. Carbon-14, for example, forms high in the atmosphere when cosmic-ray neutrons hit the ordinary nitrogen-14 that makes up most of our air:

$$^{14}_{7}\mathrm{N} + ^{1}_{0}\mathrm{n} \rightarrow ^{14}_{6}\mathrm{C} + ^{1}_{1}\mathrm{H}$$

A proton ($^{1}_{1}\mathrm{H}$) also results. Although carbon-14 decays with a half-life of 5,730 years, it's continually replenished through cosmic-ray interactions. Thus there's a nearly constant level of carbon-14 in Earth's atmosphere, and some of this radioactive carbon finds its way into living things. We'll see in the next chapter how this carbon-14 is used to date ancient objects.

Finally, there are many radioactive substances that we humans produce with our nuclear technology. Some are byproducts of the nuclear power and nuclear weapons industries. Of particular concern are the products of nuclear fission; many of these are intensely radioactive yet have half-lives long enough that they'll remain dangerous for centuries.

Background Radiation

The existence of natural and artificial radioactive materials means that we're unavoidably exposed to radiation. Figure 3.8 shows the sources of this **background radiation** exposure. For the average world citizen (figure 3.8a), the single most important source of radiation is radon-222, at 42 percent of the yearly dose. This gas, a natural decay product of uranium, seeps into buildings through porous or cracked foundations and exudes from stone, brick, concrete, and other building materials. Residents of brick and cinder-block houses, for example, receive more radiation than their neighbors in wooden houses. Other decay products in rocks and soils account for an additional 16 percent. In areas where uranium or thorium are prevalent, exposure to radon and other uranium decay products may increase substantially.

Cosmic rays are a second natural radiation source, contributing another 13 percent to the average dose, and much more for those who spend a lot of time in airplanes.

A third source is more surprising: our own bodies. We're all slightly radioactive. A typical human body contains about 5,000 Bq of potassium-40, meaning that 5,000 decays occur every second within our bodies. Potassium-40 and other radioactive substances found in air, water, and soil are incorporated into the food we eat, then into our own tissues. The

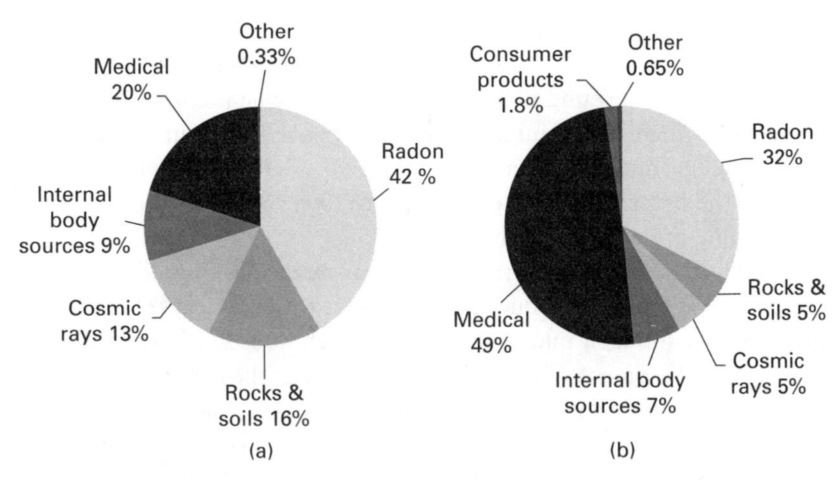

Figure 3.8
(a) Sources of background radiation for the average human. Natural sources, especially radon-222, dominate. The average yearly dose is about 3 millisieverts. (b) In the United States just over half the background radiation is from artificial sources, especially medical procedures. The U.S. average yearly dose is some 6 mSv, twice the global average. (Data source: UN Scientific Committee on the Effects of Atomic Radiation)

radiation we give ourselves is significant—about 9 percent of our total dose. In all, about 80 percent of the global average radiation exposure comes from natural sources, for an annual natural radiation dose of about 2.4 millisieverts (mSv) or 0.24 rem.

Of the remaining 20 percent of global average radiation exposure, nearly all comes from medical procedures. Less than 1 percent is attributable to nonmedical sources, including consumer products as well as nuclear power and weapons activities. Worldwide, the average exposure from all sources amounts to about 3 mSv (0.3 rem) per year.

Figure 3.8b shows a different picture for the United States, where medical procedures account for nearly half the average radiation dose. Consumer products, including smoke detectors, tobacco (which incorporates radioactive phosphorus from fertilizers), pottery and antique glassware, and granite countertops, account for some 2 percent of U.S. radiation exposure. Together, artificial sources provide slightly over half the average U.S. resident's total radiation exposure, which, at 6 mSv/year, is twice the global average.

The radiation doses in figure 3.8 are, of course, averages. Actual doses vary significantly and depend on choices you make. If you choose to live in Denver or to work on airplanes, your higher altitude results in greater cosmic-ray exposure. Should you work in a U.S. nuclear power plant, you may legally receive up to 50 mSv (5 rems) per year—nearly 10 times the average U.S. background dose. Sand in the coastal state of Kerala, India, is especially high in thorium; should you choose to live in Kerala, your yearly background radiation dose will be considerably higher than average. On the other hand, you can choose to lower your medical exposure by avoiding procedures involving radiation.

But should you avoid X-rays or nuclear medicine? Should you move out of the mountains to reduce your exposure to cosmic rays? Should you give up your job as a pilot, or sell your beach house in Kerala? Should you install a basement ventilation system to lower the radon level in your home? And what about the smoke detector that warns you of fire but contains a microcurie of americium-241? Should you get rid of it to avoid radiation exposure? These and similar questions force us to make nuclear choices. In the next chapter, we'll explore effects and uses of radiation that bear on these choices.

Summary

Unstable nuclei inevitably decay, emitting alpha, beta, or gamma radiation. This nuclear radiation is invisible but highly energetic, and readily detectable with instruments. The rate of decay of a radioactive sample is its activity, measured in becquerels or curies. Each radioisotope has a characteristic half-life, the time it takes half the nuclei in a sample of that isotope to decay; half-lives range from fractions of a second to billions of years. The effect of radiation on materials or living things is measured in sieverts or rems. We're all exposed to background radiation from natural and artificial sources; worldwide, the total background exposure is about 3 millisieverts (0.3 rem) per year; in the United States, it's twice that. Radiation exposure varies greatly with occupation, geographical location, house construction, medical procedures, and other factors.

Glossary

activity The rate at which a sample of radioactive material decays, measured in becquerels or curies.

alpha decay Radioactive decay by emission of a helium-4 nucleus, also called an alpha particle. The remaining nucleus has atomic number reduced by 2 and mass number by 4.

alpha particle A helium-4 nucleus (^4_2He), consisting of two protons and two neutrons, that's emitted in radioactive decay.

background radiation Radiation from natural or artificial sources in the everyday environment.

becquerel (Bq) A unit of radioactivity, equal to 1 decay per second.

beta decay A radioactive decay in which a neutron turns into a proton and an electron. The electron is ejected from the nucleus, leaving a nucleus with atomic number increased by 1 and mass number unchanged.

beta particle A name for the electron emitted in beta decay.

contamination Radioactive material in an undesired location.

cosmic rays High-energy particles from space that constitute a significant component of background radiation.

curie A unit of radioactivity, equal to 37 billion decays per second (37 billion Bq).

decay chain A series of isotopes formed as a result of successive radioactive decays.

electron capture The process in which an atomic nucleus captures one of the atoms innermost electrons, lowering the atomic number by 1.

excited nucleus A nucleus containing excess energy, which it may give up by emitting a gamma ray.

gamma decay The process whereby an excited nucleus sheds excess energy by emitting a gamma ray.

gamma ray A bundle of energy emitted by an excited nucleus.

Geiger counter A radiation detector in which radiation strips electrons from atoms in a gas-filled tube, resulting in a burst of electric current.

gray (Gy) The standard unit for the energy an object absorbs when exposed to radiation, equal to 100 rem.

half-life The time it takes for half the nuclei in a given radioactive material to decay.

nuclear radiation High-energy particles—alpha, beta, or gamma—emitted by radioactive nuclei.

positron Antiparticle to the electron. Positrons have the same mass as electrons but carry one unit of positive electric charge.

rad A unit that measures the energy an object absorbs when exposed to radiation, equal to 0.01 Gy.

radiation See **nuclear radiation.**

radioactive Describes a substance, in particular an isotope, which undergoes radioactive decay.

radioactive decay The process in which an unstable nucleus comes apart, usually by emitting a particle.

radioisotope Short for radioactive isotope, an isotope that undergoes radioactive decay.

radon-222 A radioactive gas formed in the decay sequence of uranium-238 and constituting a significant portion of background radiation.

rem A unit of radiation dose that describes the radiation's effect on the human body, equal to 0.01 Sv.

sievert (Sv) The standard unit of radiation dose that describes the effect of radiation on the human body, equal to 100 rem.

Note

1. In fact, a third particle is involved: an electrically neutral particle of negligible mass, called a *neutrino*. The neutrino has essentially no interaction with ordinary matter, and in this book we'll neglect it.

Further Reading

European Union and EDP Sciences. radioactivity.eu.com. An authoritative website covering all aspects of radiation and radioactivity. The predecessor of EDP Sciences was founded in 1920 by Marie Curie and other prominent physicists.

Lawrence Berkeley Laboratory (LBL), radioactivity website at https://www2.lbl.gov/abc/wallchart/chapters/03/0.html. Website intended to educate the general public about radioactivity, designed in conjunction with LBL's chart of stable and unstable nuclides.

Lillie, David W. *Our Radiant World*. Ames: Iowa State University Press, 1986. A well-written primer on radiation from both natural and artificial sources, written at about the level of this book and covering in much more depth the material of chapters 3 and 4. Mostly objective, but the author's industrial background shows. Excellent bibliography. Dated but authoritative.

Redniss, Lauren. *Radioactive: Marie and Pierre Curie: A Tale of Love and Fallout*. New York: Dey Street Books, 2015. This delightfully illustrated book combines history, biography, and physics. A moderate antinuclear bias is evident.

Tuniz, Claudio. *Radioactivity: A Very Short Introduction*. Oxford, UK: Oxford University Press, 2012. A brief but thorough introduction to radioactivity as it occurs both in nature and in technological applications.

4
Radiation

People fear radiation for good reasons: It harms biological systems, including ourselves. It's invisible and undetectable without special equipment. And nuclear technologies have created new sources of potentially hazardous radiation.

But radiation's dangers aren't the whole story. Risk from radiation exposure depends on the radiation dose one receives. Furthermore, some uses of radiation are distinctly beneficial. Many people are alive today who wouldn't be had radiation not helped diagnose or treat otherwise fatal diseases. Radiation helps us in less obvious ways, too. Food safety and shelf life are improved by irradiation. Radiation-induced sterilization reduces insect populations. Sensitive detection and analysis of pollutants relies on radiation techniques. And we're enriched by the knowledge of our own past that archaeologists gain through radioisotope dating. This chapter explores the harmful effects of radiation and also samples its beneficial uses.

Biological Effects of Radiation

Radiation consists of high-energy particles, including alpha, beta, and gamma rays as well as X-rays. Their high energy enables these particles to knock electrons out of atoms—the process of **ionization**, and hence the term **ionizing radiation**. Molecules containing ionized atoms are chemically very active. In living tissue such molecules may undergo chemical reactions whose products are detrimental to life. Ionization of water, for example, leads the formation of substances that act as cell poisons. Radiation striking more complex biological molecules, such as proteins or nucleic acids, may break the molecules and prevent their proper functioning. Loss of cell vitality, decreased enzyme activity, cancer, and genetic changes are among the possible outcomes.

Genetic Effects

If radiation strikes germ cells in reproductive organs, it may alter the genes and chromosomes that determine hereditary characteristics. DNA—the molecule that carries the genetic message—is especially sensitive to radiation. When DNA in reproductive cells is altered, offspring inherit changed genetic characteristics called **mutations**. Some mutations are obvious, but most are more subtle. Yet over 99 percent of mutations are harmful. In fact, the mutations we see in living organisms represent only a small fraction of those that actually occur. Most mutations result in death before a developing organism has sprouted, hatched, or been born.

Since the mutagenic (mutation-causing) properties of radiation were discovered in the 1920s, generations of laboratory animals have been bred under controlled conditions to study the **genetic effects** of radiation. For animals exposed to high but nonfatal radiation doses, the mutation rate increases in proportion to the dose. But even without radiation above natural background, mutations still occur. These spontaneous mutations arise naturally, some by chemical agents or natural background radiation. So the important question is how many *additional* mutations result from radiation exposure above background. Studies aimed at answering these questions have involved literally millions of mice, but for humans the largest sample we have are some 120,000 survivors of the nuclear bombings of Hiroshima and Nagasaki, along with nearly 80,000 of their offspring.

Studies of the Japanese bombings are problematic, as documented by social scientist Susan Lindee and many others.[1] Serious work didn't start until the early 1950s, some five years after the bombings, and therefore missed victims who died or moved away before then. Establishing the victims' radiation doses often relied on interviews, done years after the bombings, to determine where victims were in relation to the detonations. Dosimetry was hampered by there being minimal information on radiation release from the one nuclear test that preceded Hiroshima and Nagasaki and by most information being classified. Although Japanese scientists used ingenious means to estimate doses immediately after the bombings, those data were confiscated by the United States and have never been found. Furthermore, the bombings were sudden events that caused relatively large radiation exposures. Studies based on the bombings may or may not be relevant to victims of lower-level but longer-term exposure, such as those living in areas contaminated by nuclear accidents. Nevertheless, the Japanese studies remain one of our strongest indicators of radiation effects in large populations, and they're still the basis of many radiation exposure regulations.

Following the Japanese survivors and their offspring for decades has yielded no statistically significant evidence of adverse genetic effects.[2] Two caveats here: DNA sequencing could show mutations that haven't yet been detected, and victims could carry recessive mutations that might appear in future generations. However, it's possible to estimate the likelihood of specific genetic diseases arising from radiation exposure, and these estimates coupled with animal studies suggest that a dose of 1 Gy (for our purposes, similar to 1 Sv) results in mutations occurring at twice the spontaneous rate.

Now, 1 Sv is a large radiation dose—about 300 times the yearly background dose. What about the genetic effects of low doses, such as the 5 millisieverts typical of residents in the area surrounding Japan's 2011 Fukushima Daiichi nuclear accident? Extrapolating downward from 1 Sv for a doubled mutation rate suggests an increase of only five-thousandths, or 0.5 percent, above the natural mutation rate. But such extrapolation probably overestimates the radiation-induced mutation rate. That's because living cells contain **repair enzymes**, whose job is to fix damaged genetic material. At low doses, these repair enzymes should reduce the mutation rate below what would be expected from high-dose studies.

Most scientists believe that genetic effects of public exposure to radiation from nuclear technologies are insignificant. Some, however, argue that long-term effects of even minuscule exposure may propagate far into future generations, producing a genetically weakened human race. Do we accept the benefits of nuclear technology today in return for such a remote but potentially devastating possibility?

Epigenetic Effects

The term **epigenetic** refers to changes that don't alter the DNA sequence but are nevertheless heritable. Epigenetic effects alter the expression or activation of particular genes in the absence of DNA damage, and these alterations can be passed to a cell's offspring on cell division. If epigenetic alteration affects reproductive cells, then the effects may persist in subsequent generations. The field of epigenetics is relatively new, having become widespread in biology only in the 1990s. There are few studies of epigenetic effects on large populations exposed to radiation, but laboratory work suggests a number of mechanisms whereby radiation can alter the cellular epigenome.[3] It may be that some radiation effects reported previously as genetic are actually epigenetic.

Somatic Effects

Somatic effects of radiation are effects on an individual that aren't passed on to future generations. High radiation doses of several sieverts destroy cells that normally divide rapidly, diminishing the body's ability to replenish its red and white blood cells and the cells that line the intestinal tract. Nausea and vomiting are the first symptoms of acute **radiation sickness**, typically appearing a few hours after exposure. There follows a lull of several days to a week or more during which the victim may feel fine. But then, as red blood cells die without being replaced, anemia sets in. Intestinal bleeding, complicated by the loss of blood-clotting factors, exacerbates the illness. As white blood cells go unreplaced, the body's immunity declines. Death may follow in a matter of weeks or months. At a radiation exposure of 4 Sv, about half the victims die of radiation sickness. Blood transfusions and bone-marrow transplants can improve survival by allowing the body to make new red blood cells.

Victims exposed to doses of a few Sv exhibit similar symptoms but generally recover in several months. The same is true for those lucky enough to survive higher doses. These victims seem to recover completely and, except for an increased cancer risk, go on to lead normal lives.

While doses above about 1 Sv cause radiation sickness, doses of a few tenths of a sievert may produce no obvious effects. How are we to determine whether such radiation exposures are harmful? And how can we possibly establish harmful effects at the much lower doses—typically only a few thousandths of a sievert—received by the general public in nuclear incidents or through medical procedures? Answering these questions requires statistical analysis of large populations exposed to low-level radiation.

Cancer and Radiation

The most significant result of low-level radiation exposure is increased incidence of cancer. Numerous studies have confirmed this effect. For example, X-ray technicians working in the first half of the twentieth century showed substantial increases in a variety of cancers. Watchmakers in the 1920s applied radium-containing paint to make watch hands visible in the dark. Workers painting watch hands often licked their brushes, ingesting radium in the process. Nearly all of them eventually died of bone cancer or radiation-induced anemia. Uranium miners, exposed to radon gas trapped in the mines, used to develop lung cancer at a much higher than average rate. Many early nuclear scientists—including Marie Curie and her daughter Irene—died from leukemia that was undoubtedly caused by radiation exposure. (Pierre Curie was spared this fate; he was killed in the street when a horse bolted from its carriage.)

Today, we know enough to avoid painting our watches with radium. Strict controls on uranium mines have dropped radon exposure to the point where some American homes have higher radon levels than uranium mines. Scientists and medical personnel are more conscious of radiation's dangers and take precautions to minimize exposure. Are we then safe from radiation? Or do much lower levels still pose a risk?

A look at survivors of Hiroshima and Nagasaki shows why the issue of low-level radiation is still clouded. Figure 4.1 plots the incidence of solid cancers among Hiroshima residents exposed to radiation from the bomb, as a percent of the number of deaths expected in the absence of bomb radiation. Where that percentage is substantial, we can assume that the additional cases are due to radiation. And sure enough, large doses of radiation result in greatly increased cancer incidence—nearly double the expected rate for exposures over 2 grays.[4] But for exposures under 0.1 Gy, the cancer excess is less than 2 percent. In fact, given the statistical uncertainty, it's even possible that very low doses resulted in fewer cases than expected!

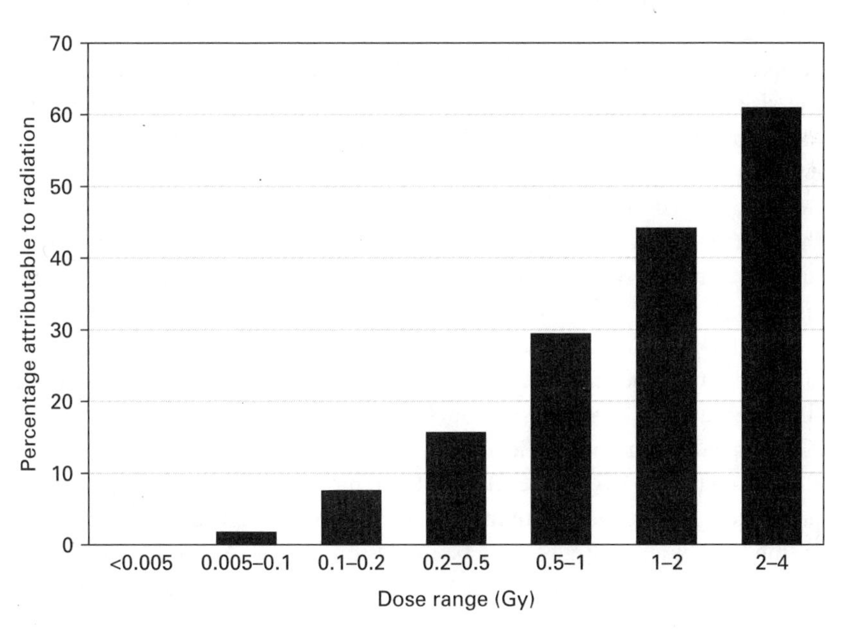

Figure 4.1
Excess solid cancers as a percentage of expected cancer incidence in Japanese bomb survivors, as a function of dose in grays (Gy). (Data source: D. L. Preston et al., "Solid Cancer Incidence in Atomic Bomb Survivors: 1958–1998," *Radiation Research* 168, no. 1 [July 2007], Table 9.)

The results shown in figure 4.1 typify an ongoing controversy over the health effects of radiation: Should effects at low doses, which are hard to measure, be extrapolated proportionately from the easier-to-measure effects at high doses? And at very low doses, should the extrapolation continue straight to zero effect at zero dose? Or are there reasons why the response at low doses might deviate from a direct proportionality? Those same repair enzymes that help ward off radiation-induced mutations at low radiation doses might also repair DNA damage that could lead to cancer. Many radiation specialists therefore suggest that risks at low doses should be lower than implied by a direct proportionality. Some would argue for a threshold dose, below which there are no harmful effects. And a few would claim **hormesis**, a response in which very low radiation doses are actually beneficial. (If that sounds absurd, note that hormesis is well established for chemicals that are toxic at high doses but beneficial at low doses.) Alternatively, it might be that low radiation doses are actually more harmful than a proportional extrapolation would suggest. Figure 4.2 shows **dose-response curves** corresponding to these possibilities.

Absent unambiguously solid evidence for radiation effects at low doses, many scientific bodies and regulatory agencies take a conservative approach, adopting the **linear no-threshold (LNT) model**. This model assumes that the risk of radiation-induced cancer is directly proportional (*linear*, in math-speak) to the dose, right down to zero dose. Some LNT models for cancer risk use a so-called *dose and dose-rate effectiveness factor*, which reduces the risk at low doses by a factor of typically 1.5 or 2 relative to the measured risk at high doses. But these models are still linear, which implies that *any* radiation exposure, no matter how small, carries some risk.

Scholarly bodies recommending use of the LNT model include the United Nations Scientific Committee on the Effects of Atomic Radiation, the International Commission on Radiological Protection, and the U.S. National Academy of Sciences in its report *Biological Effects of Ionizing Radiation* (BEIR VII, phase 2). The LNT model is the basis of radiation-protection regulations in many countries, including those promulgated by the U.S. Nuclear Regulatory Commission (NRC). A notable exception is France, whose Academy of Sciences has questioned the validity of LNT at doses below 100 mSv.

The debate over LNT and other models in figure 4.2 might seem an arcane matter, best left to scientists and statisticians. But it's not; rather, the choice of which low-dose model to adopt is a nuclear choice with

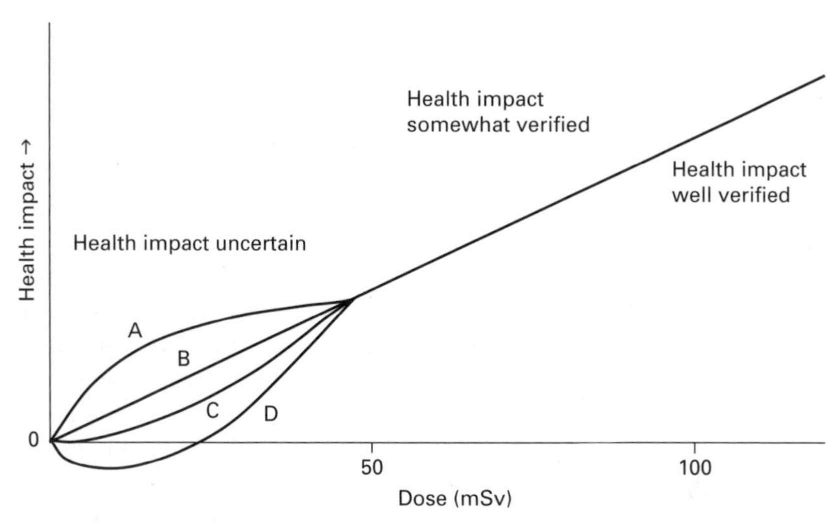

Figure 4.2
Dose-response curves corresponding to different hypotheses about the health impact of low radiation doses. Curve A indicates higher risk at low doses. Curve B is the linear no-threshold model. Curve C shows lower risk at low doses, and curve D is a hormesis model with a threshold below which radiation is actually beneficial. All are in essential agreement at higher doses (above 50 mSv), where health impacts become increasingly well verified.

enormous consequences. The size of the evacuation zone at Fukushima was based in part on the assumed risks of low-dose radiation. Adopt a threshold model instead of the LNT and many people would be spared the stress of evacuation—which brought its own health consequences. Were those consequences worse than the danger from radiation? Quite possibly—but the answer depends on assumptions about radiation risks at low doses. After the Chernobyl accident thousands of pregnancies were intentionally terminated across Europe out of fear of genetic damage. Was that a sensible precaution or an overreaction? The answer requires knowing the impact of low radiation doses. A physician's decision to administer a CAT scan or nuclear medicine procedure is predicated on the calculation that the risks of the disease being diagnosed or treated outweigh the risks of low-dose radiation associated with the procedure. Protecting the public from radiation emitted by nuclear power plants requires costly shielding and other expensive measures—and thus contributes to the economic predicament that nuclear power currently faces, especially in the West. Yet nuclear power arguably offers a low-carbon source of energy that could help stave off global warming. In all these examples, the risks of radiation weigh against non-nuclear risks, and someone has to make a choice. Intelligent choice requires understanding the effects of radiation at low doses.

The BEIR VII report develops a rough rule of thumb for estimating cancer effects of low-level radiation. It suggests that a radiation dose of 0.1 Sv (100 millisieverts or some 30 times the global average annual background dose) results in a 1-in-100 lifetime chance of developing cancer. That means that if 100 people are exposed to 100 mSv each, we can expect that, on average, one of them will eventually develop cancer as a result of their radiation exposure. Expose 1,000 people and expect 10 cancers. Consider these numbers in light of the fact that, on average, 42 out of 100 people will develop cancer in their lifetimes. Consider also that 100 mSv is a large dose—at the upper limit of what's considered "low dose" and far above the doses experienced by the general public in the worst nuclear accidents.

So what's the cancer impact of more realistic and much lower radiation exposures? If the LNT hypothesis is correct, then we can simply scale down from the BEIR-VII's 1-in-100 cancer risk for a dose of 100 mSv. For example, the 1979 Three Mile Island (TMI) nuclear accident in Pennsylvania resulted in a dose of about 0.1 millisievert to the approximately 36,000 people within five miles of TMI. That's only one one-thousandth of the 100-mSv dose that gives a 1-in-100 cancer risk. So, assuming LNT, we would expect a risk of 1 in $(100 \times 1{,}000)$ or 1 in 100,000. The exposed population was only about one-third of 100,000, so our calculation suggests it's unlikely that any cancers resulted from that accident. On the other hand, the most exposed region outside the exclusion zone from the 2011 Fukushima accident saw doses averaging from 1 to 10 mSv. For an average of 5 mSv, or one-twentieth of the 1-in-100 cancer dose, our extrapolation would then suggest a lifetime cancer risk of 1 in (100×20) or 1 in 2,000—one cancer for every 2,000 people exposed. The population of the affected region is about 2 million, so we might expect some $2{,}000{,}000/2{,}000 = 1{,}000$ lifetime cancers resulting from the accident. That sounds like a lot, but the lifetime cancer incidence in Japan is 41 in 100, implying that there should be more than 800,000 cancers, not associated with the Fukushima accident, in the affected population. Cancer rates fluctuate naturally, making detection of an additional 1,000 cancers over many decades essentially impossible. Our estimate of 1,000 Fukushima cancers is consistent with more careful studies, including one from Stanford University that put the number of cancers in the range from 24 to 1,800.[5]

It's important to view estimates of cancer incidence from radiation exposure in terms of the increase above normal cancer rates—globally, about 42 cancers for every 100 people. That's why, with rare exceptions (e.g., thyroid cancer following the Chernobyl accident), it's virtually impossible to detect radiation-induced cancers in a general population exposed to low-level radiation. However, large-scale studies of nuclear industry workers have teased out a statistical link between low-dose radiation and leukemia. The INWORKS study,[6] which released results in 2015, considered some 300,000 workers in France, the U.K., and the U.S. Study participants worked an average of 15 years in the industry, and all wore radiation badges so their doses are well known. The average annual occupational dose was under 2 mSv—about two-thirds of the global average background dose. INWORKS results are consistent with the Japanese bombing studies, and they support a linear no threshold dose-response even at low doses.

That said, the average figures we've been using here don't necessarily apply to all individuals or all cancer types. For example, the cancer risk associated with radiation exposure for a one-year-old infant at Fukushima is nearly double that for a 20-year-old. In general, young children and the unborn are more susceptible to radiation because of their rapid growth, the sensitivity of their developing organs, and the fact that they have a longer lifespan ahead in which to develop cancer. Back to Fukushima: although doses there averaged in the range of 1 to 10 mSv, some individuals received as much as 50 mSv. For them the cancer risk was 10 times our estimate based on an average of 5 mSv. But the BEIR-VII estimate of 1 in 100 cancers for a 0.1 Sv dose is just that—an estimate—and it's an average based on a population with age structure similar to the United States. Individuals face varying risks depending on age at exposure, gender, and other factors. The fact is that we don't know precisely the effect of low radiation doses in causing cancer.

Despite that uncertainty, low-level radiation exposure is not among the more serious threats to our lives. Many dangers we willingly accept in everyday life carry far greater risks than low-level radiation. If you choose to smoke, you will, on average, live 5.3 fewer years than your nonsmoking friends. Although not everyone gets killed in car accidents, enough do that the average risk amounts to a lifespan reduction of 183 days. You would be horrified at the thought of being among the people most seriously irradiated at Chernobyl, yet if you were in that crowd of some 200,000, your life expectancy would drop by 24 days—just about the same loss you face, on average, from radiation exposure due to indoor radon. The average American loses 120 days due to air pollution, much of it from the coal-fired power industry, but only 0.06 days due to radiation from the nuclear power industry (the most ardently antinuclear activists would beg to differ, but even their estimates would put the nuclear industry's contribution to your loss of life at only about 0.7 days). You'd do worse living for 40 years at the boundary of a nuclear plant, receiving an extra 1 mSv per year of radiation; that would reduce your lifespan by 11 days. If you don't like that, you can move away—but if that adds 10 miles to your round-trip commute, then the risk of this extra driving for 40 years will give an average lifespan reduction of some 28 days—more than double the risk of staying next to the nuclear plant. Radiation really is one of the lesser risks most of us face.

Although the risk from radiation is small, the number of individuals affected need not be. Even the very low chance of your personally contracting fatal cancer from nuclear power operations implies that between 25 and 300 people (depending on whether you use pro- or antinuclear estimates) in the United States will contract fatal cancer each year because of nuclear power.[7] Is that acceptable? Or do we demand that nuclear technology produce *no* excess cancers? And if we demand that of nuclear technology, why not of coal-burning technology, which kills at least 10,000 people yearly in the United States and a million in China, or of pesticides, or of artificial sweeteners? Nuclear choices, like those involving low-level radiation, cannot be divorced from a broader technological context.

Medical Uses of Radiation

You've just seen that radiation can cause health problems ranging from radiation sickness and cancer to genetic abnormalities. Yet radiation is one of modern medicine's most valuable tools for diagnosing and treating diseases, especially cancer. As figure 3.8 showed, medical procedures account for 20 percent of the global average radiation exposure and nearly half for the average American—in either case, by far the largest artificial source of radiation for the average person. Whether to use radiation for medical purposes is yet another nuclear choice we face. In most cases, the health benefits of medical radiation exposure outweigh the potential harm.

Cancer Treatment

The same properties that make radiation harmful to living cells make it a valuable tool for fighting cancer. Cells most susceptible to radiation damage are those that divide rapidly; that's why the bone-marrow cells that produce red blood cells are preferentially destroyed by high radiation doses. Since they divide rapidly, cancer cells are more easily damaged by radiation than most normal body cells. Directing a beam of radiation at a tumor can destroy enough cancer cells to shrink or eliminate the tumor, while doing less drastic damage to surrounding tissue.

Some two-thirds of all cancer patients receive **radiation therapy**. The first such therapy used X-rays, only a few years after Wilhelm Roentgen's 1895 discovery of the rays. Gamma rays, more energetic than X-rays, were widely used in early cancer treatment. A typical gamma-ray unit comprised a chunk of radioactive cobalt-60 surrounded by heavy lead shielding; a channel in the shielding let out a well-defined beam of gamma rays. Cobalt-60 units are still in use in the developing world, but in technologically advanced countries high-energy particle beams have replaced gamma rays from nuclear materials for radiation therapy applied from outside the body. Beam energy and thus penetration can be precisely controlled for maximum effectiveness against a given patient's tumor (figure 4.3). The goal in radiation therapy is to maximize radiation to the tumor itself while minimizing damage to surrounding tissues. Proton beams are especially effective at localizing the radiation dose. Some treatments use multiple beams matched to the shape of the tumor or rotate the beam about the tumor. Finally, some tumors call for implantation of radioactive material at the tumor site; radiation exposure is then limited to the tumor region. Such **brachytherapy** may be used in brief sessions where high-dose materials are placed temporarily near the tumor or by implanting radioactive "seeds" that remain in the patient's body for life. Iridium-192, with a 74-day half-life, is often used in temporary high-dose brachytherapy. Although Ir-192 decays primarily by beta emission, another decay mode produces gamma radiation that destroys cancer cells. Lower-dose permanent implants often use 59-day-half-life iodine-125, in which capture of an inner electron by the nucleus results in an excited nucleus of tellurium-125. The Te-125 almost immediately emits its excess energy as cancer-killing gamma radiation. For some cancers, radiation may be as effective as surgery; in others, it's used in conjunction with surgery and/or chemotherapy. Even when radiation can't cure cancer, it can provide pain relief and other palliative care in patients with advanced cancer.

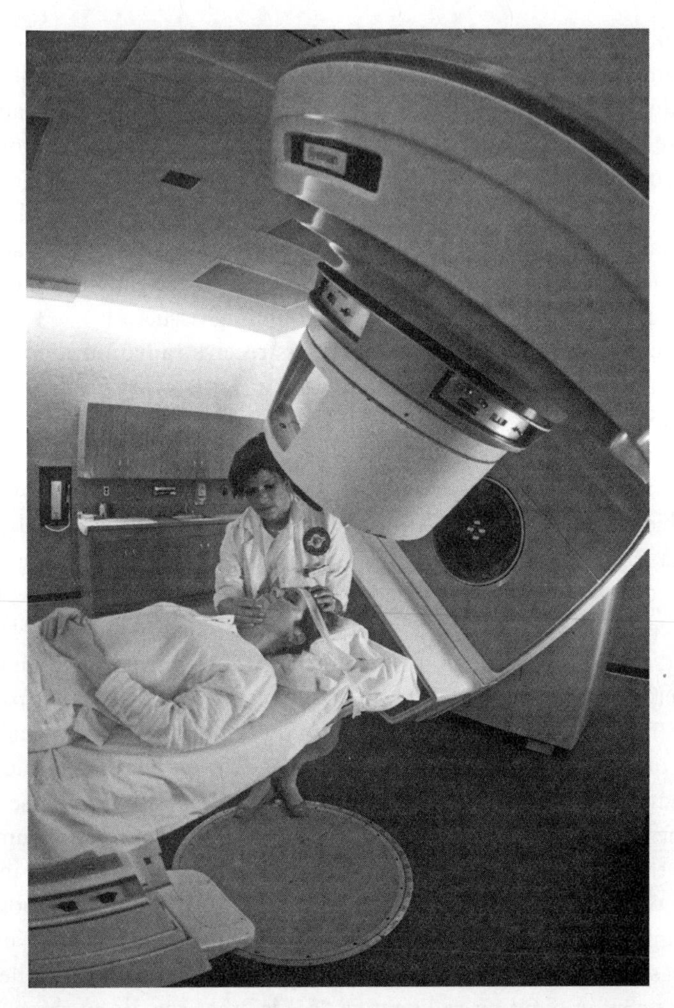

Figure 4.3
Patient being prepared for particle-beam radiation treatment. (National Cancer Institute/Science Source)

Medical Diagnosis

Nearly everyone has had diagnostic X-rays of teeth or bones. X-rays are produced by slamming high-energy electrons into metallic targets. With appropriate energy, X-rays penetrate soft tissues but not hard bones and teeth, so the hard tissues stand out in X-ray images. Applying X-ray-opaque materials to soft tissues allows them to be imaged as well. X-rays of the stomach, for example, are made after the patient swallows a solution of X-ray-opaque barium. The barium coats the walls of the esophagus, stomach, and intestine, so these organs show up in the image. X-ray imaging combined with sophisticated computer analysis is called *computerized axial tomography* (CAT). CAT scans give vividly detailed images through selected sections of the body.

Nuclear medicine uses radioactive isotopes for diagnosis. Administering radioactive versions of biologically important chemicals allows doctors to follow these chemicals through the body and into particular organs. Iodine, for example, is necessary for the functioning of the thyroid gland. Administering a radioisotope of iodine, followed by measurement of radiation from the thyroid, gives an indication of thyroid function. In general, a radioactive substance is called a **radioactive tracer** when it's used to trace the flow of material in some system—in this case, the human body.

Many radioisotopes are used as tracers. Radioactive substances can be chosen that are readily absorbed by particular organs or tumors. The radiation emitted then gives an image of the region under study. This technique can locate tumors deep within the brain (figure 4.4). Similar

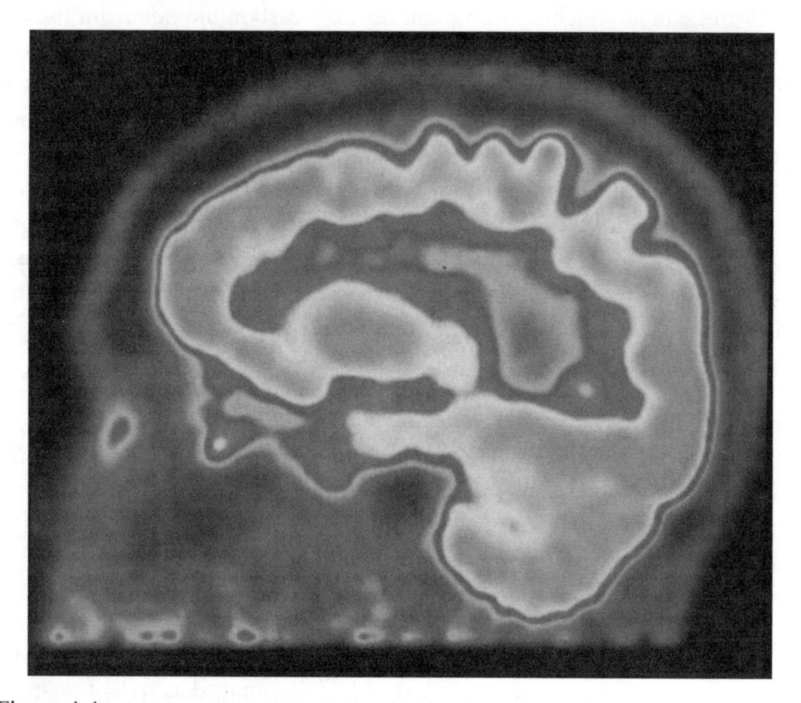

Figure 4.4
Scan of a human brain using positron emission tomography (PET). (Steven Needell/Science Source)

scanning techniques search for cancers of the liver, bones, and other organs. Oxygen-15, with a 2-minute half-life, is used in studies of blood flow and lung ventilation. The very short half-life makes this isotope especially safe, since its radioactivity decays in such a short time. But a hospital using O-15 must produce it "in house," since there's no time to transport it from elsewhere. Some major hospitals have cyclotrons or other particle accelerators to produce short-lived radioisotopes for medical diagnostics.

Nonmedical Uses of Radiation

Radiation serves a host of applications in diverse fields including industry, science, archaeology, insect control, food preservation, and security. Here we sample just a few of radiation's myriad uses.

Radioactive Tracers

You've just seen how radioisotopes serve as tracers in medical diagnosis. The same idea works in other applications. Suppose a botanist wants to know the rate at which a given plant absorbs carbon dioxide from the air. (In this age of global warming due to a buildup of carbon dioxide in the atmosphere, that's a very relevant question.) He puts the plant in a closed environment whose atmosphere contains carbon dioxide made with radioactive carbon-14. The plant absorbs the radioactive CO_2, and measurement of the plant's radioactivity tells the rate at which the plant absorbs carbon dioxide. Or suppose an engineer develops a new material for automotive engine bearings. To study how well the material wears, she introduces radioisotopes into a sample bearing. As the bearing wears, radioactive material accumulates in the engine oil. Measuring the oil's radioactivity lets the engineer infer the rate of wear.

Smoke Alarms

Smoke alarms save lives. The majority of fire deaths in the United States occur in homes that lack functioning smoke alarms. The simplest and least expensive type of alarm is the *ionization alarm*. In these devices, a small amount (about 0.3 micrograms, yielding an activity of one microcurie or 37,000 Bq) of the radioisotope americium-241 emits alpha particles that ionize air in a sampling chamber, allowing the air to carry a modest electric current. If smoke enters the chamber, it interferes with the current. An electronic circuit detects the change in current and triggers the alarm.

The americium used in smoke detectors is a transuranic element that doesn't occur in nature. It's formed in nuclear reactors when plutonium-239 absorbs a neutron and, instead of fissioning, undergoes a chain of nuclear reactions leading to Am-241.

Oil Well Logging

Radioisotope sources of neutrons or gamma rays are used to map the geology surrounding oil wells, yielding detailed records of rock density, porosity, and other characteristics as a function of depth in a well (figure 4.5). This application, known as *well logging*, probes already-drilled wells or new

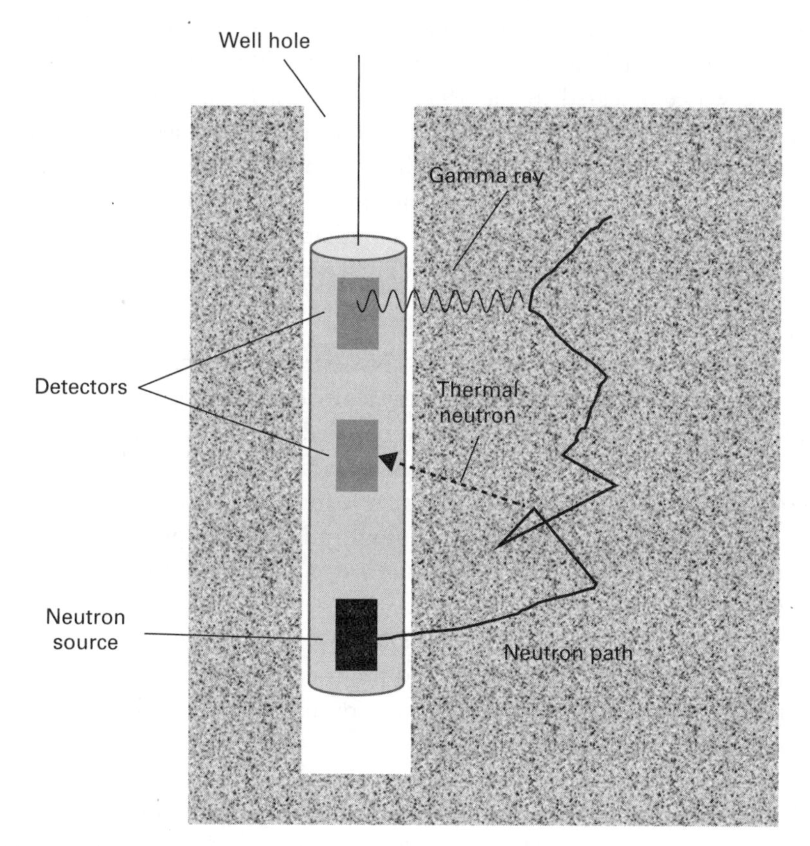

Figure 4.5
Diagram showing how a neutron source is used to probe the geology surrounding an oil well. High-energy neutrons from the source scatter off the surrounding rock, generating additional thermal neutrons and some gamma rays. Detectors sense these particles, yielding information about the rock structure. Neutrons interact most readily with hydrogen, making the technique especially useful for determining porosity of the rock—an indicator that oil might be present.

wells. For new wells, the radioactive source is mounted just behind the drilling head as the well is bored. One approach uses alpha particles from americium-241 to irradiate stable beryllium-9, which yields neutrons. Such Am-Be neutron sources are manufactured as sealed units for well logging and other applications.

Insect Control

Because radiation preferentially damages rapidly dividing cells, large radiation doses to the reproductive organs may result in sterility. In the *sterile insect technique* (SIT), used to control insect pests, a large group of irradiated male insects are released into the wild, where they mate with normal females. No offspring result, and the population declines. The screw-worm fly, a livestock pest prevalent in the southeastern United States, was eradicated in the 1950s in the first large-scale application of insect control through radiation sterilization. The medfly, a serious pest of citrus crops, has been partially controlled in the same way. Early attempts against malaria-carrying mosquitoes weren't as successful, but in 2019 a two-pronged approach using radiation sterilization and a virus nearly eradicated an aggressive mosquito species in Guangzhou, China.

Food Preservation

High doses of radiation preserve food by destroying bacteria and enzymes that cause spoilage and disease. At very high doses—thousands of grays—bacteria are completely destroyed, and sealed food then keeps for years, just as canned food does. At lower doses, bacterial populations fall to the point where the shelf life of food extends for many months. Irradiation of fruits and vegetables inhibits sprouting and overripening, again greatly extending useful shelf life (figure 4.6). Although food irradiation has been used since the 1950s, commercial irradiation through the twentieth century was largely limited to herbs and spices (and still is in Europe). The range and quantity of irradiated foods expanded in the twenty-first century, and today some 500,000 tons are sold globally each year in the 60 countries that approve the process.

Is irradiated food safe? Proponents of irradiation point to a method that's less destructive of food values than the heat treatment used in traditional canning, and which reduces the incidence of salmonella and other pathogens. They're also quick to point out that irradiation doesn't make food radioactive. Opponents agree that there's no radiation hazard to the consumer, but that ionizing radiation causes chemical changes that might result in toxic or carcinogenic substances. And some opponents of food

Figure 4.6
Food preservation by irradiation. The potato at left is eight months old and was untreated. The potato at right is the same age but received 200 Gy of radiation. (Brookhaven National Laboratory)

irradiation see an insidious connection between food and nuclear weapons, since one of the two commonly used isotopes, cesium-137, comes from spent nuclear fuel. However, health agencies such as the World Health Organization and the U.S. Food and Drug Administration have approved irradiation for a wide variety of foods.

Radioisotope Dating

As they decay, radioactive isotopes act as "clocks," enabling archaeologists, geologists, and art historians to determine the ages of ancient objects. For things that were once alive—bones, charcoal from ancient campfires, vegetable matter, textiles, and the like—the isotope carbon-14 is effective back to about 50,000 years. The box "Radiocarbon Dating" describes this process in more detail.

Carbon-14 also provides evidence that our era's troublesome increase in atmospheric carbon dioxide—the predominant cause of global warming—is a result of humankind's fossil-fuel consumption. That's because the proportion of C-14 in the atmosphere is dropping—an indication that the new atmospheric carbon may be from organic material that's been long dead. Coupled with a decrease in the stable isotope C-13, which isn't taken up in

Radiocarbon Dating

Carbon-14 forms continuously in the atmosphere as cosmic rays interact with nitrogen. Chemically identical to ordinary carbon, C-14 is incorporated into plants and then into plant-eating animals. Carbon-14 decays with a half-life of 5,730 years, but the level of C-14 in a living organism remains fairly constant as the intake of newly formed C-14 balances radioactive decay. When the organism dies, it stops taking in C-14. But radioactive decay continues, so the C-14 level drops. By measuring the ratio of C-14 to stable carbon isotopes in an ancient sample, scientists can determine how much time has passed since the sample was alive. Figure 4.7 depicts the principle of radiocarbon dating.

Figure 4.7
Radiocarbon dating. (a) Carbon-14, formed in the atmosphere by cosmic rays, is incorporated into living things through the food chain. As a result, living organisms maintain a mild but steady level of C-14 radioactivity. (b) At death, C-14 uptake ceases. (c) Much later, C-14 activity has decreased substantially. (d) Archeologists excavate the long-dead remains. By measuring C-14 activity, they infer the time since death. Note that the archeologists, with their active C-14 intake, are more radioactive than their ancient ancestor. In practice, the radioactivity measurement is done not in the field but in the laboratory. (Artwork by Robin Brickman.)

photosynthesis as readily as is C-12, the changing isotopic content of atmospheric carbon points to its source in long-dead plant matter. That's exactly what the fossil fuels are.

For longer time spans, up to the billions-of-years ages of rocks, **radioisotope dating** compares ratios of different isotopes and their decay products. For example, lead-206 is the stable end product of the uranium-238 decay chain considered in the preceding chapter. Comparing the amounts of lead-206 and uranium-238 tells how long ago the nuclei in a material formed. That's how we know that Earth and Moon are made from stuff of essentially the same age. Indeed, much of our knowledge of our own past, our planet's, and our solar system's comes from radioisotope dating.

Activation Analysis

Is my water safe, or does it contain toxic arsenic? What, exactly, is in this sample of air pollution I've collected? Is there a bomb in this airline luggage? What elements are in this meteorite? What kind of paint did Leonardo da Vinci use for the *Mona Lisa*? These and similar questions yield to the nuclear technique of **activation analysis**.

In the most common form of activation analysis, a sample is bombarded with neutrons. Nuclei in the sample absorb neutrons, producing new isotopes—many of which are radioactive. These activated nuclei decay, each type of nucleus giving off its own characteristic radiation. Measuring the intensities and energies of those radiations gives the quantities and types of radioactive nuclei present, and from this the numbers and types of the original nonradioactive nuclei can be inferred.

Activation analysis has many uses in archaeology, art history, geology, chemistry, environmental science, and possibly security. Advantages include its nondestructive nature and the ability to detect elements present only in minute quantities. A work of art, for example, can be analyzed without removing any material. In principle, neutron activation could detect explosives in airline passengers' checked luggage. Although present-day scanners don't use activation, several companies are pursuing this approach.

Summary

The high-energy particles that constitute nuclear radiation do grievous damage as they tear through biological systems, disrupting cell functions and damaging genetic material. The results may include acute radiation sickness, initiation of cancer, and undesirable genetic traits passed on to future generations. But it takes large radiation doses to produce significant effects. Fatal radiation sickness occurs at about 4 sieverts. The natural mutation rate doubles at about 1 Sv. There's considerable uncertainty about the average dose needed to cause cancer, with the best estimates suggesting that a dose of 0.1 Sv adds a 1-in-100 chance of developing cancer (against a normal cancer rate of about 42 in 100). That dose is much higher than a typical individual would receive in even the most serious nuclear accident. And it's not clear that this 1-in-100 chance scales proportionately to low doses. On the other hand, even very low doses to large populations can result in significant numbers of cancers.

The penetrating radiation emitted in nuclear processes need not be altogether harmful. The same disruptive effects that damage living things help preserve food and eliminate food-borne diseases. Medical applications of radiation have saved countless lives through diagnosis and treatment. Radioisotopes trace the flow of material in biological and industrial processes. Naturally occurring radioisotopes act as clocks, their half-lives helping us date materials from thousands to billions of years old. And the creation of short-lived radioisotopes in materials of uncertain composition results in radiation whose characteristics reveal that composition.

Is radiation bad or good? The answer has to be that it's both, and that each potential use of radiation requires a nuclear choice that weighs the bad against the good.

Glossary

activation analysis Determination of the elemental content of materials through bombardment with radiation (usually neutrons) and subsequent analysis of radioactive decays.

brachytherapy Radiation therapy involving implantation of radioactive material at a tumor site.

dose-response curve A graph that plots the health effect of a given agent, such as radiation, versus the dose received.

epigenetic effects Effects that alter the expression of genes but don't involve damage to DNA itself. These effects may be heritable if they occur in reproductive cells.

genetic effects Effects on an organism's DNA that are passed on to future generations.

hormesis A beneficial response to very low doses of an agent that would be harmful at higher doses. The possibility of hormesis with radiation is controversial.

ionization Removal of an electron from an atom. Radiation is one possible cause of ionization.

ionizing radiation Radiation of sufficient energy to knock electrons from atoms. X-rays and nuclear radiation are ionizing; light, infrared, microwaves, and radio waves are not.

linear no-threshold (LNT) model A widely accepted but not rigorously proven model in which the chance of radiation-induced cancer is directly proportional to the radiation dose, even at the lowest doses.

mutation A change in genetic characteristics brought about by alteration of DNA, the genetic material. Radiation can cause mutations.

nuclear medicine Medical diagnostic techniques making use of radioactive isotopes.

radiation sickness Illness caused by high-level radiation exposure (over 1 Sv in humans) and characterized by nausea, vomiting, intestinal bleeding, anemia and other symptoms.

radiation therapy Use of radiation to fight cancer or other diseases.

radioactive tracer A radioactive substance used to trace the movement of material in biological or other systems.

radiocarbon dating Radioisotope dating using the isotope carbon-14; useful for dating once-living materials as much as 50,000 years old.

radioisotope dating Use of radioisotopes to date ancient materials by measuring the extent to which unstable isotopes in those materials have decayed.

repair enzymes Substances in living cells that repair damage to DNA, possibly reducing the effects of low-level radiation.

somatic effects Effects on an individual organism that aren't passed on to future generations.

Notes

1. Susan Lindee, "Survivors and Scientists: Hiroshima, Fukushima, and the Radiation Effects Research Foundation, 1975–2014," *Social Studies of Science* 46, no. 2 (2016): 184–209.

2. For an up-to-date summary of these findings, see Bertrand Jordan, "The Hiroshima/Nagasaki Survivor Studies: Discrepancies Between Results and General Perception," *Genetics* 203 (August 2016): 1505–1512.

3. See Matt Merrifield and Olga Kovalchuk, "Epigenetics in Radiation Biology: A New Research Frontier," *Frontiers in Genetics* 4 (April 2013).

4. Recall that grays and sieverts are essentially interchangeable for most radiation that we're discussing, so this figure is essentially the same as 2 Sv.

5. J. Ten Hoeve and M. Jacobson, "Worldwide Health Effects of the Fukushima Daiichi Nuclear Accident," *Energy and Environmental Science* (September 2012). DOI:10.1039/c2ee22019a.

6. Klervi Leuraud et al., "Ionising Radiation and Risk of Death from Leukaemia and Lymphoma in Radiation-monitored Workers (INWORKS): An International Cohort Study," *The Lancet*, 2 (July 2015), https://www.thelancet.com/action/showPdf?pii=S2352-3026%2815%2900094-0.

7. These numbers assume that a dose of 20 sV (2,000 rems), spread among a population, will cause one fatal cancer. The antinuclear number is based on individual doses of 0.05 mSv (5 millirems) per year (somewhat more than the typical dose to persons living near a nuclear power plant) to the entire population; the pronuclear figure assumes an average dose of 0.002 mSv (0.2 mrem) per year.

Further Reading

Centers for Disease Control and Prevention. "Radiation in Your Life." An authoritative compilation of information on radiation, including artificial and natural exposures. Available at https://www.cdc.gov/nceh/radiation/sources.html, with links to detailed radiation topics.

Gale, Robert P., and Eric Lax. *Radiation: What It Is, What You Need to Know*. New York: Knopf, 2013. A thorough and rational look at radiation—both the good and the bad—for the general public.

Marra, John F. *Hot Carbon: Carbon-14 and a Revolution in Science*. New York: Columbia University Press, 2019. An environmental scientist details the discovery of carbon-14 and myriad uses for this remarkable isotope.

National Academy of Sciences. *Health Risks from Exposure to Low Levels of Ionizing Radiation* (BEIR VII, phase 2). Washington, DC: National Academies Press, 2006. This latest official study is thorough, detailed, and mathematical.

World Nuclear Association. "Radioisotopes in Medicine." Available at http://www.world-nuclear.org/information-library/non-power-nuclear-applications/radioisotopes-research/radioisotopes-in-medicine.aspx. A brief but thorough look at medical applications of radioisotopes, including a useful section on worldwide production of crucial medical isotopes.

5

Energy from the Nucleus

We've just spent two chapters considering radiation, which results when unstable nuclei emit energetic particles. Although radiation is energetic, far larger amounts of energy come from wholesale rearrangement of the nucleus. It's this larger energy that powers nuclear reactors and nuclear weapons. This chapter explores two fundamental approaches to nuclear energy.

The Curve of Binding Energy

Chapter 2 stressed the *nuclear difference*—the millionfold increase in energy from chemical to nuclear reactions. You saw how this huge increase is associated with the strength of the nuclear force that binds protons and neutrons into the atomic nucleus. Now we ask a more detailed question: just how tightly bound are the nucleons in a given nucleus?

A more precise question is this: how much energy is needed to remove one nucleon from the nucleus? This quantity is the **binding energy**. Binding energy is important because it tells not only how much energy is *needed* to take a nucleus apart but also how much energy is *released* when the nucleus forms. Here's a gravitational analogy: Lifting a bowling ball takes energy, because gravity binds the ball to Earth. Drop the ball, and you get that energy back—in the form of sound, slight heating of the floor, and perhaps permanent damage to the ball. In nuclear terms, binding energy is therefore a measure of the energy released when a nucleus forms; to be precise, it's the energy released per nucleon. Ultimately, binding energy tells how much energy we can get out of nuclear reactions.

You saw in chapter 2 that the makeup and the stability of a nucleus are determined by the interplay between the strong but short-range nuclear force and the weak but long-range electrical force. For small nuclei, the nucleons are close together, and the short-range nuclear force dominates.

On the other hand, there aren't many nucleons in a small nucleus, so each is bound to only a few others. As the size of the nucleus increases, each nucleon at first feels the attraction of additional nearby nucleons. The nucleus becomes more tightly bound, so the binding energy increases. For larger nuclei, however, most nucleons are so far apart that the nuclear force between them is insignificant. And if they're protons, those nucleons still experience the long-range electrical repulsion. The result is that the binding energy grows less rapidly with nuclear size and eventually begins to decrease—with the result that the largest nuclei are less tightly bound than those of intermediate size.

Figure 5.1 shows the **curve of binding energy**, a graph of binding energy per nucleon versus nuclear mass number. The curve shows that for light nuclei the binding energy per nucleon increases very rapidly with increasing mass. The curve exhibits a broad peak near the common iron isotope iron-56, then begins a gradual decline toward the heaviest naturally occurring isotope, uranium-238.

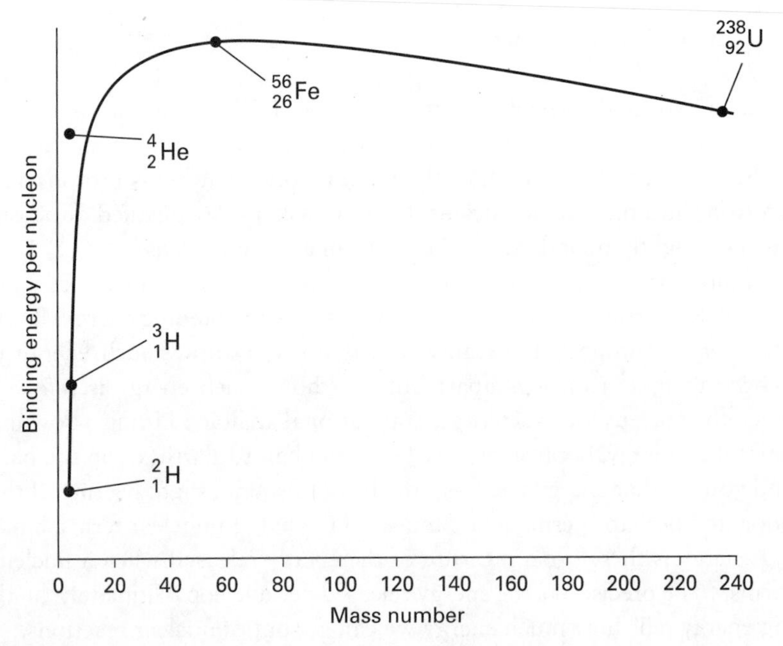

Figure 5.1
The curve of binding energy. The graph plots binding energy per nucleon versus nuclear mass number. Near the top of the curve is iron-56 ($^{56}_{26}Fe$). Several other common isotopes are also labeled. Other nuclei lie near, but not necessarily right on, the solid curve.

Nuclear Fusion

The initial increase in the curve of binding energy means that combining two light nuclei results in a more tightly bound nucleus, and therefore releases energy. Helium-4 ($^{4}_{2}He$), for example, is much higher up the binding-energy curve than deuterium ($^{2}_{1}H$). Combining two deuterons to make a single helium-4 would therefore release a substantial amount of nuclear energy. This process of combining light nuclei is nuclear **fusion.** Figure 5.2 shows two important fusion reactions. The energy released in fusion is so huge that fusion of even the minuscule amount of deuterium in ordinary water would make a gallon of water the energy equivalent of nearly 400 gallons of gasoline!

Nuclear fusion is vital to our existence: it makes the Sun shine and therefore sustains life on Earth. And we ourselves are made from the products of fusion. Carbon, oxygen, nitrogen, and other elements necessary for life were formed through nuclear fusion in the cores of ancient stars. Some of those stars underwent supernova explosions, spewing their material into interstellar space. Much later, Sun, Earth, and other constituents of our solar system formed from this cosmic debris.

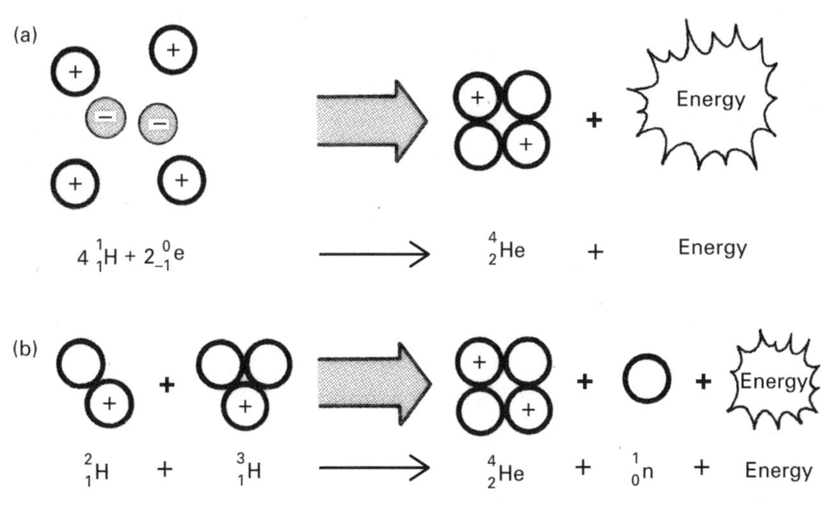

Figure 5.2
Important fusion reactions. (a) In the Sun, a sequence of reactions results in the conversion of four protons and two electrons into a helium-4 nucleus. Energy released in this reaction makes the Sun shine. (b) In thermonuclear weapons—"hydrogen bombs"—deuterium ($^{2}_{1}H$) and tritium ($^{3}_{1}H$) fuse to yield helium-4 and a neutron. This reaction may someday find peaceful use in electric power plants.

Except for hydrogen, helium, and traces of lithium, which formed shortly after the Big Bang, essentially all elements up to iron were "cooked" by fusion in the interiors of stars. Just before they go supernova, the most massive stars have iron cores, surrounded by concentric layers of ever-lighter elements. Fusion in each layer gave rise to heavier nuclei in the next layer inward.

Iron is at the top of the binding energy curve, so fusion that produces nuclei heavier than iron won't release energy—in fact, it requires energy. So where did heavier elements come from? Astrophysicists used to think they formed only in the cataclysmic supernova explosions themselves, when enough excess energy was available to make these less tightly bound nuclei. But the 2017 discovery of gravitational waves from a collision of two neutron stars showed that some heavy elements are formed in neutron-star collisions. Whether we're talking about light elements forged by fusion inside stars or heavy elements forged in supernovae and neutron-star collisions, we and virtually everything around us are products of the stars, with our constituent elements formed in nuclear reactions that occurred billions of years ago.

It's uplifting to think of ourselves as "star children," tracing our ultimate origins to fusion events inside ancient suns. But fusion also presents a sobering prospect: fusion energy released in our thermonuclear weapons has the potential to destroy us. Although fusion powers the stars, we humans have not yet learned to harness it for any purpose but destruction. We'll devote all of chapter 11 to a more detailed discussion of fusion and our attempts to harness it for energy production.

Nuclear Fission

You've seen how nuclear fusion releases energy by moving nuclei up the curve of binding energy, fusing lighter nuclei into heavier, more tightly bound ones. Now look again at the curve of binding energy, repeated in figure 5.3. You can see that it's also possible to release energy by moving up the curve from the heaviest nuclei to the more tightly bound, lighter nuclei in the vicinity of iron. That's done by splitting heavier nuclei into lighter pieces—the process of nuclear **fission**.

Some of the heaviest, least stable nuclei undergo **spontaneous fission**, in which a heavy nucleus spontaneously splits into two lighter pieces and a few neutrons. In naturally occurring isotopes, the half-lives for spontaneous fission are so long (often greater than the age of the universe) that

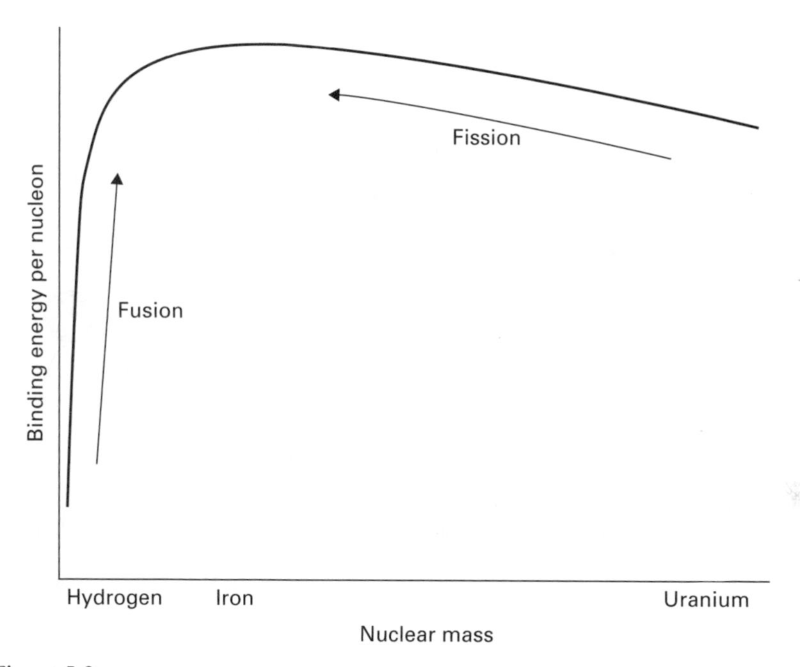

Figure 5.3
The curve of binding energy, showing energy release by fusion of light nuclei or fission of heavy nuclei.

the process is very infrequent and the rate of energy release therefore insignificant.

Of greater importance is **neutron-induced fission**, in which a heavy nucleus splits after being struck by a neutron. Isotopes whose nuclei can undergo neutron-induced fission are termed **fissionable**. Figure 5.4 shows neutron-induced fission of a uranium-235 nucleus. The U-235 nucleus absorbs the neutron and becomes a highly excited nucleus of uranium-236. The U-236 undergoes violent oscillations that distort it into a dumbbell shape. With this elongated shape, the short-range nuclear force diminishes, and the electric repulsion is then strong enough to drive the two pieces apart. The entire process takes only about a trillionth of a second.

Strictly speaking, all nuclei are fissionable, including the **uranium** isotopes U-235 and U-238. But most fissionable isotopes, including U-238, require that the incoming neutron have considerable energy. However, for uranium-233, uranium-235, and **plutonium**-239 (and a few other less significant isotopes), fission occurs with neutrons of arbitrarily low energy. These are **fissile** isotopes. Of the three listed, only U-235 occurs naturally,

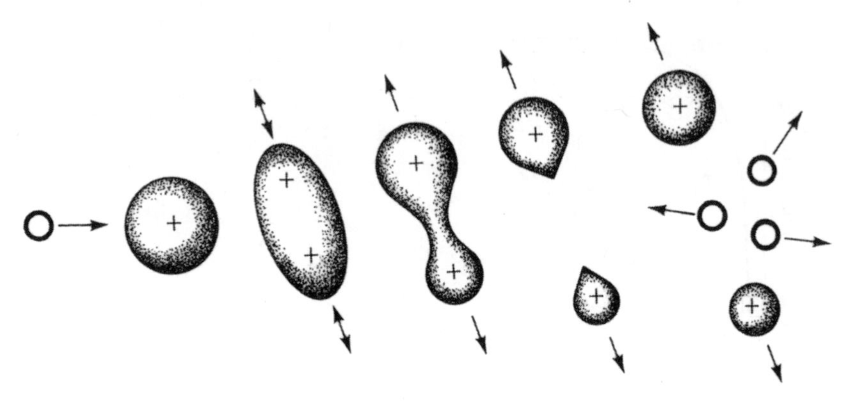

Figure 5.4
Neutron-induced fission of uranium-235. At left, a neutron strikes the U-235, forming U-236. The U-236 nucleus is highly excited and oscillates violently, assuming a dumbbell shape. Electrical repulsion drives the two ends of the dumbbell apart, resulting in two smaller nuclei. Several neutrons are released in the process (three neutrons are shown here).

and it constitutes less than 1 percent of natural uranium. You'll soon see how U-233 and Pu-239 are produced in nuclear reactors, and why Pu-239 is a particularly significant nuclear material.

The Nuclear Chain Reaction

Neutrons induce fission, and fission itself releases neutrons. Those neutrons may go on to induce more fission reactions, giving rise to still more neutrons and more fission. The result is a self-sustaining **chain reaction** (figure 5.6).

A self-sustaining chain reaction requires that, on average, at least one neutron from each fission event causes another fission. A chain reaction in which each fission event results, on average, in exactly one other fission event is said to be critical. A **critical chain reaction** releases energy at a steady rate, and is therefore desirable in a nuclear reactor. If the number of fissions from each event is less than one, it is a **subcritical chain reaction** and soon fizzles out—much as a disease epidemic fades away if each infected person infects, on average, less than one other person. If, on the other hand, each fission event triggers more than one additional fission, then the reaction is a **supercritical chain reaction**. Energy release grows exponentially and may lead quickly to a nuclear explosion—as in a fission bomb. Supercriticality in a nuclear reactor can be disastrous.

A typical fission reaction releases two or three neutrons, so it might seem easy to produce a critical or supercritical chain reaction. But imagine a mass of uranium, comprising a mix of U-235 and U-238, and consider what might happen to a neutron:

- It might leave the mass altogether, causing no further fission.
- It might be absorbed by a U-238 nucleus. Since the neutron energy required for fission of U-238 is very high, fission is an unlikely result. (However, you'll soon see that neutron absorption in U-238 is significant for another reason.)
- It might be absorbed by a U-235 nucleus, resulting in fission and the production of two or three additional neutrons.

Other outcomes are possible. For example, neutron absorption in U-235 doesn't always result in fission. The presence of other materials in or around the uranium mass may cause more neutron absorption without

Discovery of Fission

The discovery of nuclear fission was laced with international intrigue and the approaching cloud of World War II. First hints came in 1934, when the Italian physicist Enrico Fermi bombarded uranium with neutrons, producing new radioisotopes. When she heard of Fermi's results, the German chemist Ida Noddack speculated that "in the bombardment of heavy nuclei with neutrons, these nuclei break up into several large fragments." Noddack's was the first known suggestion of nuclear fission, but nuclear scientists didn't follow up on her suggestion.

In 1938, the German chemists Otto Hahn and Fritz Strassman were astonished to find an isotope of barium in uranium that had been bombarded with neutrons. They had expected to find elements heavier than uranium, and indeed they did. But the lighter barium was totally unexpected. Hahn communicated this quandary to his former colleague, the Austrian physicist Lise Meitner (figure 5.5), who by then had fled Nazi Germany for Sweden.

Meitner discussed Hahn and Strassman's result on a walk in the Swedish countryside with her nephew Otto Frisch in December 1938. They sketched a diagram like our figure 5.4, suggesting that uranium had fissioned to produce the barium that puzzled Hahn and Strassman. Meitner and Frisch calculated the enormous energy that would be released in the process. They published their work in a paper entitled "Disintegration of Uranium by Neutrons: A New Type of Nuclear Reaction," which included the first use of *fission* in the nuclear context.

(*continued*)

Discovery of Fission (*continued*)

Figure 5.5
Lise Meitner and Otto Hahn were colleagues for some 30 years in Berlin, until Meitner fled Nazi Germany in 1938. As a woman in a male-dominated field, Meitner had braved sexist policies that denied her access to laboratories when men were present. By the 1930s she had become one of the world's most respected nuclear physicists. Although Meitner is arguably the most direct discoverer of nuclear fission, she wasn't even mentioned when Hahn received the 1944 Nobel Prize for Chemistry for that discovery. She was honored in 1997 when element 109, meitnerium, was named in her honor. Inset shows the periodic table symbol for meitnerium, indicating atomic number 109 and mass number of its most stable isotope, Mt-278. (New York Public Library/Science Source)

Word of fission and its potential for enormous energy release spread rapidly through the international physics community and to the public. The military implications were obvious and ominous, and in April 1939 the *New York Times* headlined "Vision Earth Rocked by Isotope Blast; Scientists Say Bit of Uranium Could Wreck New York." Just six years later came the detonation of the first nuclear explosive in the New Mexico desert, followed by the nuclear devastation of Hiroshima and Nagasaki.

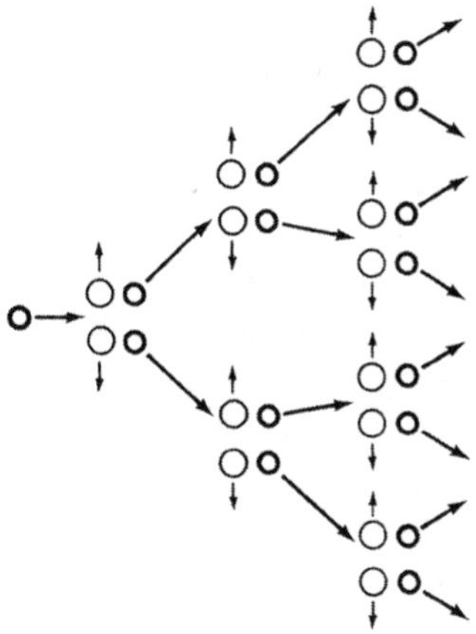

Figure 5.6
A fission chain reaction. In this case the reaction is supercritical, since each fission event results in two others. The fission rate grows exponentially, resulting in explosive energy release.

fission. But considering only the three outcomes listed will help you understand the challenge of sustaining a fission chain reaction.

For a sustained reaction, one of those two or three neutrons must be absorbed by U-235, inducing another fission event and giving rise to more neutrons. But this won't happen if too many neutrons leave the mass or are absorbed in U-238. How to minimize these losses? U-238 absorption can be reduced by **enrichment**—that is, by increasing the proportion of fissile U-235 relative to nonfissile U-238. Later you'll see how enrichment is accomplished; for now, note that enrichment to 3 to 5 percent U-235 is typical in nuclear power reactors, with much higher enrichment in some research reactors and in weapons. Enrichment technology is extremely sensitive, since a nation possessing it can produce nuclear weapons material from relatively abundant natural uranium.

Enrichment reduces losses by absorption in U-238, but it doesn't prevent neutrons from leaving the uranium mass. The easiest way to reduce that loss is to make the mass larger, increasing the likelihood that a neutron hits a U-235 nucleus and produces fission rather than escaping. Eventually a size is reached at which, on average, exactly one of the neutrons from each fission event causes another fission. At that point, the uranium constitutes a **critical mass**: it has the minimum size needed to sustain a chain reaction. A larger mass is supercritical; in it, neutrons and fission reactions multiply rapidly, resulting in catastrophic energy release—a nuclear explosion.

How big is a critical mass? That depends on the proportion of fissile material and on its shape and configuration. Surrounding the mass with a neutron-reflecting substance further reduces the size needed for criticality, by returning neutrons that would otherwise escape. So how big is a critical mass? It's alarmingly small—as the historical photo in figure 5.7 makes dramatically clear. For pure uranium-235, somewhere around 30 pounds could constitute a critical mass. For plutonium-239 the critical mass is even smaller—roughly 10 pounds, about the size of a tennis ball. These numbers are for weapons configurations, which we'll discuss in chapter 12. The critical mass in nuclear power reactors is much larger because of low enrichment, a vastly different configuration of fissile fuel, and the presence of cooling water and other materials.

Initiating a chain reaction is the easy part. Cosmic rays interacting with the atmosphere produce stray neutrons that can induce fission, as

Figure 5.7
The critical mass is alarmingly small. Here, physicist Harold Agnew carries the plutonium core of the "Fat Man" bomb that was dropped on Nagasaki in 1945. Even this crude first-generation nuclear weapon required only 14 pounds of plutonium, yet its explosive yield killed nearly 100,000 people. The plutonium itself is within the magnesium box that Agnew is holding. (Los Alamos National Laboratory)

do occasional spontaneous fission events. In a critical mass of fissile material, stray neutrons will naturally start a chain reaction. Greater reliability, especially in nuclear weapons, results from using special neutron sources to initiate the reaction. But that's not essential in a crude weapon; in fact, stray neutrons can cause the opposite problem, *preigniting* the chain reaction and blowing the device apart before fission is complete.

Fission Products

In fission, a uranium or plutonium nucleus splits into two lighter nuclei called **fission products.** The most likely outcome of a single fission reaction is two nuclei of unequal mass, one with mass number roughly in the range 85 to 105 and the other in the range 130 to 145. A common U-235 fission reaction, for example, results in molybdenum-102 and tin-131, along with three neutrons: ${}^{1}_{0}n + {}^{235}_{92}U \rightarrow {}^{102}_{42}Mo + {}^{131}_{50}Sn + 3{}^{1}_{0}n$. This reaction is significant because the tin-131 quickly beta-decays to radioactive iodine-131, a dangerous contaminant that lodges in the thyroid gland. Other important isotopes among fission products are strontium-90 (a radioisotope that mimics calcium and is therefore incorporated into bone) and cesium-137.

Fission products are highly radioactive. Figure 5.8 shows why. As you learned in chapter 2, the heaviest nuclei have more neutrons than protons, in order to overcome the electrical repulsion of protons with the nuclear force exerted by neutrons. When a heavy nucleus fissions, the fission products preserve the same ratio of neutrons to protons. But, as figure 5.8 shows, that leaves them with excess neutrons, which renders them unstable. Some of those neutrons "boil off" almost instantaneously, giving the two or three neutrons produced in each fission event. But the remaining isotopes are still rich in neutrons. They decay through a sequence of beta emissions until they achieve stability. So the fission products are inherently radioactive. Most of their half-lives range from less than a second to tens or hundreds of years.[1] Those times are much less than the roughly billion-year half-lives of uranium isotopes, so the fission products are much more intensely radioactive than the original uranium. During the refueling of a nuclear reactor, for example, fresh uranium fuel doesn't require elaborate radiation protection. But spent fuel, rich in fission products, is kept underwater to shield workers from its intense radiation.

The intense radioactivity of fission products is what makes nuclear waste a challenge and the fallout from nuclear weapons a lethal hazard.

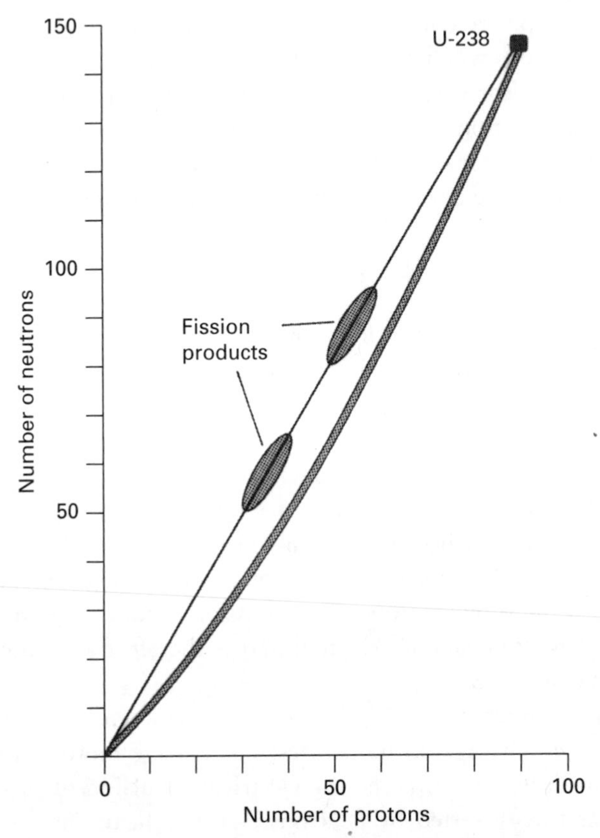

Figure 5.8
A chart of the nuclei, simplified from figure 2.11. The thick curve is the region of stability, extended to include long-lived isotopes through uranium. The thin straight line is the neutron-to-proton ratio for uranium and thus for the products of uranium fission. Gray bands mark the most probable fission products, which necessarily lie above the curve of stable isotopes and are therefore radioactive.

In a reactor, every attempt is made to contain fission products, but nuclear weapons disperse their fission products directly into the environment. Those who were young children in the mid-1950s to the early 1960s had their bones form while the United States and the Soviet Union were engaged in the atmospheric nuclear weapons tests. As a result, they have higher than normal levels of strontium-90 in their bones and can be expected to develop bone cancer at a somewhat higher than average rate.

Middle-weight fission products aren't the only radioactive isotopes formed in uranium fission. Neutrons absorbed by uranium-238—the heaviest naturally occurring isotope—give rise to still heavier **transuranic** nuclei (meaning *beyond uranium*). Many transuranics have half-lives which, although short enough that they're no longer present in nature, are long compared with those of fission products. The most important transuranic isotope is plutonium-239. Formation of Pu-239 begins when U-238 absorbs a neutron, producing unstable uranium-239. U-239 beta-decays with a 24-minute half-life, forming neptunium-239. Np-239 again beta decays, now with a 2.4-day half-life, to form plutonium-239 (figure 5.9).

With its half-life of some 24,000 years, Pu-239 stays around for a long time, yet its radioactivity is still considerable. Most important, Pu-239 is more virulently fissile than U-235, with a smaller critical mass. So neutron absorption by U-238, although antithetical to the fission chain reaction, ultimately results in a more fissile product. This **breeding** of of plutonium in specially designed reactors produces fissile Pu-239 for weapons. Using power reactors designed for breeding could extend the world's uranium

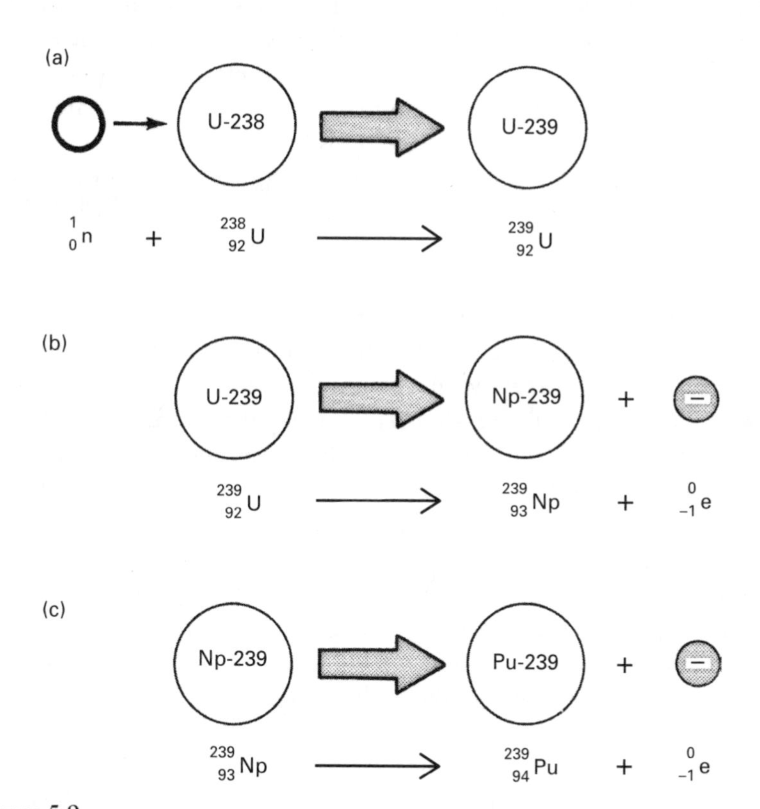

Figure 5.9
Formation of plutonium-239 begins when uranium-238 absorbs a neutron. Two subsequent beta decays yield neptunium-239 and then plutonium-239.

supplies by converting some of the 99.3 percent of natural uranium that's nonfissile U-238 into fissile Pu-239. Even ordinary reactors get some of their energy from fission of Pu-239 that was bred in the reactor.

$E=mc^2$ and All That

So far we haven't mentioned Einstein and his famous formula $E=mc^2$, which for most people is inextricably linked with nuclear energy. $E=mc^2$ does apply to nuclear energy, and in her 1939 paper on fission Lise Meitner used the formula to verify the fission energy released. But it's a common misconception to think that $E=mc^2$ is exclusively about nuclear physics, and therefore to make Einstein ultimately responsible for nuclear technology.

$E=mc^2$ expresses a remarkable unity, which Einstein was first to recognize: that matter and energy are manifestations of the same basic stuff that makes up the universe. $E=mc^2$ is the key to the interchangeability of matter and energy. It says that matter with mass m is equivalent to a quantity of energy given by the product of m times the square of the speed of light, c. Since the speed of light is huge, the energy equivalent of even a small mass is large. If we knew how to convert it entirely to energy, a single raisin could power a large city for a day. And $E=mc^2$ works both ways: a bundle of pure energy, such as a gamma ray, can suddenly turn itself into a pair of elementary particles.

This matter-energy equivalence manifests itself in nuclear reactions. If you weigh a uranium nucleus, then fission it, then weigh all the fission products, you'll find they weigh slightly less than the original uranium. The missing mass—call it m—has been converted to energy in the amount given by $E=mc^2$. Similarly, weigh two deuterons and then fuse them to make helium. The helium weighs less than the deuterons you started with, and $E=mc^2$ relates the mass difference to the energy released. Or weigh a nuclear bomb before it explodes, then collect and weigh all the resulting debris. You'll find a difference, and multiplying that by the square of the speed of light gives the energy released in the explosion. On a larger scale, the Sun, through fusion, converts 4 million tons of matter to energy every second—energy that becomes the sunlight that sustains life on Earth.

But $E=mc^2$ is a *universal* statement. It applies to *all* events that release energy. Weigh a candle and all the oxygen around it. Burn the candle and weigh the smoke and carbon dioxide that come out, along with what's left of the candle. If you could weigh with enough precision, you would find that after burning the total weight is less. Again, matter has been converted to energy. Again, $E = mc^2$ relates the mass decrease to the energy released. The difference between the candle and the bomb is that the latter converts proportionately more of its mass to energy. That's the nuclear difference, expressed in Einstein's terms. It isn't that nuclear processes convert mass to energy whereas others don't; rather, nuclear processes convert proportionately more mass—about a million times more than chemical processes. And even the nuclear processes of fission and fusion aren't all that good at converting mass to energy; in both, less than 1 percent of the mass gets converted.

So is Einstein's formula at the heart of nuclear energy? Yes, but only in the sense that it's at the heart of every energy-conversion process—burning gasoline, metabolizing food, stretching a rubber band, running a nuclear reactor. There's nothing uniquely *nuclear* about it. Was Einstein's work an essential step on the path to nuclear technology? No more than it was essential to harnessing fire or development of the steam engine.

Einstein did play a role in instigating the nuclear weapons effort of World War II. But Einstein and his famous formula are widely misunderstood; neither the man nor the equation is responsible for our nuclear dilemmas.

Summary

Nuclear energy is released when atomic nuclei rearrange to form more tightly bound structures. The curve of binding energy reveals two kinds of energy-releasing nuclear reactions. In *fusion*, light nuclei join to form a heavier nucleus. Fusion powers the stars, but the electrical repulsion of nuclei makes fusion difficult under terrestrial conditions.

Heavier nuclei release energy when they split, or *fission*. Fission occurs most readily when a neutron strikes a nucleus, and fission itself releases more neutrons. In the few so-called *fissile* isotopes—notably uranium-235 and plutonium-239—a self-sustaining fission chain reaction is possible, as neutrons from one fission event cause additional fission. A critical mass is necessary to ensure a self-sustaining reaction.

The middleweight *fission products* contain excess neutrons and are therefore highly radioactive. Absorption of neutrons by nonfissile U-238 creates other radioactive materials, the heavy transuranic elements. The most important of these is fissile plutonium-239, which, like uranium-235, can fuel nuclear reactors and weapons.

Nuclear fusion and fission convert matter to energy according to Einstein's $E = mc^2$. But nuclear processes share that equation with every other energy-releasing process, including commonplace chemical reactions. The nuclear difference lies in the millionfold increase in the proportion of matter converted to energy, not in the fact of that conversion.

You now know about the basic nuclear processes involved in radiation and energy production and are familiar with some of the materials important in those processes. In the remainder of this book you'll see applications of nuclear processes, first in energy production and then in nuclear weapons.

Glossary

binding energy The amount of energy needed to remove one nucleon from the nucleus; equivalently, the energy per nucleon released when the nucleus forms.

breeding Production of plutonium-239 (or other useful isotopes) by neutron absorption, usually in a nuclear reactor.

chain reaction A self-sustaining fission reaction in which neutrons from one fission event trigger subsequent fission.

critical chain reaction A chain reaction in which, on average, exactly one neutron from each fission event triggers another fission. A critical reaction releases nuclear energy at a steady rate.

critical mass A mass of fissile material large enough to sustain a fission chain reaction. Minimum critical masses for plutonium-239 and uranium-235 are about 10 pounds and 30 pounds, respectively.

curve of binding energy A graph showing the binding energy per nucleon versus nuclear mass. The graph peaks around iron, showing that all other nuclei are less tightly bound.

enrichment A process whereby the proportion of fissile U-235 is increased above its naturally occurring value of 0.7 percent. Enrichment to 3–5 percent is typical of fuel for power reactors. Weapons-grade uranium may be enriched to 90 percent or more.

fissile nucleus A fissionable nucleus that will undergo fission when struck by a neutron of arbitrarily low energy. The significant fissile nuclei are uranium-233, uranium-235, and plutonium-239. Of these, only U-235 occurs naturally, and it constitutes just 0.7 percent of natural uranium.

fission Splitting of a heavy nucleus to form two medium-weight nuclei, accompanied by the release of energy.

fission product Any medium-weight nucleus formed during fission of a heavy nucleus. Fission products are inherently radioactive.

fissionable nucleus A nucleus that can undergo fission when struck by a neutron. Most fissionable nuclei require a certain minimum neutron energy to cause fission.

fusion Combining of two light nuclei to form a heavier nucleus, accompanied by the release of energy.

neutron-induced fission Fission that occurs when a nucleus is struck by a neutron.

plutonium (Pu) A transuranic element usually formed by neutron absorption in uranium-238. The isotope plutonium-239 is highly fissile and is widely used in nuclear weapons.

spontaneous fission Fission that occurs spontaneously, without any initiating event. Spontaneous fission is infrequent in naturally-occurring isotopes.

subcritical chain reaction A chain reaction in which, on average, fewer than one neutron per fission event triggers another fission. A subcritical reaction soon fizzles to a halt.

supercritical chain reaction A chain reaction in which, on average, more than one neutron per fission trigger additional fission. Energy release in a supercritical reaction grows exponentially and can be explosive.

transuranic Any element heavier than uranium. The half-lives of these elements are short enough that they exist only when produced artificially, often by neutron absorption in fission chain reactions.

uranium (U) The heaviest naturally occurring element. Its two most common isotopes, U-235 (0.7 percent of natural uranium) and U-238, (99.3 percent) are fissionable, but only U-235 is fissile.

Note

1. There are a handful of very long-lived fission products, of which iodine-129 has the longest half-life, nearly 16 million years. But precisely because they're so long-lived, they don't contribute much radioactivity.

Further Reading

Garwin, Richard, and Georges Charpak. *Megawatts and Megatons: The Future of Nuclear Power and Nuclear Weapons*. Chicago: University of Chicago Press, 2002. Two eminent physicists, one of them a Nobel laureate, discuss nuclear power and nuclear weapons at about the level of *Nuclear Choices* but with a bit more math.

Glasstone, Samuel. *Sourcebook on Atomic Energy*, 3rd ed. Huntington, NY: Krieger, 1979. Originally published under the auspices of the former U.S. Atomic Energy Commission, this classic is a comprehensive compendium of basic nuclear knowledge. Old, but still relevant.

Murray, Raymond, and Keith Holbert. *Nuclear Energy: An Introduction to the Concepts, Systems, and Applications of Nuclear Processes*, 8th ed. Cambridge, MA: Butterworth-Heinmann, 2020. A textbook for undergraduate engineering majors, this one covers our material much more quantitatively and in much greater depth. Chapters 2, 6, and 7 are especially relevant to this chapter. Not for the mathematically challenged!

II

Nuclear Power

6

Energy: Using It, Making It

What *is* energy? A physicist might write an equation or mumble something about "ability to do work." But in essence, energy is what makes everything happen. All motion entails energy, whether it's the motion of a planet, a car, or a neutron. The random dance of molecules in matter involves the energy we associate with the term *heat*.[1] Take away energy, and everything would stop. Nothing would happen.

Motion isn't energy's only manifestation. Stretch a rubber band and you've stored energy. A gasoline molecule is a miniature version of the stretched rubber band, its energy stored via electric forces between its atoms. So is a uranium nucleus, its enormous stored energy associated with the interplay of electric and nuclear forces.

Other forms of energy include light and its cousins: radio waves, microwaves, infrared and ultraviolet rays, X-rays, and gamma rays. These result from atomic, nuclear, and other processes. In turn, they can transform into other forms of energy. In a microwave oven, microwave energy jostles water molecules in food, producing heat. Plants capture sunlight and store it as chemical energy. In cancer therapy, radiation bombards a tumor, its energy destroying molecules within cancer cells.

Energy is a fundamental constituent of our universe. It takes many forms, and it readily changes from one form to another. Energy is measurable, quantifiable stuff with the potential to make things happen. We humans, especially those of us in industrialized societies, use lots of energy to make lots of things happen. This chapter explores our enormous energy appetite and where we get our energy.

Energy and Power

How much energy do we use? The answer depends on whether you're asking about a year, an hour, or a second of energy use. More meaningful is this question: at what rate do we use energy? That rate—whether it's describing energy use or energy production—is called **power**.

You're probably familiar with two units for power: the watt and the horsepower. To say that you have a 1,000-watt hair dryer is to give the rate at which the dryer uses energy. To say that a car has a 200-horsepower engine is to give the rate at which the engine can produce energy of motion from energy stored in gasoline. In this book, we'll use the **watt** (W), the **kilowatt** (kW, equal to 1,000 watts), and the **megawatt** (MW, equal to 1,000,000 W or 1,000 kW) as units of power. A typical desk lamp might have a 60-watt incandescent lightbulb or a 9-W LED. An electric stove burner might use 2,000 watts, or 2 kW. That's 20 times the energy-consumption rate of a 100-watt (0.1-kW) lightbulb. One horsepower is 746 watts, or roughly 0.75 kilowatt, so a 100-horsepower car expends energy at about the same rate as 75 hair dryers or 750 100-watt lightbulbs.

You probably think of watts in the context of electrical devices, but this unit of power applies to anything that supplies or uses energy. The Sun, for example, supplies energy at the truly stupendous rate of 380,000,000,000,000,000,000,000,000 watts. At noon, solar energy reaches Earth's surface at the rate of about 1 kilowatt for each square meter (just over a square yard). Gasoline, burned at the rate of a gallon an hour, releases energy at the rate of about 40 kilowatts. The electrical power output of a typical large power plant is 1 billion watts (a gigawatt; GW), or 1,000 megawatts. Table 6.1 gives other examples of power.

Power is the *rate* of energy use or supply. If you burn a 100-watt lamp for two hours, you use twice the energy you would use in one hour. Energy is therefore the product *power* × *time*. If the power is in kilowatts and the time in hours, then energy is in **kilowatt-hours** (kWh).[2] Example: Run a 2-kW stove burner for one hour, and you use 2 kWh of energy. Run it for two hours, and you use 4 kWh. Keep it on a full 24-hour day, and you use 48 kWh. At a typical price of 10¢ per kilowatt-hour of electrical energy, that would cost you $4.80. A 1,000-MW power plant puts out 1,000 megawatt-hours (MWh), or 1 million kWh, of electrical energy each hour, for a total of 24 million kWh each a day.

You can go the other way too. A gallon of gasoline contains about 40 kWh of energy. Suppose you burn that gallon in two hours. Then you use energy at the rate of 40 kWh per two hours, or 20 kWh per hour. A kilowatt-hour per hour is simply a kilowatt (mathematically, the hours cancel in kW × h / h)—a unit expressing *rate* of energy use. So you're using gasoline energy at the rate of 20 kW.

Table 6.1
Rates of energy use: power

Energy user/supplier	Power
Cellphone (sleep mode)	0.0001 kW (0.1 W)
Cellphone (while talking)	0.006 kW (6 W)
Desk lamp (LED)	0.009 kW (9 W)
Laptop computer	0.06 kW (60 W)
TV, 55″ OLED	0.36 kW (360 W)
Sunlight on 1 square meter	1 kW
Stove burner	2 kW
Home heating, winter day	10 kW
Hot shower	25 kW
Car (60 miles/hour, 30 miles/gallon)	80 kW
Large power plant	1,000,000 kW (1,000 MW)
World energy consumption	20 billion kW (20 terawatts; TW)

Table 6.2
Energy and power: an analogy

	Distance or speed	Energy or power
Quantity & unit	distance (miles)	energy (kilowatt-hours; kWh)
Rate & unit	speed (miles per hour; mph)	power (kilowatts; kW)
Relationship	distance = speed × time	energy = power × time

Energy—measured in kilowatt-hours—is the "stuff," and power—measured in kilowatts—is the rate at which that "stuff" is used or supplied. The relation between energy and power is like the relation between distance and speed. Distance—measured in miles—is the "stuff" a car traverses; speed—measured in miles per hour—is the rate at which the car traverses that "stuff." Table 6.2 illustrates this analogy.

Energy and power are often confused. That stems in part from the fact that our unit of rate—power—is the single word *kilowatt*, while the unit of "stuff"—energy—is the compound word *kilowatt-hour*. In contrast, speed—a rate—has the compound unit *miles per hour*, whereas distance is simply *miles*. We'll almost always be talking about power, since questions about actual amounts of energy are usually meaningless unless a time is specified.

A common response to "My TV uses 360 watts" or "This power plant produces 1,000 megawatts" is to ask "Is that 360 watts in an hour?" or "Is that 1,000 MW each day?" Those responses reflect confusion between energy and power: 360 W and 1,000 MW are already *rates*. They don't need an additional time qualifier. Avoid such confusion! Power (watts, kilowatts, megawatts) always refers to a *rate* and energy (watt-hours, kilowatt-hours, megawatt-hours, megawatt-years, or whatever) to the actual "stuff" being used or supplied.

Energy, rather than power, is the useful quantity when we ask about the energy stored in fuels or the energy released in a single event such as the formation of a carbon dioxide molecule, the fission of a uranium nucleus, or the explosion of a bomb. How much energy is in a gallon of gasoline? About 40 kWh. It makes no sense to ask how much power is in that gallon. I could burn it in my car's engine for one hour, releasing energy at the rate of 40 kW, or I could light it with a match and see the whole 40 kWh released in one second—an enormous but brief burst of power.[3] Table 6.3 gives the energy contents of some important substances. Note how the nuclear difference, introduced in chapter 2, manifests itself in the huge disparity between the energy contents of chemical and nuclear fuels.

Your Energy Servants

Developed countries are prodigious energy consumers. What, exactly, does "prodigious" mean? Instead of giving an abstract number of kilowatts, let's put the question this way: Suppose our energy were supplied not by coal, oil, gas, uranium, and so forth, but by human servants working to generate the energy you use. How many such *energy servants* would you need?

For the first humans, the answer would have been "none." But as soon as people harnessed fire and domesticated animals, they began using more energy than their own bodies could supply. The number of equivalent energy servants began to rise, and it's risen more or less steadily ever since.

What's the rate at which an energy servant—that is, a human body—can produce energy? Figure 6.1 shows a person turning a hand-cranked generator; she's just able to keep a 100-W lightbulb burning. Or do deep knee bends at the rate of one each second; that, too, is equivalent to about 100 W.[4] Or figure that your average food intake amounts to some 2,000 calories per day and convert that to an energy intake rate in watts; again it's about 100 W.[5] These examples tell you what a watt means, not in abstract physics terms but in terms of what you feel in your own body and its muscles.

You can guess at the number of your energy servants by thinking about your own life. Whenever you have a 100-watt lightbulb on, that's one servant. Your car may burn an average of one gallon of gasoline each day; that's 40 kWh over 24 hours, an average of 1.7 kW, or 17 more servants. If it's winter, your heating system might consume energy at the rate of 10 kW. If there are four people in your family, you account for 2.5 kW, or another 25 servants. You cook dinner, picking up a few more. Occasionally you travel by plane; include that energy use too. And what about the energy it took to make the plane, and your car, and to process your food, and to run the refrigerator that keeps it fresh? And the gasoline to run the tractor to plow the field that grew the grain that fed the cow that got ground into the hamburger you cooked? Then there's the store where you shopped: banks of bright lights, freezer cases, trucks delivering food. All these use energy, some of it in your name.

You could get the average number of energy servants for a country by adding up all the energy used in the country each year, measured in kilowatt-hours, and dividing by the number of hours in a year. The result is the total rate of energy consumption in kilowatts. Divide by the population, and you have the per-capita energy consumption rate. Taking each servant as 100 W or 0.1 kW, you can then find the number of energy servants.

So how many servants? The answer, for the average world citizen of the twenty-first century, is approximately 20 (see figure 6.2). For Europe it's some 50 servants, and for North America it's more like 100. If the energy used in our names were supplied by human servants, each North American would have 100 servants working round the clock, for an average power of 10,000 W or 10 kW. In other words, North Americans use energy at 100 times the rate our own bodies can supply it. That's what it means to live in a high-energy society.

Your Servants' Work

What do your energy servants do for you? Figure 6.3 shows how energy use divides among four broad areas: transportation, residential, commercial, and industrial.[6] Worldwide (figure 6.3a), more than half goes to industry and a quarter to transportation. In the United States, transportation is the dominant use, with industry a close second. Because U.S. residents use an average of 10 kW or 100 energy servants, you can think of the percentages in figure 6.3b as also being numbers of energy servants. Thus, if you're in the U.S., 38 of your servants are busy moving

Table 6.3
Energy contents of fuels and other substances

Substance	Energy content
coal	7,300 kWh/ton
oil, gasoline	40 kWh/gallon
natural gas	30 kWh/100 cubic feet
sugar	0.02 kWh/teaspoon
biomass, dry (wood, garbage, etc.)	5,000 kWh/ton
uranium (fission)	
natural abundance	160 million kWh/ton
	5,000 kWh/ounce
pure U-235	20 billion kWh/ton
	700,000 kWh/ounce
deuterium (fusion)	
natural water	13,000 kWh/gallon
pure deuterium	3.6 million kWh/ounce
Hiroshima bomb	15 million kWh

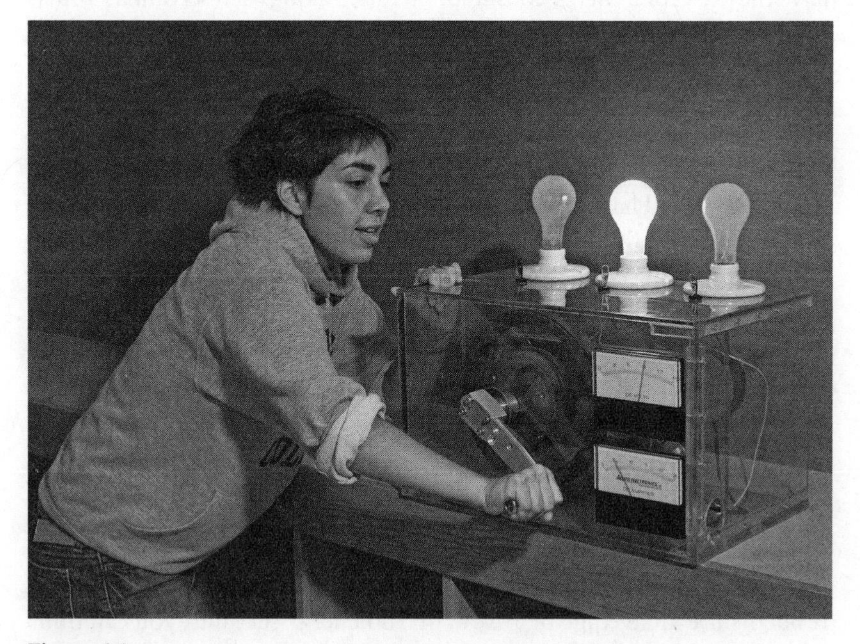

Figure 6.1
The power output of the average human body is about 100 watts. Here, using a hand-cranked generator, that power ends up lighting a 100-watt lightbulb. (Tad Merrick Photography)

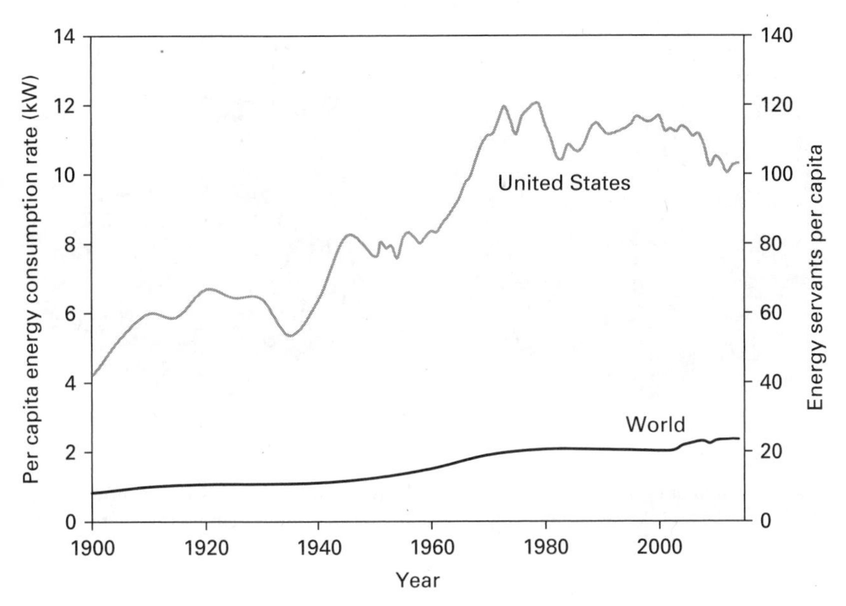

Figure 6.2
Per-capita energy consumption since 1900, given both in kilowatts (kW) and in equivalent energy servants of 100 watts each. For the United States, the Great Depression shows as the obvious dip in the 1930s, hinting at a relation between energy consumption and economic prosperity. The decline in the 1980s is a response to the oil crises of the 1970s, while recent declines reflect modest gains in energy efficiency.

you and your goods around. Of those 38 transportation servants, over half drive your private car; the rest are at work in public transportation or in moving goods that are ultimately for your use. Half of your residential and commercial energy servants work to heat and cool buildings and provide hot water; the rest run refrigerators, other appliances, lights, and electronics. Your 36 industrial servants are engaged in a variety of activities. More than three-fourths refine petroleum, synthesize chemicals, smelt metals, and make paper. The rest process food and produce the vast array of goods available in our industrial society.

Electrical Energy

Some 17 percent of the energy delivered to users worldwide is in the form of electricity, and the figure for the United States is nearly identical. But inefficiencies in electric power generation, which we'll explore shortly, mean that nearly 40 percent of our total energy consumption is associated with electricity generation and supply. Those percentages are

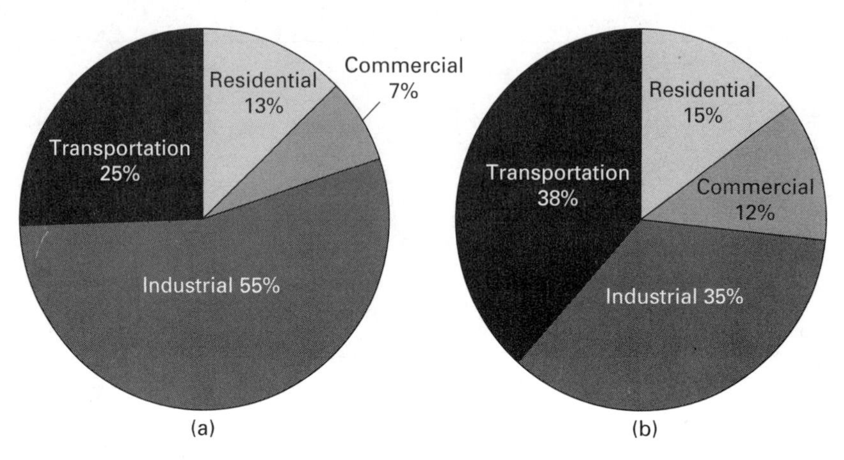

Figure 6.3
Energy consumption by sector: (a) world; (b) United States. For the U.S., the entire pie represents the 10-kW per-capita energy consumption rate, equivalent to 100 energy servants.

expected to rise rapidly as the world embraces new sources of electricity like wind and solar, and new uses for electricity, like electric vehicles and heating/cooling using heat-pump technology. We're particularly interested in electrical energy in this book because electricity generation is by far the dominant use of energy from nuclear fission.

Of the first electric power plants, built before 1900, about half were fossil fueled (coal, oil, gas) and half hydroelectric (water powered). The abundance of cheap fossil fuels, and the convenience of siting fossil-fueled plants, soon made fossil fuels the dominant source of electrical energy in most countries. Worldwide, coal retains the greatest share of electric power generation, although in the United States natural gas surpassed coal in 2014. Nuclear fission came on the commercial scene in the 1950s, and its share of world electricity generation rose through the mid 1990s, then declined to its present 10 percent (20 percent in the U.S.). Antinuclear activists might applaud that decline, but climate activists might decry the loss of a low-carbon energy source. There's a nuclear choice that splits the environmental community! Environmentalists of both stripes can cheer the rise in renewable electricity, but worldwide solar, wind, biomass, and geothermal energy still account for only 10 percent of total generation—although that share is rising rapidly. Figure 6.4 quantifies sources of world and U.S. electricity generation.

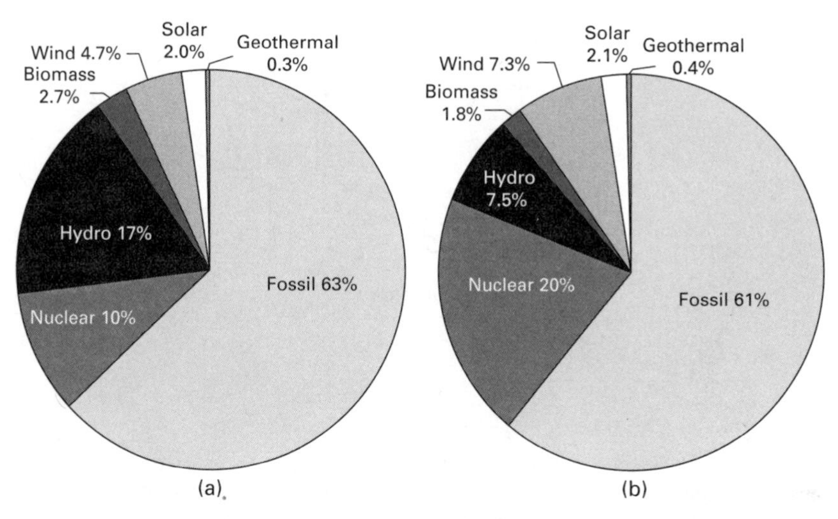

Figure 6.4
Sources of electrical energy: (a) world; (b) United States.

Making Electricity

Most electricity is produced in large, centralized power plants that send their product over a complex network of transmission lines—the **power grid**—to homes, industries, and other end users. Every outlet and switch in your home is connected to the grid and to all its implications. Flip a switch on and you may be asking coal miners to dig a little deeper, coal plants to spew a little more planet-warming carbon dioxide, or nuclear plants to produce more nuclear waste and weapons-usable plutonium. Chapter 10 explores alternatives to this very real connection between today's electrical enterprise and serious environmental problems. Here we'll focus on just how electricity generation works, especially but not exclusively in the nuclear context.

We have many ways to produce electricity. Batteries convert chemical energy to electrical energy. Solar panels convert sunlight energy to electricity. Spacecraft traveling to the outer planets make electricity from heat, generated by the radioactive decay of plutonium-238. But worldwide, electricity generation is predominantly by the conversion of mechanical energy to electricity in **electric generators**.

Electric generators use a fundamental principle of physics that links electricity and magnetism. Known since the early 1800s, this principle asserts that moving an electrical conductor in the presence of a magnet

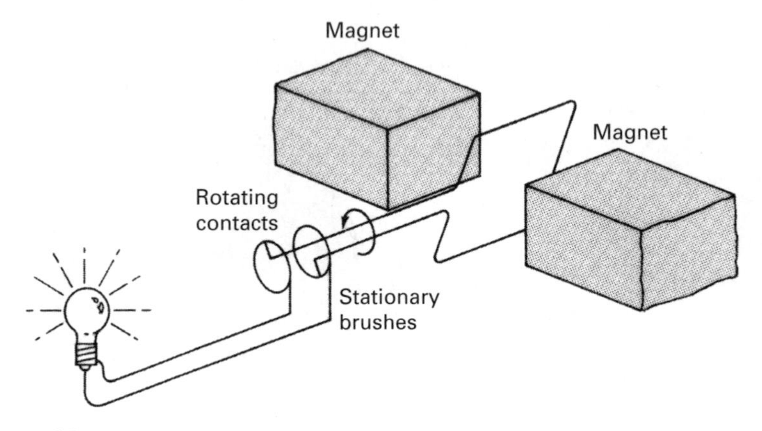

Figure 6.5
An electric generator. A wire loop rotates between magnetic poles, giving rise to an electric current. Conducting brushes contacting the circular rings convey current from the rotating loop to wires that take it where it's needed—in this case to a lightbulb.

generates electric current. Figure 6.5 shows how this is applied in an electric generator. The generator consists of one or more coils of wire that rotate between magnetic poles. So-called *brushes* contact the rotating conductors and carry the electricity to stationary wires for delivery to an end user like the lightbulb shown in figure 6.5.

To make electricity with the generator of figure 6.5, you have to spin the coil. And that's not easy, because you're supplying the mechanical energy that gets converted to electricity to light the lightbulb. Recall figure 6.1, which reinforces the fact that the generator doesn't *make* electricity but only *converts* energy from mechanical to electrical form. If you've had the opportunity to turn a hand- or pedal-cranked electric generator, then you've felt in your muscles the work you need to do to supply mechanical energy. Demand more electricity of the generator, perhaps by turning on an additional lightbulb, and it gets harder to turn.

The same effect applies to large generators in power plants. If it didn't, power companies could set their generators spinning and they would happily produce electricity forever without burning coal or fissioning uranium. So what turns the generators?

Hydro and Wind Generators

Hydroelectric power plants use moving water to spin fanlike turbine blades (figure 6.6a). Hydro plants include some of the smallest as well as the largest facilities in the electric-power industry (China's Three Gorges, at 22,500 megawatts, is the world's largest power plant). Hydropower

(a)

(b)

Figure 6.6
(a) A hydro turbine being installed at Wheeler Dam in Alabama (Tennessee Valley Authority); (b) close-up of a wind turbine. A 2-MW electric generator occupies the housing to right of the blades. (Alex Bartel/Science Source)

is the dominant source of electrical energy in Brazil, Canada, Norway, and many developing countries. Hydro accounts for 17 percent of the world's electricity. The corresponding figure for the U.S. is only 7 percent, although hydro dominates in the Pacific Northwest.

Wind turbines are similar to hydro turbines, except that moving air is the driving force (figure 6.6b). Wind is one of the fastest-growing sources of electrical energy, and in 2020 accounted for 6 percent worldwide and 9 percent in the U.S.

Thermal Power Plants

Figure 6.4a showed that 73 percent of the world's electricity is generated by fossil-fueled and nuclear power plants. These, along with a handful of solar power plants that concentrate sunlight to produce heat, constitute **thermal power plants**. Thermal plants are huge steam engines in which a heat source boils water to produce high-pressure steam. The steam spins a turbine connected to an electric generator. The heat source can be nuclear fission, concentrated sunlight, or combustion of coal, gas, or oil. The basic operation of thermal power plants, and even many of the details, are the same whether the plant is fossil fueled or nuclear.

Figure 6.7 shows the essential features of a thermal power plant. The boiler might be a large vessel, as shown, or it might consist of many thin pipes carrying water through the heated region. Steam exits the boiler at high pressure. It hits the small turbine blades, then expands, cools, and slows through a sequence of larger blades engineered to extract maximum energy (figure 6.8).

Before it returns to the boiler, the steam must be turned back into liquid water. That's the job of the **condenser**, in which steam pipes contact cool water pumped in from a river, lake, or bay. To prevent ecological disturbance, the heated water is cooled before being returned to its source. This is usually done in the enormous **cooling towers** that, for many people, symbolize nuclear power. Actually, though, those towers are as likely to be found at fossil-fueled plants because any thermal power plant needs a condenser and cooling system (figure 6.9).

A power plant's cooling system dumps a lot of energy into the environment—energy wrung from expensive coal, gas, oil, or uranium. That's a huge waste! *A typical thermal power plant dumps about two-thirds of the energy released from its fuel into the environment as waste heat!* Only one-third ends up as useful electricity. A 1,000-MW power plant is really a 3,000-MW plant, but 2,000 MW go up the cooling

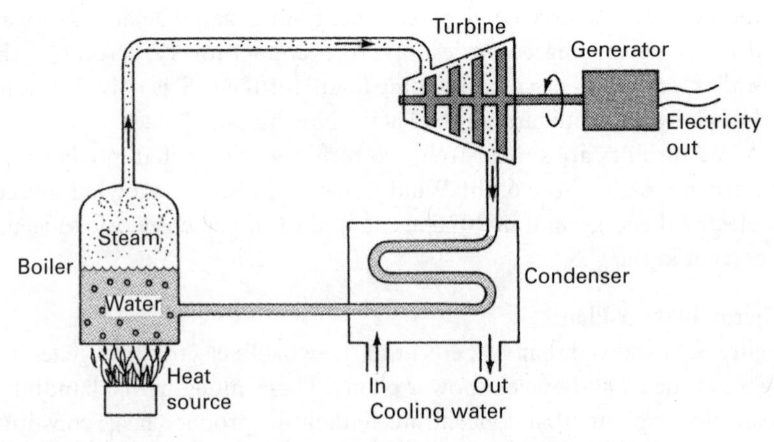

Figure 6.7
Simplified diagram of a thermal electric power plant.

Figure 6.8
Turbine assembly from an 820-MW turbine-generator. Note the smaller blade diameter at center, where high-pressure steam enters. (Talen Energy)

towers as waste heat.[7] So when you turn on your 100-W lightbulb, it isn't just one servant you press into service—it's three, one to light the lamp and two to make waste heat.

Why can't we do better? Are our engineers too dumb to make an efficient power plant? The problem lies deeper than that, in a fundamental principle of physics that precludes our converting heat to electricity with 100 percent efficiency.

Energy Quality

If I offer you a kilowatt-hour of energy, would you rather have it in the form of electricity or a bathtub full of hot water? Either will do if you want to take a bath, since you can run an electric heater that turns the

Figure 6.9
Clouds rise from 414-foot-high cooling towers in this archival photo of the Martins Creek power plant in Pennsylvania. Inside the towers, updrafts of air cool water from the plant's condenser before it's returned to the Delaware River. At the time of this photo, the plant was burning coal or oil; today it burns natural gas or oil. (Talen Energy)

electricity to heat without, in principle, any energy loss. But suppose you want to run your computer. You can do that with electricity but not with hot water. In that sense electricity is a more useful and therefore higher-quality form of energy.

Energy quality is the subject of one of the most famous laws of physics: the **second law of thermodynamics**. Fundamentally, the second law says something very simple and familiar: left to themselves, things get more chaotic, not more organized. Scramble an egg, mixing yolk and white; they stay mixed and won't spontaneously separate no matter how much you stir. Drop a pile of carefully organized papers and you've got a random mess that won't reassemble by itself into its original organized state. Mix hot and cold water and you've got lukewarm. No matter how long you wait, the water won't spontaneously separate into hot and cold.

Nuclear-to-Fossil and Fossil-to-Nuclear

The similarity between fossil-fueled and nuclear power plants is sufficiently close that on occasion plants of one type have been converted into the other. In 1984, construction was halted on a nuclear plant in Midland, Michigan, that was 85 percent complete. But the troubled plant got a new lease on life when it was converted to a gas-fired plant. Much of the nuclear plant's equipment—$1.5 billion worth, including the turbine-generators—was used in the gas-fired plant. The gas-fired plant became operational in 1990 and today it continues to supply some 10 percent of lower Michigan's electricity. It also operates as a cogeneration facility, providing steam for nearby Dow Chemical.

A similar conversion occurred in Platteville, Colorado, where the Fort St. Vrain power plant used an unusual high-temperature gas-cooled nuclear reactor to produce electricity from 1979 to 1989. Technically but not commercially successful, Fort St. Vrain's reactor was decommissioned in the 1990s. In its place rose a gas-fired power plant that utilized the original turbine-generator from the nuclear plant in a combined-cycle operation.

Today we're beginning to see conversions the other way, as China develops high-temperature nuclear reactors to replace coal-fired boilers. The new reactors are designed to be compatible with steam turbines and electric generators now installed in China's advanced coal plants, so the reactors can be used to convert coal plants to nuclear. The first conversions are planned for regions with severe air pollution. Later, coal-to-nuclear conversion may be used to help China meet its commitments to cut carbon emissions under the Paris climate agreement.

In this water example, the total energy stays the same, but the quality of that energy has diminished. Another way of stating the second law is to say that energy quality generally decreases and can't spontaneously increase.

Electricity can be converted into any other form of energy with, in principle, 100 percent efficiency. Thus electricity is energy of the highest quality. So is the energy associated with the motion of everyday- and larger-sized objects (baseballs, people, cars, planets, and so forth). But the molecular motion we call, loosely, *heat* is of lower quality. You can't do much with that lukewarm water, but you can do anything you want with the same amount of energy taken as electricity. And there's more versatility in your initial hot and cold water than when they're mixed. So the energy associated with separate hot and cold things is of higher quality than when those things have come to a common temperature. The greater the temperature difference—in particular, the hotter the higher temperature—the greater the energy quality. Yet another way of stating the second law is to say that higher-quality energy can be 100 percent converted to a lower-quality form, but that you can't go the other way: you can't convert 100 percent of lower-quality energy to higher quality.

Energy Quality and Power-Plant Efficiencies

What's this got to do with thermal power plants? A thermal plant starts with heat from burning or fissioning fuel, and ends up with electricity. But because that heat is of lower quality than electricity, the plant can't convert 100 percent of the heat energy to electricity. How much can it convert? That depends on the quality of the heat—which, as you've just seen, depends on the associated temperature difference. The lowest temperature available to a power plant is typically that of the ambient environment, so the quality of the heat energy depends on the highest temperature available from burning or fissioning fuel. As a practical matter, materials used in power-plant construction set an upper limit of about 1,100°F for the thin-tube boilers used in most fossil-fueled plants, and 600°F for the large pressure vessels in the most widely used nuclear plant designs. Coupled with losses due to friction and other imperfections, this limits large fossil and nuclear plants to efficiencies in the range of typically 30 to 35 percent. That's why some two-thirds of the energy released from fuels ends up as waste heat. Natural gas plants can do better, with efficiencies from 40 to 60 percent, especially if they use *combined cycle* technology where high-temperature jet-engine-like gas turbines are coupled with the conventional steam cycle shown in figure 6.7. But given that coal still dominates world electricity production, our estimate that two-thirds of the energy released in thermal power plants gets wasted is still a good rule of thumb.

Couldn't we do something with that waste heat? After all, it's still useful for low-energy-quality applications like heat and hot water. Increasingly, we *are* using waste heat. In Europe, whole cities are heated by circulating water carrying the waste heat from electric power plants. Many industries and institutions generate their own electricity and use the waste heat for space heating or industrial processes. Use of waste heat from power generation is termed **cogeneration** or **combined heat and power** (CHP). Worldwide, however, fewer than 10 percent of electric power plants employ cogeneration.

Summary

Citizens of modern industrialized societies use energy at enormous rates—many times the 100 watts a typical human body can supply. That statement, like many others about energy, is expressed in terms of *power*, or *rate* of energy use. In this book, we usually measure power in watts, kilowatts, or megawatts, and energy itself in kilowatt-hours. Worldwide, humankind uses energy at the average rate of about 2 kW per capita; in North America it's 10 kW—the equivalent of 100 energy servants working round the clock to supply the energy we use directly or that's used for our benefit. Nearly 40 percent of that energy goes to making electricity, and about 10 percent of the world's electricity comes from nuclear fission. But electrical energy generation in fossil and nuclear power plants is inefficient, due largely to limitations imposed by the second law of thermodynamics. In fact, about two-thirds of the energy released from burning or fissioning fuel in large power plants is dumped to the environment as waste heat.

Glossary

cogeneration or **combined heat and power (CHP)** Generating electricity while utilizing the waste heat for heating or industrial processes.

condenser Part of a thermal power plant that converts steam back into liquid water.

cooling tower A device that extracts waste heat from the cooling water of a thermal power plant and discharges it, usually to the atmosphere.

electric generator A device that converts mechanical energy into electrical energy by moving electrical conductors in the presence of a magnet.

energy quality A characteristic of energy that determines how useful it is for performing a variety of tasks. Electricity and motion, the highest-quality forms of energy, can be used fully for any task requiring energy. High-temperature heat energy is lower in quality, and low-temperature heat is lowest. Lower-quality energy cannot be converted to higher quality with 100 percent efficiency.

kilowatt (kW) A unit of power, equal to 1,000 watts.

kilowatt-hour (kWh) A unit of energy. A device using energy at the rate of 1 kilowatt uses 1 kilowatt-hour of energy each hour.

megawatt (MW) A unit of power, equal to 1,000,000 watts or 1,000 kilowatts.

power The rate at which energy is used or supplied; measured in watts or kilowatts.

power grid The interconnection of power plants, transmission systems, and end users of electricity.

second law of thermodynamics The statement that things tend toward more chaotic states. Its implications include the inability of energy to convert spontaneously from lower-quality to higher-quality forms, with the highest energy quality in the form of electricity or motion. The second law sets fundamental limits on the efficiency of thermal power plants.

thermal power plant A power plant in which a heat source—usually burning fuel or nuclear fission—boils water to produce steam that turns a turbine connected to an electric generator.

watt (W) A unit of power. (In physics terms, 1 W is 1 joule per second.)

Notes

1. We'll be using the term *heat* somewhat loosely in the sense it's commonly understood. To a physicist, though, *heat* is energy that's in transit specifically because of a temperature difference. Once that energy is in something—for example, hot water—then strictly speaking it should be called *internal energy*.

2. Other energy units you may know of include the calorie, used to describe the energy in foods and chemical reactions; the joule, official energy unit of the international scientific community; the electron-volt (eV), used in atomic and nuclear physics; the British thermal unit (Btu), used by the U.S. engineering community and building trades; and the megaton, a unit for the explosive energy of nuclear weapons. All these units measure the same thing, namely energy.

3. How much power? One second is 1/3,600 of an hour, so the power would be 40 kWh/(1/3,600 hour) = 144,000 kW, or 144 MW. That's about 15 percent of the output of a large power plant!

4. If you've had high school physics, you can easily estimate this power. As I do knee bends, I raise the upper part of my body about 10 inches, or about 0.25 meters. Suppose about 2/3 of my 155 pounds (70 kilograms) is involved. Then the energy involved in raising that mass m a distance $h = 0.25$ meters against the $g = 9.8$ newtons/kilogram force of gravity is mgh, or $(2/3) \times (70\text{kg}) \times (9.8\ \text{N/kg}) \times (0.25\ \text{m}) = 114$ joules. I do this once a second, and 1 watt is 1 joule per second, so the power is 114 watts.

5. A food Calorie (Cal or kcal) is actually 1,000 calories (cal), and 1 cal is 4.184 joules, so 1 kcal is 4,184 joules. A 24-hour day contains 86,400 seconds, so 2,000 kcal/day is $(2{,}000\ \text{kcal}) \times (4{,}184\ \text{J/kcal})/(86{,}400\ \text{s}) = 97\ \text{W}$.

6. Figure 6.3 shows what's called *end-use* energy—the energy actually delivered to users. Accounting for *primary energy*—the total energy consumed, including energy lost to inefficiencies in electric power generation—would give somewhat different pictures.

7. Sometimes you'll see the abbreviations MW_e and MW_{th} for *megawatts electric* and *megawatts thermal* to describe, respectively, the electric power output of a power plant and the total power released from burning or fissioning fuel. The latter is usually about three times larger than the former to account for waste heat.

Further Reading

BP. *BP Statistical Review of World Energy*. London: BP P.L.C., published annually. Available for download at https://www.bp.com/en/global/corporate/energy-economics/statistical-review-of-world-energy.html. Produced by one of the world's largest energy companies, this report contains a thorough listing of worldwide energy statistics, organized by world regions and individual countries. Also available as an app for mobile devices.

Paige, J., et al. *Electrification Futures Study: End-Use Electric Technology Cost and Performance Projections through 2050*. Golden, CO: National Renewable Energy Laboratory NREL/TP-6A20–70485, 2017. Available at https://www.nrel.gov/docs/fy18osti/70485.pdf. This authoritative report examines the implications of increasing the electricity portion of our energy supply, considering energy used in transportation, buildings, and industry.

U.S. Energy Information Administration. Form EIA-860 detailed data, available for download at https://www.eia.gov/electricity/data/eia860/. Fascinating information on electric power generators in the United States. The spreadsheet 3_1_GeneratorYyyyy is especially interesting; it shows locations, rated power output, and fuel or other energy source. You can sort by state, type of plant, power output, and other categories. A must-read for energy geeks!

U.S. Energy Information Administration. *International Energy Statistics*. A website where the user can customize maps, graphs, and data tables for a host of world energy statistics, including by individual countries. Search on "EIA International Energy Statistics" for the latest URL.

U.S. Energy Information Administration. *Monthly Energy Review*. Recent and historical data on U.S. energy usage and production, going back to 1949. Available at https://www.eia.gov/totalenergy/data/monthly/pdf/mer.pdf, with updates monthly.

Wolfson, Richard. *Energy, Environment, and Climate*, 3rd ed. New York: Norton, 2018. One of your authors explores energy and its environmental implications in an undergraduate textbook. Chapters 2–4 cover the basics of energy in more depth than does *Nuclear Choices*.

7

Nuclear Reactors

In a nuclear power plant, fission provides the heat that boils water to drive the turbine-generator. Most of the energy released in fission takes the form of rapid motion of the fission products. These collide with particles in the surrounding material, producing the heat that ultimately turns the generator.

The goal in running a nuclear power plant is to maintain a steady fission chain reaction. You saw in chapter 5 how each fission event releases several neutrons, some of which may strike fissile nuclei and cause additional fission—hence the term *chain reaction*. In a power plant, we want the chain reaction to be just critical, meaning that an average of exactly one neutron from each fission causes another fission; that way, the rate of energy release remains steady. A supercritical chain reaction, illustrated in figure 5.6, would be disastrous in a power plant. Energy release would grow exponentially, resulting in violent disruption of the power plant. That's exactly what happened in the Chernobyl accident.

At the heart of a nuclear power plant is its **nuclear reactor**, a system engineered to sustain a controllable chain reaction. Understanding the many controversies surrounding nuclear power requires a close look at the workings of nuclear reactors.

Neutrons, Fast and Slow

When a neutron hits a uranium nucleus, will it always cause fission? You've already seen part of the answer: for low-energy neutrons (also called **slow neutrons** or **thermal neutrons**), only the fissile isotope U-235 will fission. Nuclei of the more common U-238 may absorb a neutron, eventually becoming plutonium-239. High-energy neutrons (**fast neutrons**) will also fission U-235 and, if fast enough, may even fission U-238. But the likelihood of fission varies drastically with neutron energy. For

Table 7.1
Fission probability versus neutron speed

	Neutron speed		
Isotope	Slow	Fast (as emitted in fission)	Very fast (much faster than emitted in fission)
U-235	Extremely likely	Possible	Possible
U-238	Impossible	Unlikely	Possible

U-235, slow neutrons are several hundred times more likely than fast neutrons to cause fission. Thus it's much easier to achieve a chain reaction if the fast neutrons emitted in fission can be slowed. In contrast, U-238 fission requires extremely fast neutrons—even the fast neutrons from U-235 fission are unlikely to fission U-238. This's one reason why it's not possible to sustain a chain reaction in U-238. Table 7.1 summarizes the likelihood of fission with different neutron speeds.

Because slow-neutron fission of U-235 is so much more likely, a chain reaction using slow neutrons can be sustained even with a great deal of nonfissile U-238 present. Even though the U-238 absorbs many neutrons, those that do strike U-235 are very effective in causing fission. Slow-neutron chain reactions are possible even in uranium with U-235 at its natural abundance of only 0.7 percent, although most reactor designs require enrichment to 3–5 percent U-235. Use of natural or slightly enriched uranium in power reactors is important for two reasons: first, enrichment is expensive; second, natural or slightly enriched uranium cannot be used directly in nuclear weapons.

The fact that most power reactors use slow neutrons provides an answer to a common question: can a nuclear reactor blow up like a bomb? The answer, for a slow-neutron reactor, is a definitive no. A bomb gets its destructive power not only from the sheer amount of energy released but also from the suddenness of that release; a typical nuclear explosion is over in a millionth of a second. In a bomb, the chain reaction is sustained by fast neutrons that spend only about 10 billionths of a second between being released in one fission event and striking a nucleus to cause another fission. In a reactor, neutrons take 10,000 times longer between fission events, so even a reactor that was badly out of control could not undergo a bomb-like explosion. That's not to say that a nuclear reactor couldn't fail catastrophically or even explosively; however, even the most violent reactor disaster would not have an explosive effect approaching that of a nuclear weapon.

In most power reactors, slow neutrons sustain the chain reaction. But the neutrons released in fission are fast neutrons. Slowing them is the job of the **moderator**, a substance that absorbs neutrons' energy. At the microscopic level, collisions between neutrons and nuclei in the moderator are what slow the neutrons.

When objects collide—whether subatomic particles, billiard balls, or cars—the most effective energy transfer occurs if the objects have similar masses. A good moderator is, therefore, a substance whose nuclei have about the same mass as a neutron. That makes hydrogen (${}^{1}_{1}H$), with its single proton, a good moderator. Water (H_2O) is full of hydrogen, so it's an appropriate moderator substance. Indeed, water is the moderator in most of the world's power reactors. But hydrogen has one failing: it often absorbs neutrons. A moderator should absorb neutron *energy* but not neutrons themselves; otherwise it reduces the number of neutrons available to sustain the chain reaction. This neutron loss means that reactors moderated with ordinary water can't run on unenriched, natural uranium. So fuel for most power reactors must be enriched, to about 3–5 percent U-235, to overcome the effect of neutron absorption in the moderator.

The next heaviest nucleus after hydrogen is deuterium (${}^{2}_{1}H$ or D), the hydrogen isotope whose nucleus contains a proton and a neutron. Since its mass is twice that of the neutron, deuterium isn't as efficient at absorbing neutron energy. But it makes up for that by its very low probability of absorbing neutrons themselves. Neutrons in a deuterium-moderated reactor are therefore so abundant that the reactor can run on natural, unenriched uranium despite its low (0.7 percent) proportion of fissile U-235. Since it's chemically similar to ordinary hydrogen, deuterium also combines with oxygen to make water—in this case **heavy water** (${}^{2}_{1}H_2O$, or D_2O). Heavy water is present in minute quantities in natural water and can be separated and used as a reactor moderator. The Canadian nuclear program emphasizes **heavy-water reactors** (HWRs). Reactors using ordinary **light water** are, in contrast, **light-water reactors** (LWRs). Heavy water has been considered a sensitive material since the dawn of the nuclear age, because a nation possessing it can make plutonium for bombs without needing enriched uranium. During World War II, a daring raid by Norwegian commandos set back the German nuclear bomb program by destroying what was then the world's only commercial heavy-water production facility, in Nazi-occupied Norway.

Some reactor designs use solid graphite—the common form of carbon used in pencil "lead"—as moderator. Carbon nuclei in graphite are 12 times more massive than neutrons but, like deuterium, exhibit very little neutron absorption. Graphite-moderated power reactors operate today only in the U.K. and Russia. Graphite-moderated reactors were widely used in the early days of the nuclear age to produce plutonium for nuclear weapons, but most of those so-called *production reactors* are now obsolete.

People are often confused about the role of the moderator in a nuclear reactor. The name suggests that it moderates—that is, controls or tempers—the nuclear reaction. In fact, the opposite is true: the moderator makes the nuclear reaction go. Without the moderator, only fast neutrons would be present. In reactor fuel, with its low proportion of fissile U-235, those fast neutrons couldn't sustain a chain reaction. What the moderator does moderate is the speed of the neutrons, yielding slow neutrons that are more effective in causing fission.

Controlling the Chain Reaction

A steady chain reaction requires maintaining that delicate average of exactly 1 neutron from each fission causing another fission. If that number—called the **multiplication factor**—drops below 1, the reaction fizzles; if it goes over 1, the reaction goes out of control.

A rise in the multiplication factor can mean disaster. Consider a reactor running at its rated power level of 3,000 MW. Suppose the multiplication factor rises from 1 to 1.01. Then, in the time that it takes neutrons from one fission to cause another—called the **generation time**—the fission rate increases by a factor of 1.01, increasing the reactor power to 3,000 MW × 1.01 or 3,030 MW. After two generations, it's 3,030 MW × 1.01 or 3,060 MW. Carrying this procedure through 10 generations gives a power level of 3,314 MW (still tolerable) but after 100 generations the power has nearly tripled, exceeding 8,000 MW. And after 500 generations, the reactor power has risen to nearly 150 times its design value. Since the generation time is only a fraction of a second, a multiplication factor even a little over 1 can quickly lead to disaster.

Controlling a nuclear reactor means keeping the multiplication factor at *exactly* 1. This is done by inserting neutron-absorbing material, in the form of **control rods**, amidst the uranium fuel. If the multiplication factor rises above 1, the control rods are inserted further to absorb more neutrons and thereby lower the multiplication factor. Should the multiplication factor drop below 1, the rods are withdrawn to provide more fission-causing neutrons and thus maintain the chain reaction. Control rods also adjust the reactor's power output and, when inserted fully, halt the chain reaction. That happens in an emergency shutdown, or **scram**. Reactors are designed to scram automatically if a potentially dangerous situation arises (see figure 7.1).

If a reactor does get out of control, will the control rods have time to react? Even a slow neutron takes only 1/10,000 of a second from its release in one fission event until it causes another fission. At that rate, the 500 generations of neutrons that blew up our 3,000-MW reactor a

Figure 7.1
A reactor scram occurs when control rods are inserted fully in response to a potentially dangerous situation. The operating mechanisms for the rods are therefore very sensitive, as this sign in the Vermont Yankee reactor building suggests. (Photo by Dexter Mahaffey.)

few paragraphs ago would be over in less than one-tenth of a second. No mechanical system can possibly respond in that time. Fortunately, reactor control *is* possible, thanks to the fact that a very few neutrons are released not in the fission event itself but in the subsequent decay of fission products. The effect of these **delayed neutrons** is to make the average time for one generation of neutrons about one-tenth of a second. That means it could take a minute or more for a reactor's power to increase significantly if the multiplication factor rises slightly above 1, so there's enough time for mechanical control rods to operate.

Even with delayed neutrons, care in reactor engineering and operation is essential. The reactor needs its multiplication factor so close to 1 that the chain reaction wouldn't continue without the delayed neutrons (here "so close" means less than 1.0065). Then slight variations result in slow changes in reactor power, and the control rods can respond. But if the multiplication factor goes over 1.0065, then **prompt neutrons** alone sustain the reaction. The generation time becomes dramatically shorter and the reaction goes rapidly out of control. That's precisely what happened at Chernobyl, where operators inadvertently put the reactor into this situation of **prompt criticality**. The reactor power then soared to 500 times its design value in 5 seconds.

Keeping Cool

Fission in a nuclear reactor produces heat. Unless that heat is removed, the temperature will continually rise, possibly resulting in a meltdown. In a power reactor, of course, we want to get heat out of the reactor in the form of steam to turn the turbine-generator. Removing heat is the job of the **coolant**, a fluid that circulates through the reactor and transports heat away, ultimately to the steam turbine.

Coolants, like moderators, vary with reactor design. Light-water reactors use ordinary water as the coolant. The same water, in fact, serves as both coolant and moderator. This is an important safety feature, since loss of coolant also means loss of moderator—and that stops the chain reaction. There's still the danger of meltdown, since decaying fission products generate considerable heat, but the power level is well below that of the chain reaction itself. In addition to ordinary water, coolants used in commercial slow-neutron reactors include heavy water and carbon dioxide gas.

Reactor Designs

All slow-neutron reactors share the three common features we've just discussed: moderator, control rods, and coolant. Working with those three elements, nuclear engineers around the world have developed very different reactor designs. The design details aren't just for engineers, since they have direct bearing on public debate over the safety of nuclear power. Here we give a brief overview of the reactor designs used in today's nuclear power plants.

Light-Water Reactors

Asked to design a nuclear reactor, you might consider putting uranium fuel in a container of water. The water would slow neutrons to maintain a chain reaction and would simultaneously absorb heat. The water would boil, and the resulting steam could run a turbine-generator. That, in a nutshell, is a **boiling-water reactor**, or BWR, perhaps the simplest design for a power reactor. Your BWR-based power plant might look something like figure 7.2.

Roughly one-fifth of the world's light-water power reactors are BWRs. The uranium fuel in these reactors takes the form of uranium-dioxide (UO_2) **fuel pellets** sealed into **fuel rods**—long tubes of corrosion-resistant

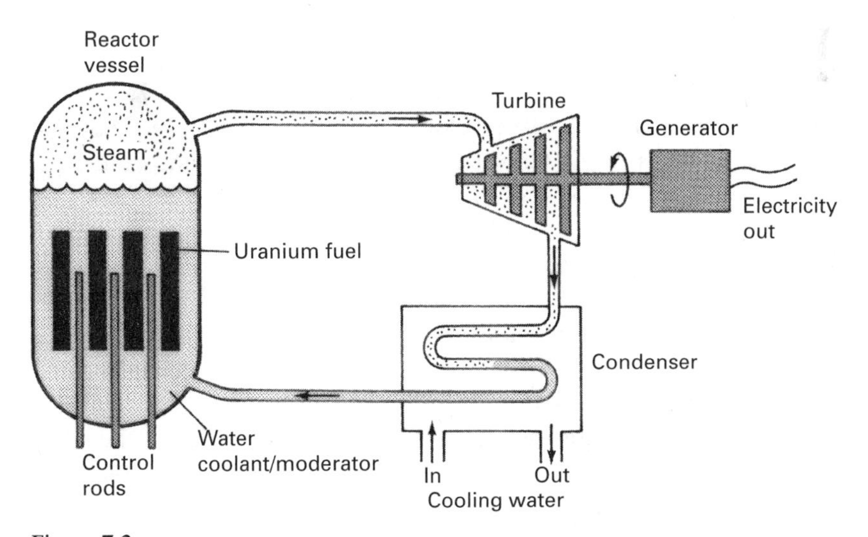

Figure 7.2
Diagram of a nuclear power plant using a boiling-water reactor. The turbine is driven by steam produced in the reactor vessel itself.

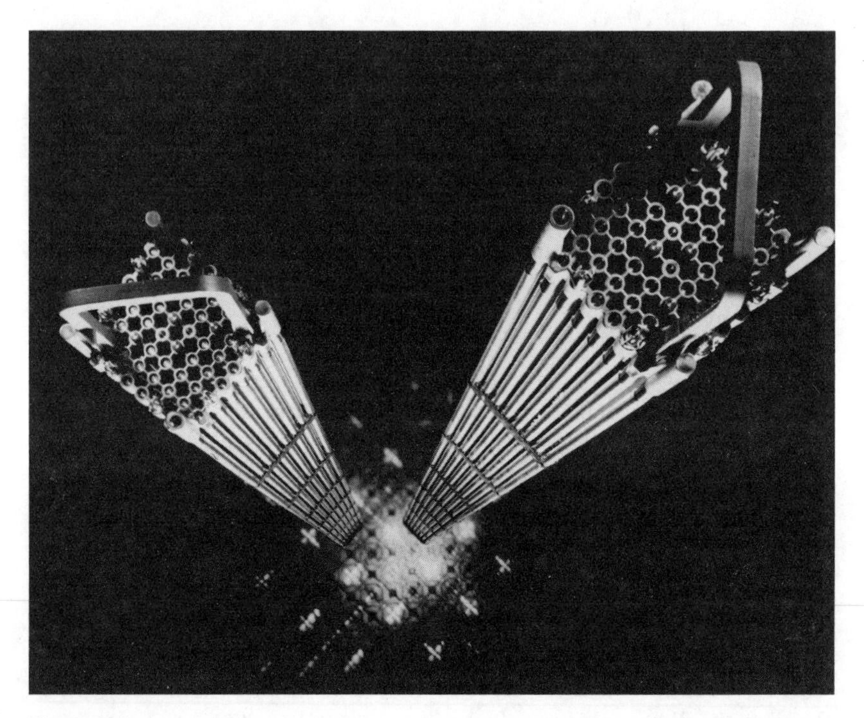

Figure 7.3
A pair of fuel bundles being lowered into a reactor core. (Courtesy of GE Hitachi Nuclear Energy Americas LLC.)

zirconium alloy. The tubes are joined into **fuel bundles** (figure 7.3). Many bundles, mounted inside the reactor's heavy steel pressure vessel, comprise the reactor's **core**. The water that serves as coolant and moderator circulates among the fuel rods. In normal operation, most radioactive fission products remain encased in the fuel. But some gaseous fission products work their way into the water, and neutron absorption in the water gives rise to radioactive tritium. For these reasons, water and steam circulating through the turbine are radioactive. This means that a break in a steam pipe will release radioactivity, and it also requires radiation protection for workers servicing the turbine.

The **pressurized-water reactor** (PWR) is a close cousin of the BWR and is by far the most common type of power reactor. In a PWR, the reactor vessel is held under such high pressure that the water isn't able to boil, despite its high temperature. This superheated water goes from the reactor vessel to **steam generators**, where it flows through pipes in contact with other water that is allowed to boil. The resulting steam turns

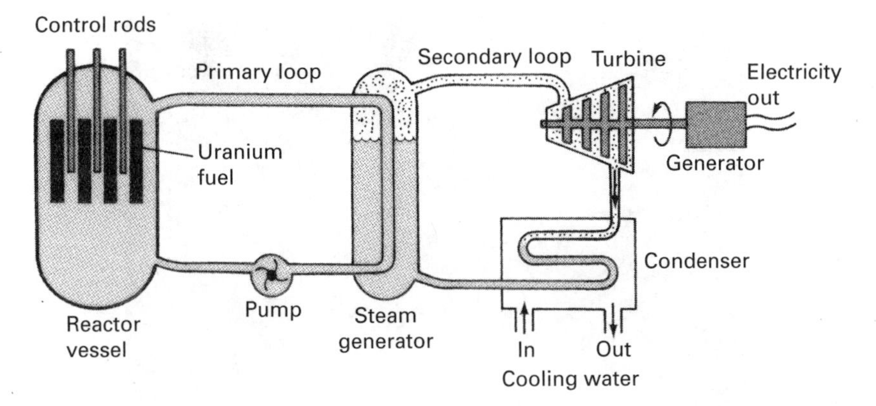

Figure 7.4
Diagram of a nuclear power plant using a pressurized-water reactor, showing primary and secondary coolant loops that share a steam generator.

the turbine-generator, is condensed back to water, and then returns to the steam generator to complete the cycle. The turbine, condenser, and boiling-water part of the steam generator constitute the **secondary loop** of the cooling system. Water in the secondary loop never enters the reactor core, so it's not radioactive. Radioactive water is confined to the **primary loop**, consisting of the reactor vessel and the primary coolant pipes that flow through the steam generators. Steam generators are mounted close to the reactor vessel, within the thick containment structure, reducing the chance of radiation release due to a break in the primary coolant loop. Figure 7.4 shows the essential features of a pressurized-water reactor.

PWRs and BWRs share some important features. In both, the reactor core is enclosed in a heavy steel pressure vessel. To refuel the reactor, it must be shut down and the lid of the pressure vessel removed (figure 7.5). Spent fuel bundles are extracted, a process that's carried out underwater because of the intense radioactivity, and then replaced with fresh ones. Typically, a reactor is refueled about every 12 to 18 months, with one-third of the fuel bundles being replaced. Refueling lasts six to eight weeks, a time that utilities seek to minimize because of the high cost of replacement power. The long interval between refuelings makes spent fuel from light-water reactors a less-than-ideal source of weapons-grade plutonium.

You've seen that use of a common moderator and coolant is a safety feature of light-water reactors, because loss of coolant means loss of moderator, which halts the chain reaction. Using water as moderator has an additional safety effect that occurs because the density of water decreases as its temperature rises. A slight increase in reactor power raises the water's

Figure 7.5
Removing the lid of the pressure vessel before refueling a boiling-water reactor. (Talen Energy)

temperature, lowering its density and therefore—since there are fewer hydrogen nuclei in a given volume—reducing the water's effectiveness as a moderator. As a result, fission slows and the temperature again drops. This feature of LWRs provides some stability against runaway chain reactions.

Heavy-Water Reactors

Canada, a country with considerable nuclear expertise, developed the commercial heavy-water reactor design known as **CANDU** (for Canada deuterium uranium). Recall that heavy water, because of its low neutron absorption, allows reactors to use unenriched uranium. In CANDU reactors in service today, bundles of unenriched uranium fuel are mounted in individual coolant channels, through which pressurized heavy-water coolant circulates. The entire assembly of fuel bundles and coolant channels is immersed in a tank of unpressurized heavy water, which serves as the neutron moderator. Thus the CANDU design is cooled and moderated by heavy water but with the two functions physically distinct. The pressurized coolant carries heat to steam generators, where it boils ordinary light water, which drives the turbine-generators, as in a pressurized-water reactor.

Because the CANDU's coolant flows within individual pressurized channels, there's no need for the large, heavy pressure vessel found in light-water reactors. The lack of a single pressure vessel makes catastrophic failure of the high-pressure coolant system less likely than in an LWR. Furthermore, individual channels can be removed and refueled without shutting down the reactor. In practice, CANDU refueling is done almost continuously, with an average of 15 fuel bundles replaced each day. Continuous refueling has several important implications. First, it makes reactor control easier, because conditions within the reactor core remain essentially the same over time. (In contrast, depletion of U-235 and buildup of fission products over the long refueling interval make for substantial changes in the core of a light-water reactor.) As a result, the CANDU design requires fewer control rods. A second implication of continuous refueling entails the ever-present connection between nuclear power and nuclear weapons. Fuel that's spent a long time in a reactor—as happens in light-water reactors—is a poor source of bomb-grade plutonium. But in a reactor that can be refueled frequently, it's easier to "cook" fuel just long enough to optimize plutonium for use in nuclear weapons. With a continuously refueled reactor, it's also less practical for international inspection teams to monitor diversion of nuclear materials. On the other hand, the CANDU's unenriched uranium fuel obviates the need for enrichment facilities, which can provide highly enriched weapons-grade uranium.

Graphite-Moderated Reactors

The first reactor of the nuclear age was a 400-ton pile of graphite embedded with nearly 50 tons of uranium compounds. Built under the stands at the University of Chicago stadium, it first went critical on December 2, 1942 (figure 7.6). The successful demonstration at Chicago led quickly to the construction of huge graphite-moderated reactors at Hanford, Washington, which produced plutonium for early nuclear weapons.

Graphite has also proved a workable moderator in commercial power reactors. Graphite-moderated reactors cooled with carbon dioxide gas

Figure 7.6
Sketch of Chicago Pile I, the first reactor to achieve a self-sustaining chain reaction. The man on the floor in front of the pile is manually removing a control rod to start the chain reaction. Because of wartime secrecy, no photographs were taken of the completed reactor. (Argonne National Laboratory)

dominate the United Kingdom's commercial reactor fleet and provide nearly one-fifth of the country's electricity. The former Soviet Union developed a graphite-moderated, light-water-cooled design called the **RBMK.** Developed during the Cold War, these reactors produced both electricity for the civilian population and plutonium for weapons. The RBMK design had some serious flaws, and you'll soon see how these contributed to the Chernobyl accident. RBMKs were modified after the accident, and more than 30 years later some 10 of these reactors are still operating in Russia.

The choice of graphite as a moderator has important safety implications. First, loss of coolant—whether gas or liquid—does not mean loss of moderator. Thus, the fission chain reaction won't automatically halt in the event of a cooling system failure. In a water-cooled RBMK design there's an additional dangerous possibility. Because the hydrogen in light water absorbs neutrons, water in the RBMK acts more like a control rod than a moderator. Overheating and excessive boiling then enhance the rate of fission by reducing neutron absorption. The enhanced fission rate exacerbates overheating, leading to a further increase in the fission rate. In contrast to the stabilizing effect of water in a light-water reactor, water in the RBMK design is therefore destabilizing. Complex systems are necessary to keep the RBMK under control. These features contributed to the Chernobyl accident, as you'll see in the next chapter.

A second issue with graphite is that it's flammable. Both at Chernobyl and in the 1957 Windscale accident—the U.K.'s worst nuclear accident—graphite moderators caught fire, and the smoke from those fires helped spread radioactive fallout over large areas.

Breeder Reactors

The nuclear reactors we've considered so far have been slow-neutron reactors, using a moderator to reduce neutron energy and enhance fission. Those slow-neutron reactors use natural or slightly enriched uranium at far below the enrichment necessary for weapons. Their relatively long generation time from one fission event to the next ensures that these reactors, even if wildly out of control, can't blow up with the explosive power of a nuclear weapon.

Slow-neutron reactors have one great inefficiency: they make little use of the 99.3 percent of uranium that's U-238. Yet all that U-238 could be converted by neutron absorption to fissile plutonium-239, providing far more nuclear fuel than U-235 alone. Slow-neutron reactors do, in fact, produce Pu-239 but at somewhat less than one Pu-239 nucleus per fission event. That plutonium "burns"[1] in the reactor, ultimately producing about one-third of its total energy. But even so the vast majority of the U-238 remains useless.

Enter the **breeder reactor**, in which fission neutrons "breed" plutonium-239 from uranium-238. Breeders have been around for a long time; in fact, the first nuclear-generated electricity came from an experimental breeder in 1951. Many countries began developing breeder reactors, and France deployed several commercial breeders as part of its policy of achieving energy independence through nuclear power. However, they proved economically unviable and were eventually shut down. Although breeder research continues, today the only breeders in commercial operation are two units in Russia, including an 800-MW plant that come online in 2016. A true breeder is distinguished from other reactors in that it produces more fissile fuel than it consumes; that is, for each fission event, more than one neutron gets absorbed in U-238 to produce Pu-239. Breeding therefore requires lots of neutrons. With one neutron needed to sustain fission, and more to breed new fuel, there isn't much leeway, since only two or three neutrons result from each fission event. It turns out that more neutrons result when fission is induced by fast neutrons, with the most neutrons obtained from fast-neutron fission in Pu-239. Successful breeders therefore use a plutonium-uranium mix for fuel, and they have no moderator to slow the neutrons—hence the name **fast breeder**. Since fission is less likely with fast neutrons, fast breeders need more highly enriched fuel (around 20 percent U-235 in today's breeders) than slow-neutron reactors. The breeder's fast neutrons also mean a much lower generation time between fissions, and therefore an out-of-control reaction can grow much more rapidly.

With no moderator, the core of a breeder can be very compact. A small core produces too much heat to permit the use of water as coolant. Furthermore, the need to be parsimonious with neutrons rules out neutron-absorbing coolants. And the coolant must not act as a moderator, since the breeder requires fast neutrons. Breeders operating today use liquid sodium metal coolant. Sodium has enormous ability to carry heat without the need for high pressure, and it exhibits very low neutron absorption; on the other hand, sodium burns spontaneously on contact with air, and it reacts violently with water. Contact between radioactive sodium and water would be especially disastrous; for this reason, breeder reactors have three separate cooling loops. There's a plus side to liquid sodium, though: a breeder's sodium coolant loops don't operate at high pressure and can therefore be less prone to leakage and failure.

A typical breeder surrounds the reactor core with a blanket of uranium-238, in which fission neutrons produce plutonium-239. The irradiated uranium-plutonium mix is occasionally removed and sent to a reprocessing plant for extraction of plutonium and fabrication of new reactor fuel. That poses a significant proliferation risk, because the reprocessed plutonium can be used directly in nuclear weapons.

Which Reactor Is Best?

There's no clear answer. Safety, economics, technology, fuel availability, enrichment, waste issues, and weapons proliferation all weigh into this nuclear choice. In some of its design features, CANDU has a safety edge over LWRs. But a CANDU research reactor provided India with its first bomb plutonium. The flammable moderator of graphite reactors poses serious issues in the event of a fire, but gas-cooled graphite reactors are immune to steam explosions. The nuclear accidents at Three Mile Island, Chernobyl, and Fukushima involved, respectively, a PWR, an RBMK, and multiple BWRs. Breeders were involved in some early nuclear accidents, but the few modern breeders, despite significant incidents, have not experienced major radiation-release accidents.

In one sense, though, the nuclear marketplace has answered the "best reactor" question. Today, pressurized light-water reactors dominate the

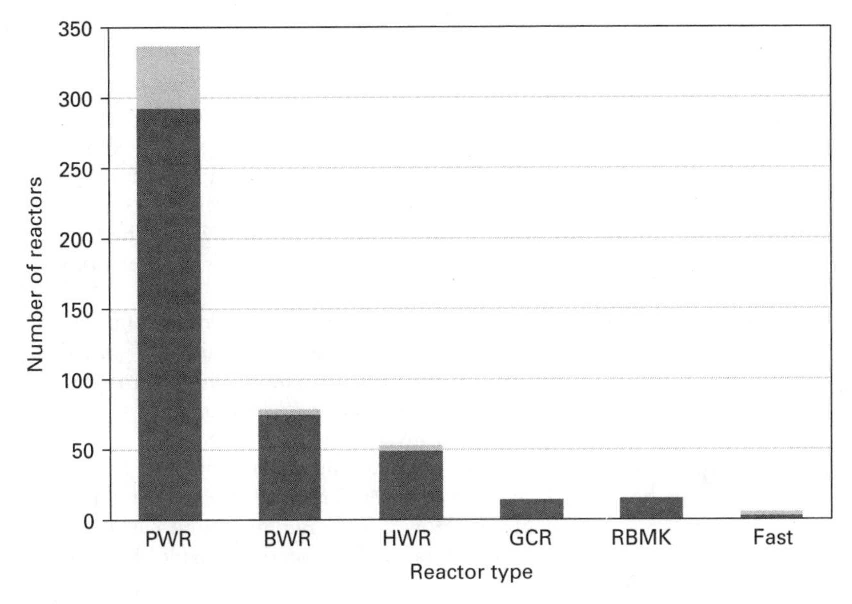

Figure 7.7
Worldwide distribution of reactor types, showing both those currently operating (dark) and under construction (light gray). Only 4 BWRs, 4 HWRs, and 2 fast reactors are under construction, as compared with 45 PWRs. HWRs are CANDU or related designs and GCRs are the U.K.'s graphite-moderated gas-cooled reactors.

world's commercial nuclear reactor fleet (see figure 7.7). Most of the other reactor designs operating today are legacies from the past. Of the 55 power reactors under construction as of 2020, 45 are PWRs. The rest include four each of advanced BWRs and heavy-water reactors derived from the CANDU design, one breeder, and one pebble-bed reactor.

Evolving Reactor Designs

Two-thirds of the approximately 450 commercial power reactors operating worldwide today are more than 30 years old. Most of those came online in the 1970s and 1980s, and their designs date even earlier. What has changed since then?

The first power reactors, deployed in the 1950s and 1960s, were one-of-a-kind units designed to show that nuclear fission could work as a source of electrical energy. The last of these **Generation I** reactors, at the Wylfa Nuclear Power Station in Wales, U.K., shut down in 2015. The majority of today's operating reactors are **Generation II** models, deployed in the late 1960s through early 1990s with designs improved by the nuclear industry's

A Natural Reactor

In 1972, a worker at a French nuclear fuel plant discovered a curious thing: samples of uranium arriving from a mine at Oklo in the Gabon Republic of West Africa contained less fissile uranium-235 than the normally low 0.7 percent, with some samples as low as 0.44 percent U-235. This was particularly baffling because the ratio of U-235 to U-238 is the same throughout the solar system, as confirmed by measurements on meteorites and Moon rocks.

Further analysis showed that the Oklo samples contained an unusual blend of isotopes that would be expected among the stable "offspring" formed in the decay of nuclear fission products. The conclusion was inescapable: a natural fission chain reaction had occurred at Oklo some 2 billion years ago. Humankind didn't invent the fission reactor!

You've just seen the challenges in sustaining a chain reaction, such as uranium enrichment and carefully engineered reactor systems including a moderator to slow the neutrons. How could random natural events put a reactor together? Several circumstances conspired to enable the Oklo chain reaction. First, the ore at Oklo is rich in uranium; that's why it was developed for a mine. Second, the ore body at the time of the reaction was saturated with groundwater that could serve as a moderator. Today, the neutron-absorbing properties of light water preclude its use as a moderator in a reactor fueled with natural uranium. But 2 billion years ago, things were different. The half-life of uranium-235 is 700 million years; that of U-238 is 4.5 billion years. So U-235 has decayed more rapidly than U-238, and that implies a greater proportion of U-235 in the past. Two billion years ago, in fact, the proportion of U-235 in natural uranium was about 3.7 percent—consistent with today's enriched light-water reactor fuels!

Eventually, 16 natural reactor zones were identified at Oklo. The reactors probably ran for several hundred thousand years, with total power output between 10 and 100 kW. The need for moderating water kept the chain reactions under control: if a reaction ran too fast, water boiled away and the reaction slowed. The very low power level precluded any danger of meltdown.

The fossil reactors at Oklo are more than scientific curiosities. They've served as natural laboratories for studying the long-term behavior of nuclear fission products. Analysis shows very modest migration of fission products from uranium-bearing regions into adjacent clay; plutonium decay products, on the other hand, show no migration—an indication that plutonium remained fixed at the sites where it formed for at least its 24,000-year half-life. These results are encouraging to those who advocate underground storage of nuclear wastes—although they could be fortuitous consequences of Oklo's particular geological setting.

experience with Gen-I units. Gen-II reactors were to have 40-year lifespans, but many have had their operating licenses extended.

Generation III reactors appeared in the 1990s. These were evolutionary improvements on Gen-II designs, and they featured greater efficiency, passive safety systems that reduced reliance on pumps and other mechanical systems, and standardized designs with modular construction for better economics. Few Gen-III reactors were built due to a worldwide slump in the nuclear power industry. Gen-III was superseded by **Generation III+**, a further evolution with more passive safety features, including gravity-driven emergency cooling systems, more efficient use of nuclear fuel, and decreased nuclear waste production. Reactors under construction today are nearly all Gen-III+ models. Figure 7.8 is a cutaway diagram of the Westinghouse AP1000 Gen-III+ reactor.

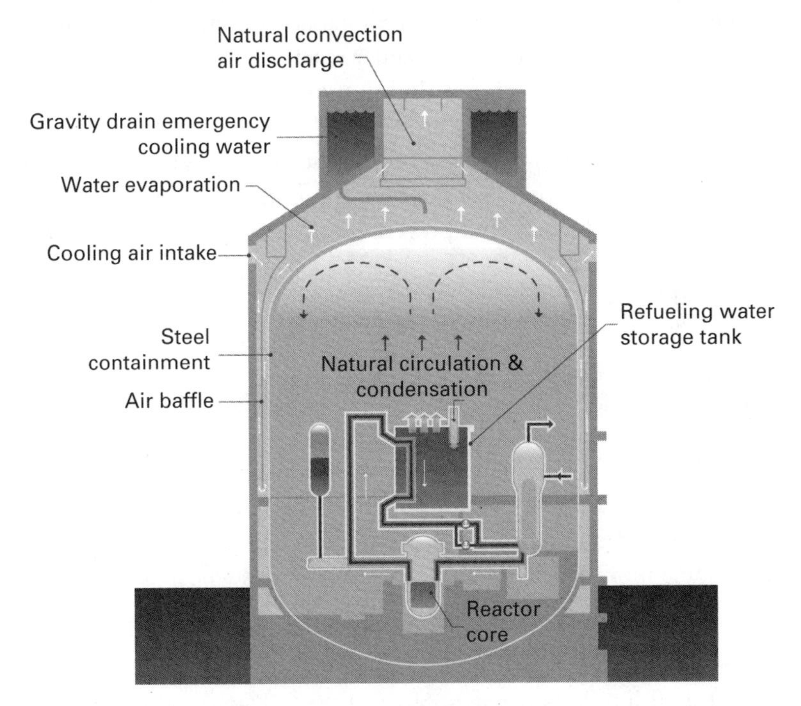

Figure 7.8
Cutaway diagram of the Westinghouse AP1000 reactor emphasizing its passive emergency cooling features. A number of these reactors are under construction around the world, including in the United States. The first completed AP1000 went online in 2018 at China's Sanmen nuclear plant. (Courtesy of Westinghouse Electric Company.)

Small Modular Reactors

Although most Gen-III+ reactors are large units with electric power outputs in the 1,000-MW range, a few are **small modular reactors** (SMRs) rated from tens to a few hundred megawatts. SMRs are often factory-built systems that are delivered to sites where their power is needed. Some are sealed units with enough fuel to last the reactor's lifetime. Russia has pioneered barge-based SMRs, scaled-down versions of nautical propulsion reactors, that can be permanently anchored offshore to deliver power to a city or other end user; the first of these went online in 2019. Argentina is constructing a 25-MW SMR with a natural-circulation cooling system contained entirely within the reactor vessel. SMRs should offer greater safety due to their lower power and therefore reduced inventory of radioactive waste, self-contained integral construction, possible underground siting, and design simplicity. They should also offer economic benefits associated with modular design and factory production. A number of companies around the world are currently preparing to market small modular reactors.

This brief history of reactor evolution shows that most reactors operating or under construction today are either of decades-old design or represent only evolutionary improvements on those designs. Where are the truly revolutionary reactors that could deliver dramatic improvements in safety, efficiency, waste reduction, and prevention of weapons proliferation?

Gen-IV and Beyond

In 2001, a group of countries that rely significantly on nuclear power joined to form the **Generation IV International Forum** (GIF), with the goal of collaborating on advanced reactor designs. Since then GIF has grown to 14 countries and has identified seven reactor concepts to pursue. The hope is that some of these reactors will be in commercial operation by 2030. Here we describe concepts under consideration by GIF, then discuss some more speculative reactor possibilities whose deployment, if ever, is likely farther off.

GIF Reactor Concepts

In contrast with today's reactors, most of the GIF concepts are unmoderated fast-neutron reactors. Unlike true breeder reactors, though, these are designed to produce plutonium within the reactor core itself and to burn that plutonium as part of regular reactor operation. This so-called closed fuel cycle greatly reduces the proliferation risk associated with removing irradiated fuel from the reactor and processing it elsewhere to extract plutonium (more on fuel cycles in chapter 9). GIF's concept reactors also tend to operate at higher temperatures than today's light-water designs, leading to greater thermodynamic efficiency (recall the second law efficiency limitations introduced in chapter 6) and gives some reactors the capability of producing hydrogen from thermal dissociation of water. Since hydrogen can serve as a transportation fuel, this would expand nuclear energy's usefulness beyond the electricity sector.

Among the GIF ideas for fast-neutron reactors is the **gas-cooled fast reactor** (GFR), using helium gas as its coolant. The hot gas can directly drive a jet-engine-like gas turbine, with the turbine exhaust still hot enough to boil water to steam and drive a conventional steam turbine. The resulting power plant would be thermodynamically similar to gas-fired combined-cycle power plants described in chapter 6 and would exhibit similar high efficiency. The **lead-cooled fast reactor** (LFR) is a different design, using liquid lead or lead alloy as its coolant. The LFR could burn a variety of nuclear fuels, including transuranic waste from light-water reactors, thus reducing the longest-lived component of today's nuclear waste. Another concept using liquid-metal coolant is the **sodium-cooled fast reactor** (SFR). This scheme is related to early breeder designs and is closest of the GIF concepts to realization. Russia's BN-800, which became operational in 2016, is a prototype SFR. Figure 7.9 diagrams the SFR concept. Advocates claim that the SFR will reduce the burden of nuclear waste by burning long-lived transuranics, but critics point to serious problems in the U.S. Department of Energy's attempts to separate highly reactive sodium from nuclear waste in its experimental breeder program.

A very different concept is the **molten salt reactor** (MSR), the only reactor with fuel in liquid form. In its fast-neutron version the fissile fuel—uranium-233 bred by neutron bombardment of thorium-232—is dissolved in a molten salt composed of lithium fluoride (LiF) and thorium fluoride (ThF_4, the breeding stock for fissile U-233). This liquid system obviates the need to manufacture solid fuel and gives the MSR lower inventories of fissile fuel than other reactor designs—a safety advantage. Other fissionable isotopes, including transuranic waste from light-water reactors, could be mixed into the liquid, where fast neutrons would transmute them to shorter-lived isotopes (more on transmutation in chapter 9). A further advantage is that U-233 is less likely than U-238 to absorb neutrons, and when it does it produces much less long-lived transuranic waste. Removal of fission products and transuranics can occur continuously in the circulating liquid. The liquid salt serves not only to carry the fuel but also acts as coolant; since this is a fast reactor there's no moderator.

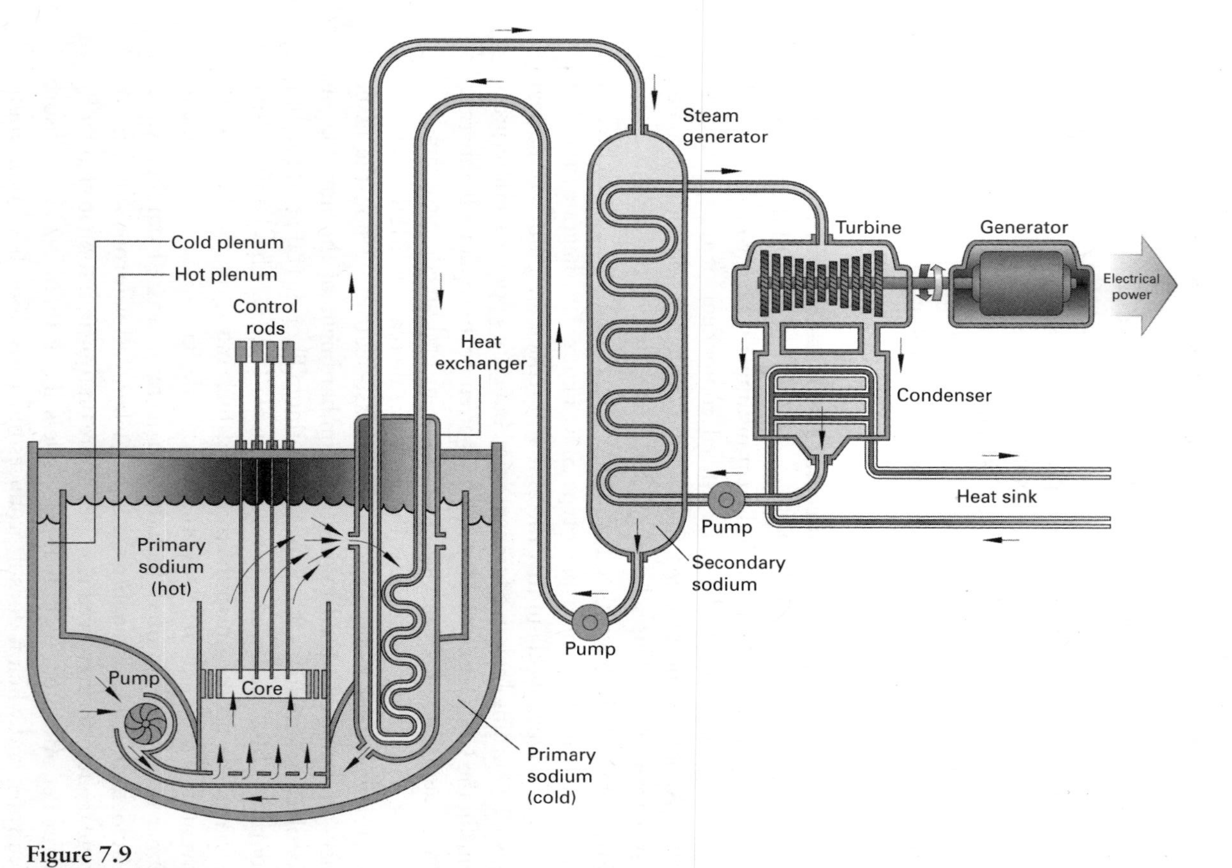

Figure 7.9
Diagram of a sodium-cooled fast reactor. The reactor core and liquid-sodium primary coolant occupy the container at left, which is at essentially atmospheric pressure. The secondary cooling loop is also sodium filled, while the tertiary loop contains water that boils in the steam generator. (Adapted from U.S. Department of Energy, Idaho National Laboratory.)

A second MSR variant is the *advanced high-temperature reactor*, a slow-neutron version with a solid graphite moderator. Both MSRs could produce hydrogen in addition to electricity. The basic molten salt reactor concept isn't new; demonstration slow-neutron MSRs were built in the 1950s but were never commercialized. Despite MSR's long history, the Gen-IV International Forum considers the concept in need of substantially more research and design, with the R&D phase likely to continue through the 2020s.

The two final Gen-IV Forum concepts are slow-neutron reactors, although one has a fast variant. In the **supercritical water-cooled reactor**, the term *supercritical* refers not to the nuclear reaction but to cooling water that's under such high temperature and pressure that it's in a state where liquid and gaseous phases become indistinguishable. Supercritical technology is used in modern coal-fired plants, where the high temperature and the fact that energy isn't needed to change liquid water to gas contribute to efficiencies that are some 30 percent higher than in conventional coal plants. A similar efficiency increase should accrue to supercritical nuclear plants. A fast-neutron variant would share a closed fuel cycle with other Gen-IV fast reactors. In contrast to the prevalence of water cooling in today's operating reactors, the supercritical reactor is the only Gen-IV water-cooled concept. The second slow-neutron concept is the **very high temperature gas reactor**, a graphite-moderated system cooled with helium gas. This design could produce hydrogen or drive a highly efficient gas turbine combined-cycle generating system. Its enriched uranium fuel takes the form of millimeter-sized pellets embedded in high-temperature ceramics, ensuring that fission products remain contained within the pellet. One variant, the *pebble-bed reactor*, incorporates these pellets into billiard-ball sized graphite pebbles (figure 7.10); another puts them into graphite blocks. Either way, the nuclear fuel and nearly all its waste products are locked into what are, in effect, miniature containment systems. The pebbles or bricks would themselves be disposed of as nuclear waste, increasing the waste volume over conventional designs but making waste handling considerably easier. Thousands of pebbles or blocks constitute the reactor core, with helium coolant flowing among them. Proponents claim that the concept is intrinsically safe, and experimental pebble beds have been operated with the cooling system turned entirely off—yet they've sustained no damage. The Chinese reactors featured in chapter 6's box "Nuclear-to-Fossil and Fossil-to-Nuclear" are pebble-bed units.

The Gen-IV reactor concepts described here are in various stages of research and development. Some, such as lead-cooled and sodium-cooled fast reactors, are based on proven nuclear technologies. Others, including

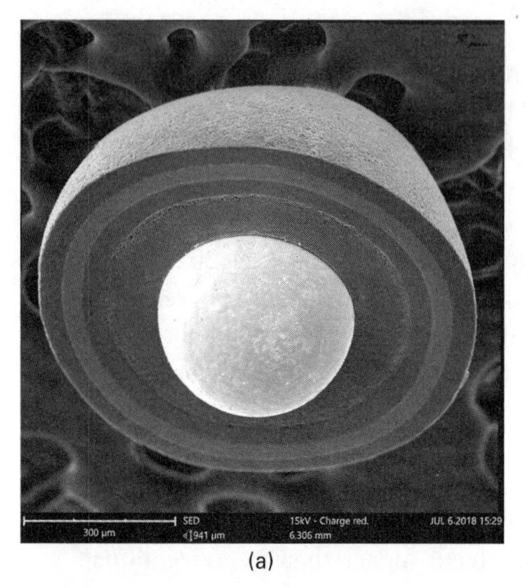

(a)

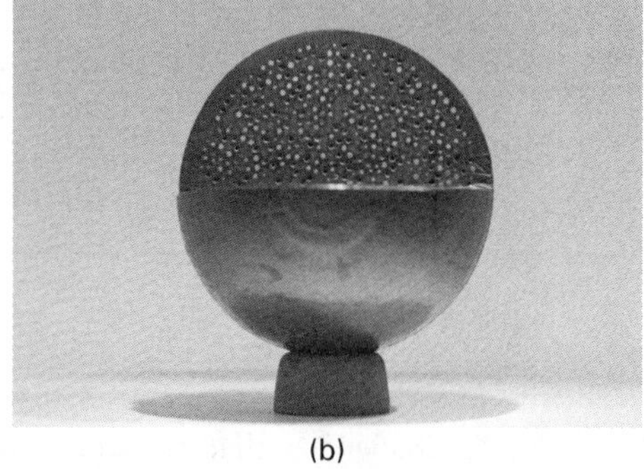

(b)

Figure 7.10
(a) Cutaway image of a single fuel particle developed by X-energy for their XE-100 high-temperature pebble-bed reactor. The particle is just under 1 millimeter in diameter and consists of an inner sphere of uranium carbide/oxide encapsulated in carbon and silicon carbide. (b) A fuel pebble, which measures 60 millimeters (about the size of a billiard ball) in diameter, contains tens of thousands of fuel particles. Individual particles are visible in this cutaway image, mixed with graphite for neutron moderation. Some 224,000 pebbles comprise the reactor core. (Both images courtesy of X-energy.)

the molten salt reactor and the pebble bed, have long histories as ideas but are still far from commercial realization. Even a few successful Gen-IV concepts could revolutionize the nuclear industry and indeed our entire energy production system, but we can't count on that and in any event we're talking decades in the future.

Far-Out Reactor Ideas

The Gen-IV Forum's reactor concepts have been thoroughly vetted by an international group of experts, with the expectation that all are worth exploring. Other advanced reactor designs are more speculative. One example is TerraPower's traveling-wave reactor, a fast-neutron system cooled with liquid sodium. The innovation here is a reactor that needs no refueling over its 40-year lifetime, thus reducing uranium mining and enrichment needs. Its core contains a mix of fuel rods, some with uranium enriched to nearly 20 percent U-235 and the others containing depleted uranium—mostly U-238 left over from the enrichment process. The reaction starts in U-235, whose fission neutrons then breed Pu-239 from U-238 in the depleted-uranium rods. The Pu-239 then fissions, continuing the process in a wave that sweeps from the center of the core outward—hence the name "traveling wave." Every few years the fuel rods are reshuffled, as the original enriched uranium rods have now become depleted and are ready as breeding stock, while the originally depleted rods now contain actively fissioning Pu-239. Technical and commercial viability of the traveling-wave reactor, if ever achieved, are a long way off. In 2015 TerraPower and the China National Nuclear Corporation agreed to work jointly on traveling-wave development, but that soon ended with the Trump administration's ban on export of nuclear technology to China. However, TerraPower enjoys the support of Microsoft founder Bill Gates, who chairs its board.

The **nuclear battery** would be an extreme variant of the small modular reactor concept, with power output of a few megawatts to about 25 MW and fast neutrons to breed plutonium that's then consumed in the reactor. The reactor could run for decades without refueling, at which point the entire reactor vessel would be replaced with a new one containing fresh fuel—much like replacing a battery. Nuclear batteries are envisioned as more economical replacements for diesel generators in remote locations.

A third speculative design is one that doesn't rely on a fission chain reaction. In the **accelerator-driven subcritical reactor** (ADSR), a high-energy proton beam slams into a lead target, releasing neutrons that both breed U-233 from thorium-232 and induce fission in the U-233. The mass of fissile U-233 remains subcritical, which works because accelerator-produced neutrons take the place of some fission neutrons in keeping fission going. The reaction stops immediately if the proton beam is turned off, and because it's subcritical there's no way it can go out of control. The accelerator-based reactor shares the molten salt reactor's advantage of producing far less waste from U-233 fission than in a reactor burning U-235 or Pu-239. The ADSR would also be especially useful for burning transuranic waste. That's because putting too much transuranic material in an ordinary reactor makes it hard to control the delicate balance of the just-critical chain reaction. Since the ADSR is subcritical, that isn't an issue.

Summary

Nuclear reactors come in many types. Common to nearly all reactors are *control rods* that regulate the number of neutrons and therefore the rate of fission, and a *coolant* to carry off heat. Slow-neutron reactors make use of the fact that uranium-235 undergoes fission more readily when struck by slow neutrons, and therefore employ a *moderator* to slow neutrons from the high speeds they have when they emerge in fission events. The most common moderators are water, heavy water, and graphite. Because slow neutrons are highly efficient at causing fission, most slow-neutron reactors operate with natural or only slightly enriched uranium. Slow-neutron reactors exhibit a variety of pros and cons relating to safety, economics, and weapons proliferation.

Fast reactors, in contrast, use fast neutrons to sustain the chain reaction. They require highly enriched fuel—often a uranium-plutonium mix—to overcome fast neutrons' lower effectiveness in causing fission. True *breeder reactors* produce more fuel than they consume, by converting nonfissile uranium-238 into fissile plutonium-239. (In this context U-238 is termed a **fertile isotope** because, although it isn't fissile, it can be used to breed a fissile isotope.) Breeder reactors are more complex than conventional designs, and they raise the specter of international plutonium trade. Advanced fast reactor designs would avoid this issue by burning the plutonium they produce right in the reactor itself.

Most of today's operating reactors are based on decades-old designs, while reactors under construction are evolutionary advances on those designs, with greater efficiency, safety, and proliferation resistance. Future reactor designs, particularly those endorsed by the Gen-IV collaboration, would feature mostly fast-neutron designs and include very efficient high-temperature schemes and liquid-fuel molten salt reactors.

Glossary

accelerator-driven subcritical reactor (ADSR) A possible future reactor in which neutrons are produced in reactions driven by a high-energy proton beam. The ADSR does not use a critical mass and therefore is intrinsically safe.

boiling-water reactor (BWR) A reactor in which water boils within the core to produce steam that directly drives a turbine-generator.

breeder reactor A reactor designed to produce more fissile material than it consumes, by converting nonfissile uranium-238 into fissile plutonium-239.

CANDU A Canadian reactor design using heavy water for moderator and coolant and fueled by unenriched uranium. The acronym stands for Canada deuterium uranium.

control rod A rod made of neutron-absorbing material that's inserted into a nuclear reactor to control the fission rate and therefore the reactor's power.

coolant A fluid that circulates through the core of a nuclear reactor to remove heat and, in a power reactor, ultimately deliver it to a turbine-generator that produces electricity.

core The heart of a nuclear reactor, containing nuclear fuel, moderator, and control rods.

delayed neutrons Neutrons that emerge in the decay of short-lived fission products, rather than in fission events themselves. Delayed neutrons help maintain reactor control by increasing the time for significant changes in power.

fast breeder A breeder reactor in which fast neutrons sustain the chain reaction.

fast neutrons High-energy neutrons, generally with speeds comparable to the speed at which they're ejected in fission events.

fertile isotope An isotope that, although not itself fissile, can be used to breed a fissile isotope. An example is U-238, which can breed Pu-239.

fuel bundle An assembly of fuel rods.

fuel pellet A small (~ ½ inch) cylinder of uranium dioxide (UO_2) that is the basic fuel unit in light-water reactors.

fuel rod A zirconium-alloy tube containing uranium fuel pellets.

gas-cooled fast reactor A Gen-IV fast-neutron design cooled with gaseous helium. Its high temperature allows the helium to drive a gas turbine.

Generation I reactors Early reactors deployed in the 1950s and early 1960 as proofs-of-concept for nuclear power.

Generation II reactors Reactors deployed from the 1960s to early 1990s and comprising the majority of reactors operating today.

Generation III and III+ reactors The few reactors deployed in the 1990s and early 2000s were Gen-III reactors, then superseded by Gen-III+ designs that are under construction today. Gen-III and Gen-III+ feature evolutionary improvements on the older Gen-II designs, including passive safety systems and higher efficiency.

Generation IV International Forum (GIF) A collaboration of 14 countries organized to pursue research and development of Gen-IV reactor concepts that are considered to have a reasonable chance of commercial success. Most GIF reactor concepts are fast-neutron designs radically different from today's operating power reactors. They generally offer higher temperature and efficiency, the ability to produce hydrogen as well as electricity, proliferation resistance, capability of burning transuranic waste, and other advanced features.

generation time The average time between emission of a neutron in a fission event and that neutron's causing a subsequent fission event.

heavy water Water whose hydrogen nuclei are those of the heavy hydrogen isotope deuterium (^2_1H).

heavy-water reactor (HWR) A nuclear reactor that uses heavy water as its moderator. Heavy water absorbs fewer neutrons than ordinary (light) water, permitting the use of unenriched uranium in heavy-water reactors.

lead-cooled fast reactor A Gen-IV fast-neutron reactor using molten lead or lead alloy as its coolant. Derived from Russian nautical reactors, the LFR should be able to burn a variety of fuels including transuranic waste from LWRs.

light water Ordinary water, whose hydrogen nuclei consist of a single proton (^1_1H).

light-water reactor (LWR) A nuclear reactor using ordinary (light) water as its moderator. Common boiling-water and pressurized-water reactors are LWRs.

meltdown Melting of a reactor core due to loss of coolant.

moderator A substance placed in a nuclear reactor in order to slow neutrons. Common moderators include water, heavy water, and graphite.

molten salt reactor (MSR) A Gen-IV fast-neutron design in which U-233 fissile fuel, derived from neutron breeding in thorium-232, is dissolved in a mix of molten salts, including thorium fluoride to breed the U-233. The only reactor featuring liquid-based fuel.

multiplication factor The average number of neutrons from each fission event that go on to cause another fission. In a reactor operating at steady power, the multiplication factor must be exactly 1.

nuclear battery A very small reactor, still on the drawing board, that would run for decades without refueling and then have its entire pressure vessel replaced by a new vessel containing fresh fuel.

nuclear reactor A system, containing nuclear fuel and other materials, designed to sustain and control a fission chain reaction.

pressurized-water reactor (PWR) A reactor in which cooling water circulating through the core is kept under enough pressure that it doesn't boil. This primary cooling water then transfers its heat to a secondary system in which water does boil, producing steam to drive a turbine-generator.

primary loop The part of a reactor's cooling system that brings coolant into direct contact with the core.

prompt criticality The dangerous reactor condition when the chain reaction is sustained by prompt neutrons alone, which risks rapid loss of control.

prompt neutrons Neutrons emitted during a fission event, as opposed to delayed neutrons that are emitted a brief time later.

RBMK A Russian reactor design using a graphite moderator and water cooling. The Chernobyl reactor was an RBMK.

scram Emergency shutdown of a reactor, accomplished by fully inserting the control rods.

secondary loop A loop of coolant fluid that does not come into direct contact with the core but picks up heat from the primary loop.

slow neutrons Low-energy neutrons moving much more slowly than neutrons ejected in fission events. Slow neutrons are far more effective than fast neutrons in causing fission of uranium-235.

small modular reactor (SMR) A small reactor with power output from tens of megawatts to a few hundred MW, often designed as a manufactured unit that's delivered by land or sea to its destination.

sodium-cooled fast reactor A Gen-IV fast-neutron reactor using liquid sodium metal as its coolant. Derived from early fast breeder designs.

steam generator A device in which nonboiling coolant transfers its heat to water, which then boils to produce steam. Used in most reactor systems except boiling-water reactors.

supercritical water-cooled reactor The only water-cooled reactor among the GIF Gen-IV concepts. Uses supercritical water to achieve high thermal efficiency. Likely to be a slow-neutron reactor, but fast-neutron designs are also under consideration.

thermal neutrons Same as slow neutrons; "thermal" refers to neutrons whose energies are consistent with the temperature in a reactor.

very high temperature gas reactor A graphite-moderated helium-cooled Gen-IV reactor design, one of whose variants is the pebble bed reactor.

Note

1. Although the term "burn" is generally associated with chemical burning (combining with oxygen), it's also used to describe the consumption of nuclear fuel in a reactor.

Further Reading

Bascomb, Neal. *The Winter Fortress: The Epic Mission to Sabotage Hitler's Atomic Bomb*. Boston: Houghton Mifflin Harcourt, 2016. A carefully researched and gripping account of the Norwegian raid that kept heavy water out of Nazi hands.

Kadak, Andrew. *A Comparison of Advanced Nuclear Technologies*. New York: Columbia University Center on Global Energy Policy, 2017. PDF download at https://energypolicy.columbia.edu/sites/default/files/A%20Comparison%20of%20Nuclear%20Technologies%20033017.pdf. An accessible review of current and next-generation nuclear technologies, including Gen-IV and beyond.

Murray, Raymond, and Keith Holbert. *Nuclear Energy: An Introduction to the Concepts, Systems, and Applications of Nuclear Processes*, 8th ed. Cambridge, MA: Butterworth-Heinmann, 2020. A textbook for undergraduate engineering majors, this one covers our material much more quantitatively and in much greater depth. Chapters 16–20 are especially relevant to this chapter. Not for the mathematically challenged!

United States Nuclear Regulatory Commission. *2011 Essential Guide to Nuclear Power Plants and Nuclear Energy*. Washington, DC: U.S. Government Printing Office, 2011. Also in Kindle edition.

World Nuclear Association. *Generation IV Nuclear Reactors*. http://www.world-nuclear.org/information-library/nuclear-fuel-cycle/nuclear-power-reactors/generation-iv-nuclear-reactors.aspx. Describes the GIF collaboration and the reactor concepts it's pursuing. A nice table lists the proposed designs by neutron speed, coolant, temperature, pressure, fuel and fuel cycle, typical power output, and proposed use.

World Nuclear Association. *Nuclear Power Reactors*. http://www.world-nuclear.org/information-library/nuclear-fuel-cycle/nuclear-power-reactors/nuclear-power-reactors.aspx. A brief review of reactor types, emphasizing those in operation today. Separate links cover advanced reactor concepts.

8

Reactor Safety

Are nuclear reactors safe? We raised that question repeatedly in the preceding chapter, comparing the pros and cons of each reactor type. Here we look further at reactor safety, considering both normal operation and the potential for nuclear accidents. The events at Three Mile Island, Chernobyl, and Fukushima illustrate how sequences of errors or equipment failures can get nuclear power plants into big trouble.

Reactor safety is far from an academic issue. Concern for public and environmental safety is one reason for the worldwide decline in nuclear power that we'll discuss in chapter 10. Yet safety issues are complex and subtle, and intelligent debate requires knowledge about the operation of nuclear reactors as well as the safety implications of alternatives to nuclear power.

Normal Reactor Operation

Most nuclear power plants operate day after day without anything newsworthy happening. Even the second reactor at Three Mile Island continued to operate quietly for four decades after the famous accident involving its twin, and the last of the Chernobyl reactors churned out electricity into the twenty-first century. What effect does normal reactor operation have on public health and safety?

Advocates of nuclear power argue that nuclear plants are among the cleanest of energy sources. In terms of sheer quantity of effluents, that's certainly true. A 1,000-megawatt coal-burning power plant produces some 600 pounds of carbon dioxide gas, 20 pounds of ash, and a pound of acid-rain-causing sulfur dioxide *every second*. In contrast, most products of nuclear fission remain locked within the fuel rods of a properly operating reactor. A very few fission products—mostly radioisotopes of the gases krypton and xenon, and small amounts of radioactive iodine—work their way through the fuel-rod cladding material and into the cooling water, and neutrons absorbed in that water give rise to radioactive tritium. These substances are routinely released to the environment—tritium in the plant's liquid discharges and gases to the atmosphere.

How serious are these normal radiation releases? Chapter 3 introduced the curie and becquerel as units of radioactivity and noted that the amounts of radioactive iodine-131 released at Three Mile Island and Fukushima were 15 curies (500 billion Bq) and 15 million curies (500 quadrillion Bq), respectively. These accidental releases occurred over a few days. In contrast, a 1,000-MW nuclear power plant normally releases less than 0.03 curies (100 million Bq) of I-131 per year. Much larger quantities of radioactive krypton and xenon are also released, but these chemically inert gases pose little health threat. The U.S. Nuclear Regulatory Commission limits the maximum public exposure from a nuclear power plant's effluents to 0.5 millisieverts per year. In practice, individuals living in the immediate neighborhood of a nuclear plant typically receive about 0.01 mSv per year, negligible compared with the 6-mSv average background dose in the United States. That 0.01 mSv is about the same dose you would get from cosmic rays on a flight from New York to Chicago. Incidentally, coal-burning power plants also release radiation—in the form of uranium, thorium, and their decay products, which occur naturally in coal—and the radiation dose from a coal plant may, depending on the coal source, exceed that from a nuclear plant of comparable size.[1]

It sounds as if normal emissions from nuclear power plants are pretty benign. On the other hand, you saw in chapter 4 that the effects of low-level radiation on children and the unborn are underemphasized in the official standards. While reports of bizarre deformities among creatures born near nuclear power plants are certainly hype, nagging suggestions of increased human infant mortality, cancer, and other maladies in the vicinity of nuclear plants surface occasionally. But the numbers are usually so small that it's impossible to reach statistically valid conclusions. What's clear is that people aren't dying in droves from the effects of nuclear power—like they are from smoking, drunk driving, and handguns, or, for that matter, from coal-burning power plants, whose emissions probably cause some 10,000 deaths each year in the United States and a million in China. A recent Florida study[2] suggested that living near a fossil- or waste-fueled power plant is associated with an increased risk of adverse birth outcomes. No such association was found when the nearest power plant was nuclear, and in fact premature births were actually lower in the vicinity of nuclear plants.

When Things Go Wrong

A nuclear power plant is a complicated system, in which a lot can go wrong. Abnormal incidents occur frequently at nuclear plants, but most are minor and don't result in radiation release or other danger. Occasional errors or equipment failures do result in radiation emission exceeding health and safety standards—which, in the United States, are set by the **Nuclear Regulatory Commission** (NRC). Other failures—for example, in backup safety systems—raise the potential for serious accidents, but in most cases those accidents don't materialize. Only rarely does a sequence of failures compound into a major reactor accident.

The core of a nuclear reactor contains an enormous amount of highly radioactive material, whose dispersal to the environment is probably the worst thing that can happen at a nuclear plant. Numerous safety systems exist to prevent accidents that could damage the core and to contain radiation in the event that an accident does occur. How likely are such major accidents? How effective are the safety systems?

Overheating of nuclear fuel is the most probable cause of a major reactor accident. That could occur either through an increase in the fission rate or by a loss of coolant. You've already seen how different reactor designs differ in their potential for such occurrences. The Russian RBMK reactor is prone to a runaway chain reaction—as happened at Chernobyl. On the other hand, the RBMK's massive graphite moderator absorbs heat, preventing rapid temperature rise in the event of coolant loss. Light-water reactors are susceptible to very different safety problems. Since the coolant is also the moderator, the chain reaction in a light-water reactor stops automatically in the event of coolant loss. But even after shutdown, the reactor core continues to generate heat from radioactive decay of fission products. Immediately after shutdown, decay heat amounts to some 7 percent of the reactor's normal power output, dropping to 1 percent after several hours. Even a fraction of a percent of full power can be tens of millions of watts, a rate of energy production that requires active cooling to prevent damage to the core. For this reason, the worst-case accident scenario in a light-water reactor is a **loss-of-coolant accident** (LOCA).

Numerous events could lead to a LOCA. Pipes could leak or break, dumping cooling water. Circulation pumps could fail or lose power (as happened at Fukushima), allowing the water in the reactor vessel to boil away. Obstructions in piping could block coolant flow. Less likely is a rupture of the reactor vessel itself, accompanied by catastrophic coolant loss. These failures could occur because of defective materials, pressure surges, human error, corrosion, or external factors such as earthquakes, airplane crashes, terrorism, or acts of war.

In the event of a LOCA, part or all of the reactor core may be exposed to air. The temperature of the uranium fuel then rises rapidly, melting the zirconium alloy cladding that surrounds it. At that point, radioactive fission products spread freely throughout the reactor vessel and the primary cooling loop. They may also escape through whatever breach in the cooling system caused the loss of coolant in the first place. For that reason, a heavy concrete **containment** structure surrounds the reactor vessel (and, in a PWR, the entire primary cooling loop; in a boiling-water reactor, the primary coolant loop extends to the turbine-generator, outside the containment.) The containment is designed to contain not only radioactive material but also pressure from hot gases and steam. On the other hand, containment structures may not withstand either the most violent events possible within a reactor or energetic external impacts like airplane crashes, so there's a small chance that a LOCA could lead to breaching the containment and widespread dispersal of radioactive fission products.

A further complication during a LOCA is the possibility of a chemical reaction between the zirconium cladding and high-temperature steam, producing explosive hydrogen gas (H_2). As you'll see, this occurred in all three of the major nuclear power accidents.

In a serious LOCA, not only the fuel-rod cladding but also the uranium fuel itself may melt. In a **meltdown**, 100 tons of molten uranium could melt its way through the reactor vessel and the containment floor, possibly resulting in a large steam explosion when molten uranium reaches groundwater. The term *China syndrome*—popularized as the title of a 1979 film thriller about an accident at a U.S. nuclear plant—describes such a meltdown, in which molten uranium melts its way into the ground, heading for China. (Today, with nuclear power booming in China but not in the U.S., perhaps things would go the other way.) In reality, the uranium would be cooled and contained within some tens of feet, and this type of containment failure might be less serious than an explosive breach of the containment with direct dispersal of radioactive material to the atmosphere. Another meltdown scenario has molten uranium regrouping in a critical configuration, starting an uncontrolled chain reaction. There's some evidence that such a *criticality event*, although brief, may have happened at Fukushima—a possibility uncovered by one of your authors (FDV).[3]

Loss of coolant need not lead to meltdown, provided replacement water reaches the reactor core. That's the job of the **emergency core-cooling system**

(ECCS). Emergency core cooling is the most important safety system in a nuclear plant, and advocates of nuclear power point to the ECCS as a prime example of engineered safety. Critics, on the other hand, note that there's never been a full-scale test of an emergency core-cooling system. Advocates counter by pointing to ECCS tests at small experimental reactors, to computer simulations that indicate ECCS failure is unlikely, and to successful ECCS operation in a number of accidents that stopped short of meltdown. Critics point to accidents where the ECCS has been disabled, either through equipment damage or human error. Operators at both Three Mile Island and Chernobyl turned off their emergency core-cooling systems, for reasons we'll examine shortly, while at Fukushima the backup generators for powering ECCS pumps were flooded and inoperable.

Nuclear Accident Analyses

Regulatory agencies, academic nuclear professionals, antinuclear groups, and others have performed studies to assess the chances of nuclear accidents and the risks they pose to public health. These generally use **probabilistic risk assessment** (PRA), a technique that seeks to answer three questions: (1) What can go wrong? (2) How likely is it? and (3) What are the consequences? A typical PRA uses a *fault tree* to follow the consequences of any one failure, as it might be compounded by subsequent failures. A successful PRA therefore requires identifying possible failures and assigning each a probability of occurrence.

An early study, and the first to use PRA in the nuclear context, was the 1975 WASH-1400, or Rasmussen Report, by the then U.S. Atomic Energy Commission. WASH-1400 suggested that dispersal of radioactive material after a containment breach could kill several thousand people immediately and nearly 50,000 more with eventual cancers. Subsequent studies have reduced those projections substantially. The latest report, the U.S. Nuclear Regulatory Commission's 2012 State-of-the-Art Reactor Consequence Analysis (SOARCA),[4] examined the consequences of six hypothetical accident scenarios at two actual U.S. nuclear plants, one BWR and one PWR. SOARCA considered that operator response and safety equipment might mitigate accidents, but they also modeled scenarios in which mitigation doesn't occur. In evaluating the public health impact, SOARCA assumed that successful evacuations would take place. A quick read of SOARCA's summary results suggests there's almost no chance of anyone developing fatal cancer as a result of radiation exposure due to the serious accident scenarios considered. For the unmitigated scenarios, the annual risk of fatal cancer for an individual within 10 miles of the power plant range from 1 in 1 billion to 1 in 100 billion. For all but one of the mitigated scenarios, the claimed risk is zero; for the one case where it's not, the risk is 1 in 10 billion.

But a more careful read of the SOARCA report shows that those odds factor in the very small chance of an accident actually occurring. So that cancer risk of 1 in 10 billion isn't your risk if an accident were to occur; rather, it's your risk if you live for a year within 10 miles of the nuclear plant—during which time an accident is very unlikely. SOARCA's summary report doesn't state your odds if an accident should occur, but they're available in the full report. One SOARCA accident scenario considered for Pennsylvania's Peach Bottom nuclear plant leads to some 1,000 fatal cancers among the 5.5 million people living within 50 miles of the plant. That compares with nearly a million cancer deaths normally expected among that population.

Given the SOARCA results, can those of us living near light-water reactors sleep easier? Is SOARCA too optimistic, or are even its low numbers still unacceptable? Nuclear skeptics, especially the Union of Concerned Scientists (UCS), have challenged SOARCA's methodology and its assumptions about mitigation procedures and evacuations. For example, SOARCA assumes that mitigation procedures (some of them using new equipment mandated since the September 11, 2001, attacks) would be in place and would work flawlessly—and that personnel would know how to use it. The UCS also points out that the SOARCA modeling uses average weather conditions, and that health effects could be a factor of 10 worse under unfavorable conditions. UCS is also troubled that none of the accident scenarios involved multiple reactors—despite the fact that the power plants studied each have two reactors onsite. One scenario even considered an earthquake, which could easily damage both reactors. At Fukushima, which had similarly not planned for multiple-reactor incidents, five of six reactors were crippled. Several SOARCA scenarios envision bringing in outside equipment to replace damaged pumps and other safety systems—but they don't consider the difficulties of doing so in the aftermath of a major natural disaster. Again, Fukushima experienced just such difficulties. Critics also note that SOARCA doesn't consider the most severe conceivable accident scenarios because they're deemed extremely unlikely.

The Big Ones: TMI, Chernobyl, and Fukushima

The nuclear accidents at Three Mile Island in the United States, Chernobyl in Ukraine, and Fukushima in Japan aren't the world's only serious nuclear reactor accidents. But they're the most severe among commercial nuclear power plants and most familiar to the public. A close look at these accidents shows a complex interplay of nuclear physics, chemistry, reactor design, safety systems, government regulations, and human behavior. Follow-up studies of Three Mile Island, Chernobyl, and Fukushima also show how these accidents could have been prevented—and also how they might have been worse.

Three Mile Island

The 1979 accident at Metropolitan Edison's Three Mile Island (TMI) nuclear power plant in Pennsylvania (figure 8.1) ranks as the most serious commercial nuclear power incident in the United States. At 4 A.M. on March 28, 1979, for reasons that have never been fully explained, pumps tripped off in the condenser system of TMI's Unit 2 reactor, halting the flow of cooling water in the secondary loop. A bypass valve was supposed to open to keep the water flowing, but it failed. Immediately the turbine tripped off and backup pumps came on to keep the cooling water flowing—both actions that happened automatically, as they were supposed to. Unfortunately, valves on the backup pumps had inadvertently been left closed, and a paper maintenance tag on the reactor control panel

Figure 8.1
The Three Mile Island nuclear power plant. The containment structures for reactor Units 1 and 2 are the cylindrical buildings at right center. The accident occurred in the Unit 2 reactor, at right. (Nuclear Regulatory Commission)

obscured the lights indicating the status of those valves. With no way to shed heat, temperature and pressure in the reactor vessel began to rise. What had been a minor failure in the condenser system now threatened the reactor itself.

Eight seconds after the start of the accident, the reactor automatically scrammed, inserting control rods to halt the chain reaction. But radioactive decay of fission products continued to produce heat, and pressure in the reactor rose enough to open a pressure relief valve at the top of the reactor vessel. The pressure then fell. Already in these first few seconds, one minor equipment failure had occurred, one safety system had started as designed but had been foiled by closed valves, and several other systems had acted perfectly to mitigate the pressure buildup. All would have been well but for another equipment failure.

As the reactor pressure dropped, the relief valve at the top of the reactor vessel was supposed to close. But it stuck open—despite a control-panel light suggesting it had closed. As a result, there was a breach in the primary coolant system through which water and steam could escape into the containment structure (figure 8.2). TMI was experiencing a loss-of-coolant accident, but the operators didn't realize it.

Less than a minute into the accident, the operators turned on pumps to inject additional water into the primary cooling loop, as they were trained to do in order to maintain pressure. But the pressure continued to fall, due to the stuck valve that the operators couldn't know about. This confused the operators, who didn't grasp that a LOCA was underway. Fortunately, the emergency core-cooling system came on automatically, as it was designed to do. But the operators, confused by conflicting alarms and signals, were most concerned that water levels in parts of the coolant loop outside the reactor vessel were rising. So, four minutes into the accident, they cut the emergency cooling water flow to almost nothing.

Confusion prevailed for the next two hours as operators struggled with a situation they still thought was caused by too much water (figure 8.3). What was happening, in fact, was that steam in the reactor vessel was pushing up water levels in other parts of the system. But the control panel didn't have a direct indication of water level in the reactor itself, so the operators were understandably misled, and they continued to believe they had only a minor incident on their hands. All the while, though, the water level in the reactor vessel was dropping.

Soon the primary-loop coolant pumps began to vibrate excessively, due to the buildup of steam in what was supposed to be only liquid water. To avoid pump damage, operators shut off one pump and throttled

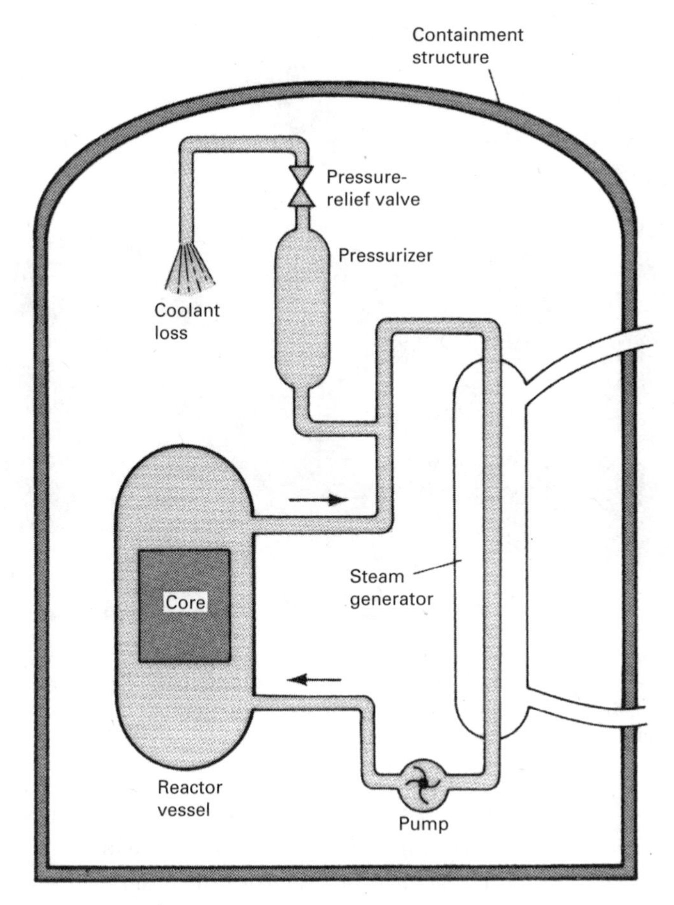

Figure 8.2
Simplified diagram of the TMI Unit 2 reactor, showing coolant escaping from the pressure relief valve.

back another. Now the reactor core had only minimal cooling. Steam and hydrogen, the latter generated by chemical reaction of steam with the zirconium fuel-rod cladding, blocked the remaining coolant flow. Years later, remote cameras fed into the reactor vessel showed that half the core had melted in these early hours—although at the time no one had any idea that things were this serious.

As the fuel cladding failed, radioactive fission products escaped through the stuck relief valve into the containment, setting off alarms in the control room. Radioactive water soon overflowed into an auxiliary building outside the containment, triggering further alarms. At 6:45 A.M., less than three hours into the accident, supervisory personnel declared a "site area

Figure 8.3
Discussions in the control room during the TMI accident. Normally four operators are on duty, but other personnel joined them as the accident progressed. (Nuclear Regulatory Commission)

emergency," based on the possibility that radioactive material would be released in the immediate vicinity of the plant. Forty-five minutes later, the situation was elevated to "general emergency," since significant radiation might now escape the plant site. Still no one believed the core was uncovered, and operators dismissed instrument readings suggesting abnormally high core temperatures.

By 9 A.M., five hours into the accident, the Nuclear Regulatory Commission had been notified, and by then at least some operators and NRC commissioners felt that the core was, in fact, uncovered. Gradually, through the morning, other federal and state agencies were notified of the emergency at TMI. But information was incomplete and often misleading. Local emergency officials complained that the information they received from the Pennsylvania Emergency Management Agency was either so vague or so technical as to be useless; often radio and television proved their best source of information. The mayor of Harrisburg got a call from a Boston radio station asking him about the nuclear emergency at nearby TMI and replied, "What emergency?" No one really knew what had happened, and agencies responsible for public safety couldn't act effectively in the confusing situation.

Meanwhile, contaminated water overflowed holding tanks outside the containment, releasing low levels of radiation into the environment. Further releases accompanied attempts to relieve pressure inside the reactor. By noon—seven hours into the accident—officials began to discuss the possibility of evacuation. Confusion reigned again, as it was unclear what agency had ultimate responsibility.

Just before 2 P.M., instruments recorded a sudden, brief rise in pressure within the containment structure. Most in the control room thought this was an instrument malfunction. Actually, there had been a minor explosion of hydrogen gas, not strong enough to damage the containment.

Within 10 hours, the reactor was effectively under control. But releases of radioactive material, both intentional and inadvertent, continued for several days. Schools closed, people living near the plant were urged to stay indoors, and the question of evacuation hung in the air. Actual measured radiation levels outside the plant weren't high, and it's unlikely that any member of the public received more than 1 mSv (100 mrem), or one-third of what was then the yearly background dose in the United States. But general fear and ignorance about radiation, coupled with confusing and often contradictory reports from the media, the utility company, and federal, state, and local officials, fostered a sense of panic and dread among the populace. Eventually an advisory was issued for voluntary evacuation of pregnant women and preschool children within five miles of TMI. Despite no large-scale evacuation order, a total of 115,000 people within 15 miles of TMI evacuated at some point during the crisis.

The accident had begun on a Wednesday. By Saturday, a new worry occupied officials: a further buildup of hydrogen in the containment structure might lead to an explosion that could rupture the containment and result in widespread dispersal of radioactive material. On Saturday afternoon, the chairperson of the Nuclear Regulatory Commission told reporters that an evacuation of the nearly 1 million people living within 20 miles of TMI might prove necessary. By Sunday afternoon, for reasons not fully understood, the hydrogen bubble began to disperse. By late that day, it was clear to the NRC that an explosion was impossible, but not until Monday morning was this good news made public. Although the danger was over, the damage had been done—damage to the reactor, damage to the public psyche, damage to the credibility of the Nuclear Regulatory Commission and other government agencies, and damage to the nuclear power industry.

Cleanup of TMI Unit 2 was a lengthy, expensive process. The last of the radioactive debris was removed in 1990, by which time it was clear that TMI had experienced a significant meltdown, with some 20 tons of molten uranium at 5,000°F (about 3,000°C) flowing into the bottom of the reactor pressure vessel. Nevertheless, the pressure vessel itself appeared to have come through with its integrity uncompromised—despite not having been designed to withstand a meltdown. Engineers later recognized that the first batch of melted uranium had cooled and formed a solid protective coating that prevented additional uranium from melting through.

What caused the Three Mile Island accident? It started with a minor pump malfunction. But that was compounded by many factors: additional equipment failures, valves left in the wrong position, faulty instrument readings, operators' errors and confusion, miscommunication between government agencies and the public, a control panel and operating procedures that were poorly designed for emergency situations, and failure of the operators to take seriously some indications of what was actually happening. There were broader causes, too: Nuclear Regulatory Commission licensing procedures that allowed plants with known safety problems to continue operation, inadequate training of plant operators, and communication failures within the nuclear industry. In short, the TMI accident had many causes—mechanical, human, and institutional. Improvements in any of these areas might have prevented it.

How serious was TMI? In its consequences for public health, it wasn't at all serious. No one was killed outright, and most follow-up studies have not identified increases in cancer in the affected population. Some, in fact, show lower cancer incidence than expected normally, although a few others, including a 2017 study, show increases in thyroid cancer but stop short of claiming a causal connection to TMI.[5] The consequences for nuclear power in the United States were more serious. The accident dealt a devastating blow to an industry already plagued by public opposition and declining orders for nuclear plants. In the years following TMI, 61 previously ordered reactors were canceled, and it was 33 years—2012—before the NRC granted licenses for new reactors. We'll explore the history and future of the nuclear power industry further in chapter 10.

TMI could have been much worse. Only a minuscule fraction of the radioactive fission products escaped to the environment. Breaching of the containment—which could have occurred had operators not acted soon enough to halt the loss of coolant—might have spewed radioactive material onto the surrounding countryside, resulting in death and illness among the population. Nuclear critics argue that TMI came within 60 minutes of such a disaster, while nuclear advocates counter by pointing to safety systems and operators' actions that ultimately worked to prevent a full meltdown and containment breaching, despite numerous malfunctions and human errors.

Chernobyl

Seven years after Three Mile Island, the world got its first taste of a really serious reactor accident. At the Chernobyl nuclear power plant in what was then Soviet Ukraine, a runaway nuclear reaction led to an explosion and fire that spread radioactive material over much of Europe.

Ironically, the Chernobyl accident occurred during a reactor safety test. Officials wanted to be sure that the spinning turbine could provide temporary power to run the emergency core-cooling pumps in the event of a reactor shutdown.

A remarkably candid report by the Soviet government showed how the test procedure, operator errors, and reactor design all contributed to the accident. Figure 8.4 shows the sequence of events. Starting on April 25, 1986, operators prepared for the test by reducing the reactor's thermal power from its normal 3,200 MW to 1,600 MW. They then turned off the emergency core-cooling system. Although the test procedure called for this action to prevent the ECCS from interfering with the test, it constituted a clear violation of the reactor's operating procedures.

As operators resumed lowering the reactor power, one of them neglected to set a control to maintain power at the test level of around 700 MW. Instead, the reactor power plunged to a mere 30 MW. It's very difficult to restart a reactor after such a rapid decline in power. That's because the isotope xenon-135 absorbs neutrons voraciously. Xe-135 forms from the decay of the fission product iodine-135. In an operating reactor, Xe-135 is destroyed by neutron absorption as rapidly as it forms. But when reactor power drops, Xe-135 builds up because fewer neutrons are present. Attempts to increase power then fail because Xe-135 "poisons" the reactor by absorbing so many neutrons that increased fission is impossible. The xenon-135 decays with a nine-hour half-life, so the reactor will restart easily after about a day.

Officials at Chernobyl were impatient to continue their test. Instead of waiting for xenon-135 to decay, they compensated for its neutron absorption by removing more control rods than regulations allowed. By 1:19 A.M. on April 26, this had raised the reactor power to 200 MW. At about the time that they withdrew the control rods, the operators turned on two additional cooling-water pumps, as the test procedure dictated.

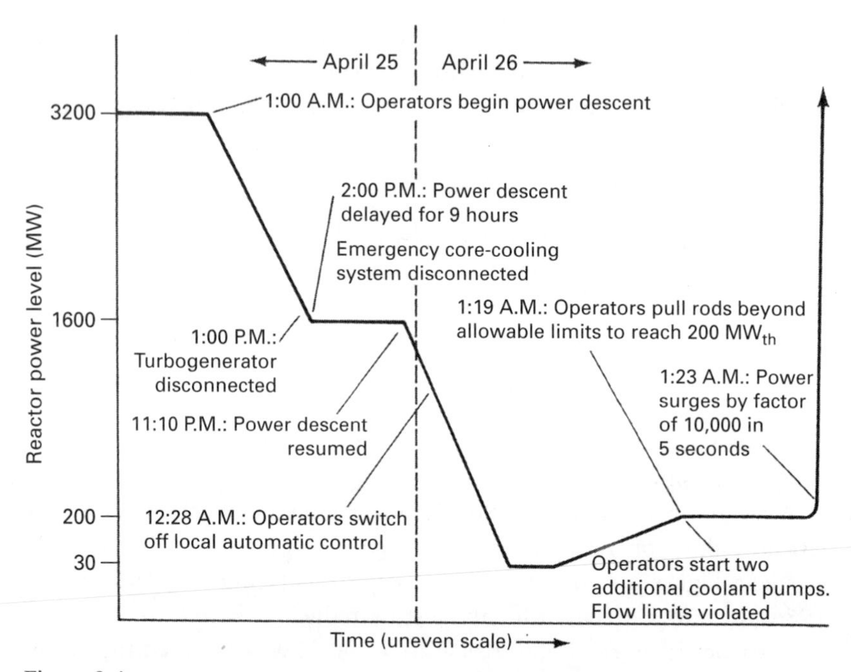

Figure 8.4
Sequence of events in the Chernobyl Accident, shown on a plot of reactor power versus time. (Adapted from *Physics Today*, December 1986, with permission of the American Institute of Physics.)

In a graphite-moderated reactor such as Chernobyl, the primary effect of water is to absorb neutrons and therefore inhibit the chain reaction. This is the exact opposite of water's role in a U.S. light-water reactor, where its moderating properties make water essential to the reaction. So Chernobyl was in a situation where loss of water would increase the fission rate.

The Chernobyl operators found there was too much water flowing through the reactor, and at 1:22 A.M. they reduced the flow. They did not immediately reinsert the control rods, however, and 30 seconds later a computer warned that the reactor was in a dangerous condition requiring immediate shutdown. But the operators proceeded with the test. They diverted steam from the turbine-generator, lowering the load on the reactor and causing more water to boil. The decrease in liquid water led to less neutron absorption and more rapid fission. The chain reaction went supercritical from prompt neutrons alone, and the power level soared to 500 times its design value in a mere five seconds. Two explosions followed. The first, which blew the heavy concrete lid off the reactor,

Figure 8.5
A technician in a helicopter checks radiation levels above the damaged Chernobyl reactor. (Novosti Press Agency, Moscow)

is generally interpreted as resulting from cooling water having flashed instantly to steam. The second explosion was probably due to hydrogen generated as water reacted with the fuel-rod cladding. An alternative scenario was proposed in 2018: it has the first explosion due to intense nuclear reactions within a few fuel rods, followed by a steam explosion.[6] In any event, the graphite moderator caught fire and radioactive fission products were carried aloft with the smoke. The ensuing damage to the reactor is clearly evident in figure 8.5.

What caused the Chernobyl accident? As with Three Mile Island, many factors conspired. Human error and folly played a larger role than at TMI, but the graphite-moderated, water-cooled reactor design contributed as well. The bureaucratic pressure to complete the test had no immediate counterpart at TMI; on the other hand, TMI happened amidst a long tradition of neglect for safety by the industry and the Nuclear Regulatory Commission. In both accidents operators became confused, made mistakes, and failed to heed warnings. Although the details of the two accidents are very different, both ultimately involved complex and potentially unstable systems operated by fallible human beings.

Two people died in the explosions at Chernobyl. Another 28—mostly power plant workers and firefighters—died of acute radiation sickness. Radioactive emissions continued for 10 days, spreading measurable fallout over much of Europe. The reactor itself was hastily entombed in a concrete sarcophagus, which was replaced in 2017 by a huge, hermetically sealed shelter called the New Safe Confinement, funded in part by international donors (figure 8.6). The accident contaminated thousands of square miles of Ukrainian agricultural land, and ultimately more than 300,000 people living within 20 miles of the reactor were evacuated. Authorities decided not to resettle the nearly 100 towns in the evacuation zone—which, ironically, is now a burgeoning tourist attraction. A casual visit suggests a thriving wilderness ecosystem, although scientific studies of long-term effects of nuclear accidents on nonhuman species show measurable effects persisting for many decades.[7]

Health studies by the United Nations Scientific Committee on the Effects of Atomic Radiation, the World Health Organization (WHO), and the International Atomic Energy Agency (IAEA) examined both the immediate impact of Chernobyl and its effect decades after the accident. A 2006 WHO report documented a large increase in thyroid cancer among those who were below age 18 at the time of the accident. Most of this increase is attributed to radiation. Radiation-induced thyroid cancer was exacerbated by iodine deficiency in the local diet, which results in greater uptake of radioactive iodine. Studies also show modest evidence of increased leukemia and breast cancer in the most heavily contaminated areas. Among the groups most exposed—including 240,000 "liquidators" who worked on the disabled plant, as well as evacuees and residents of controlled zones around the reactor—WHO estimates that there will be an excess of about 4,000 cancer deaths due to Chernobyl radiation. This estimate assumes linear non-threshold dose-response and is in addition to the expected 120,000 cancer deaths from other causes. Expanding the affected population to 5 million people with lower exposures suggests an additional 5,000 cancer deaths—but this estimate is very rough because of uncertainties in doses and, as you saw in chapter 4, because of our poor knowledge of the impact of low radiation doses. Others contest these numbers, in part because of the studies' connections

Figure 8.6
Chernobyl's New Safe Confinement structure, completed in 2017, completely covers the damaged Reactor #4. In the foreground is one of several monuments to the Chernobyl disaster. (Photo by Ferenc Dalnoki-Veress.)

to the IAEA, whose charge includes promoting nuclear energy. For example, the respected physicist and nuclear expert Richard Garwin has used the BEIR VII cancer dose of 0.01 cancers per 100 mSv that we introduced in chapter 4 to suggest that Chernobyl's toll will include another 20,000 fatal cancers beyond the "official" estimates.[8] In addition to cancer, recent studies show evidence of congenital abnormalities among those exposed to Chernobyl radiation.[9]

So how serious was Chernobyl? Essentially all the reactor's gaseous fission products escaped to the environment, along with about 3 percent of the solids. Nevertheless, many nuclear experts see it as a worst case, and something that couldn't happen to a light-water reactor. They take comfort in an ultimate death toll of "only" 9,000 and compare that 50-year figure with the estimated 10,000 deaths caused *each year* in the United States alone (and a million in China) by emissions from coal-burning power plants. Nuclear critics—including Richard Garwin, cited in the preceding paragraph—are less sanguine, with estimates of tens of thousands of Chernobyl deaths. Nuclear advocates contrast Chernobyl with light-water reactors to emphasize the safety advantages of LWRs, while critics stress that unexpected problems can arise with *any* reactor design. Proponents note that even the rupture of the containment at Chernobyl caused relatively few deaths; opponents argue that a comparable accident at a reactor near a major city—for example, the Indian Point nuclear plant, 26 miles north of New York—could take a far greater toll. And that comparison with coal? It's valid—but, as we've noted before, coal and nuclear aren't our only energy choices.

Fukushima

Unlike TMI and Chernobyl, the 2011 disaster at Japan's Fukushima Daiichi nuclear power plant began with a natural event—Japan's strongest earthquake ever measured (magnitude 9) and a subsequent tsunami. Four nuclear power plants on Japan's northeast coast experienced sufficient earth movement that the 11 reactors operating at the time shut down automatically, exactly as they were supposed to do. These shutdowns included reactor Units 1, 2, and 3 at Fukushima Daiichi. Fukushima's remaining Units 4, 5, and 6 were down for routine maintenance.

The earthquake did little obvious damage at the Fukushima nuclear plant but knocked out the local power grid. This caused a **station blackout**, a worrisome condition in which a power plant has no external power source. No problem in this case: all of Fukushima's backup diesel generators came on, supplying power for coolant pumps, instrumentation, and control systems. All would have been well were it not for the tsunami.

Forty-one minutes after the first earthquake tremors, a 13-foot-high wave hit the seawall installed to protect the Fukushima plant from just such occurrences. No problem: the seawall could deflect up to a 33-foot wave. But seven minutes later an unplanned-for 50-foot wave struck (waves that height were expected once in 500 years, which was deemed sufficiently rare not to warrant protection against). The monster wave flooded the lower levels of the nuclear plant, disrupting electrical distribution panels and knocking out basement-housed emergency generators. Even generators above water were useless because of the damaged electrical distribution system. So now Fukushima was without AC power (alternating current, the standard type of electric power supplied by the grid and backup generators).

No problem: like all nuclear plants, Fukushima had backup batteries supplying DC (direct current) power capable of running cooling pumps, instruments, and control systems for up to eight hours. The assumption was that eight hours was plenty of time to restore external power. But that assumption didn't account for widespread damage and chaos beyond the plant. Not only was the electric power grid wiped out throughout the region, but whole towns had been demolished and some 19,000 people had perished. And roads that might have allowed emergency equipment to arrive were clogged with debris.

Getting power restored to Fukushima also meant distributing power within the plant once it was reconnected to the grid—but at that point the plant's electrical distribution systems were inoperable. Even backup batteries relied on those distribution systems, so they weren't much use. And the batteries for Unit 1 had themselves been flooded, taking out the instruments in the control room shared by Units 1 and 2. At that point all reactors but Unit 6 were without AC power, and Units 1 and 2 had no DC power either—a dire situation that hadn't been anticipated in emergency plans.

The reactors at Fukushima were a widely used type of boiling-water reactor designed by General Electric in the 1960s and put into service at Fukushima in the 1970s. Units 1 through 5 used the so-called *Mark I* containment system, which employed a less expensive and less robust concrete-and-steel containment augmented by active systems designed to condense steam and prevent dangerous pressure buildup. Unit 1, the oldest reactor, had operated for 40 years and was scheduled to be shut down permanently in early 2011 but had been granted a 10-year extension only weeks before the earthquake. Unit 1 featured an auxiliary cooling system called an *isolation condenser*, whose job was to condense steam and feed the resulting water back into the reactor if the primary cooling system failed. Units 2, 3, and 4 had steam-powered auxiliary cooling called the *reactor core isolation cooling system* (RCIC) that didn't require electric power for pumping—but that did need electricity for control and monitoring. Since there was no electric power whatsoever at Units 1 and 2, operators had no easy way to know whether these systems were operating. Furthermore, the operators on duty weren't familiar with Unit 1's isolation condenser. In a desperate attempt to power up the Unit 1 and 2 control room, plant personnel scrounged batteries from undamaged vehicles. But finding the right electrical connection points in the dark, and sometimes in standing water, was arduous, time-consuming, and dangerous. All the while the water level in the reactor pressure vessels of Units 1, 2, and 3 was dropping.

Plant operators knew that cooling water was essential. So they engaged onsite fire engines—whose presence had been mandated after a 2007 earthquake-induced fire at another Japanese nuclear plant—to pump water into the reactors. This was challenging, since the reactors weren't designed to be cooled this way, and because high pressure inside reactor vessels would prevent the relatively low-pressure fire pumps from pushing water in. Reducing the pressure required venting the reactors—which could release radiation to the environment. So venting was essential, but it was also politically fraught because it required evacuation of the population in the plant's immediate vicinity. And, given the lack of electric power, the vent valves would have to be opened by hand. Workers attempting this operation were driven back by high radiation levels and the venting was, at best, partially successful. A few hours later, backup power and compressed air were brought to the scene and the valves were successfully opened, relieving pressure in Unit 1. However, simulations done in 2012 suggest that high pressure had caused bolts on the containment to stretch, also relieving pressure by releasing gas into the reactor building.

Then, just 24 hours after the tsunami, an explosion blew the roof off Unit 1's reactor building (figure 8.7)—a structure that was not designed to withstand explosive events but that could serve as backup radiation containment in the event that the primary containment leaked or failed. Now, with its reactor building's innards exposed to the sky, any radiation that escaped Unit 1's primary containment would go directly into the atmosphere.

It's now known that Unit 1's fuel began to be uncovered as early as three hours into the disaster, and that most of Unit 1's core melted during the first day. Hydrogen was produced as the hot, exposed fuel cladding reacted with steam, and it was mixed with the gases and steam that were vented from Unit 1. It's also possible that hydrogen had escaped the reactor vessel earlier through leaks opened by the extreme pressure. In any event, it was this hydrogen that exploded.

With Unit 1 vented, fire engines were able to deliver a steady flow of cooling water. At first they used freshwater but soon switched to seawater as the freshwater ran short—although the explosion at Unit 1 disrupted the cooling preparations for several hours. Seawater would quickly destroy the reactors' innards, so using it was an admission by the plant's owner, Tokyo Electric Power Company, that the reactors would never operate again.

Over the next few days, the backup cooling systems at Units 2 and 3 also failed, and those reactors, too, experienced substantial core melting. Three days after the tsunami, a hydrogen explosion shattered part of the Unit 3 reactor building, injuring workers, strewing radioactive waste around, and damaging emergency cooling and venting lines that were being set up for Unit 2. Nevertheless, the seawater injection soon resumed.

Figure 8.7
Workers inspect damage following a hydrogen explosion at Fukushima. (Public Health England/Science Source)

Then, on the fourth day, came a real surprise: another hydrogen explosion, this one at Unit 4. This reactor had not been operating at the time of the earthquake, and four months earlier its fuel had been removed from the core to a cooling pool within its reactor building. In boiling-water reactors, cooling pools sit high in the reactor building, connected to the reactor itself by channels that allow highly radioactive spent fuel to remain continuously underwater for radiation shielding (figure 9.10 will show this). A frightening new possibility loomed: one or more cooling pools might have been damaged in the earthquake, or perhaps water was boiling away because the pools' active cooling systems couldn't operate without electric power. Overheated fuel rods would generate explosive hydrogen, and were they exposed to air they might catch fire and spew radiation into the environment through the now-roofless reactor buildings.

To combat the new threat, fire hoses, riot-control water cannons, and helicopters were used, without much effect, to deliver water to the cooling pools of reactors 1, 3, and 4, where water access through the missing roofs was possible. Later, large concrete-pumping trucks were brought in and their nozzles positioned above the spent-fuel pools to deliver water directly. A week after the earthquake, danger from the reactors themselves had diminished, and the highest priority was cooling the spent-fuel pools. After another week external power had been largely restored and the Fukushima plant was, if not fully under control, then headed toward the safety of cold shutdown.

How serious was Fukushima? In terms of radiation release and public health, it was intermediate between TMI and Chernobyl, with Fukushima releasing one-tenth the radioactivity of Chernobyl. In terms of myriad and often unexpected challenges associated with simultaneous meltdowns at multiple reactors, it was probably the most harrowing of the three accidents. A 2012 Stanford University study estimated the number of fatal cancers from Fukushima radiation at between 15 and 1,100 cases, with the most probable number around 130.[10] Another authoritative study estimates fatal cancers at 1,000,[11] and a study by an antinuclear expert puts it at 5,000.[12] In any event, the toll at Fukushima would have been worse had prevailing winds not blown some 80 percent of the radioactive releases out over the Pacific Ocean. At the same time, whatever estimate you accept for deaths from Fukushima radiation, you should consider it in the context of 19,000 killed in the earthquake and tsunami, and compare it with estimates of some 2,200 deaths associated with stress and the rigors of evacuations.

TMI, Chernobyl, Fukushima: Lessons Learned?

Three Mile Island, Chernobyl, and Fukushima were very different events, but all were characterized by flawed reactor and power plant designs, by human error, and by chaos and confusion associated with unexpected and unplanned-for events. Following TMI, President Jimmy Carter appointed a commission chaired by mathematician and Dartmouth College president John Kemeny to investigate the accident and make recommendations. The Kemeny Commission noted that equipment failures contributed to the accident but placed most of the blame on humans, including the Nuclear Regulatory Commission itself. Later the NRC issued a lengthy list of upgrades required of operating nuclear plants, including personnel and training as well as equipment and, especially, control and instrumentation upgrades. Incidentally, the TMI Unit 1 reactor, twin to the damaged Unit 2, was allowed to restart six years after the accident and continued operating until 2019.

Worldwide, public response to the Chernobyl accident was widespread and largely reinforced antinuclear sentiments. Because Chernobyl involved the relatively rare RBMK reactor with no counterpart in the West, few specific safety improvements to the global nuclear industry resulted. A U.S. NRC study of Chernobyl concluded that there was no need for immediate changes in U.S. nuclear plants or their operating procedures. Several planned RBMK reactors were canceled following the accident, although the three undamaged units at Chernobyl continued to operate, one of them until the year 2000. In 1988, the Soviet Union issued a tightened set of nuclear safety regulations that took Chernobyl into account, and work has continued to reduce the instabilities inherent in the RBMK design.

Fukushima, too, elicited worldwide antinuclear response and prompted several other countries to reconsider their nuclear programs. Germany permanently closed 8 of its 17 reactors and announced that it would phase out all nuclear power within a decade. Two months after Fukushima, Italian voters, in a binding referendum, canceled all plans for new nuclear plants. In the U.S., the Nuclear Regulatory Commission issued orders for nuclear plants to deploy additional emergency equipment such as pumps and generators, instrumentation for monitoring spent-fuel pools, and improved emergency venting systems for BWRs similar to those at Fukushima. The NRC also asked U.S. nuclear plants to reassess their earthquake and flood preparedness, and in 2019 the NRC issued formal regulations with a two-year implementation time. Japan's response to Fukushima was to shut down all 56 of its reactors within a year of the disaster—a difficult move for a country that got nearly one-third of its electricity from nuclear power plants. Twenty reactors closed permanently, while others undergo stress tests and safety improvements. By 2020, 9 of Japan's 36 remaining operational reactors had been restarted, while others awaited approval. Japan is likely to need those reactors to meet its carbon-emissions reductions under the Paris Climate Agreement.

Can Reactors Be Safer?

Light-water reactors like those at TMI and Fukushima evolved as scaled-up versions of submarine reactors developed in the 1950s. Graphite-moderated reactors like Chernobyl are relatives of the original graphite "piles" first built in the 1940s to produce plutonium for nuclear weapons. Gen-II light-water and graphite reactors operating today incorporate safety features not found in their ancestors, and more safety features have been retrofitted in response to TMI, Chernobyl, Fukushima, and a host of lesser but potentially serious reactor incidents. Yet fundamentally, most

The International Nuclear and Radiological Event Scale

In 1990, the International Atomic Energy Agency introduced an eight-point scale, running from 0 to 7, rating potentially dangerous events at nuclear plants and related facilities. This **International Nuclear and Radiological Event Scale** (INES) is, like the Richter scale used to quantify earthquakes, logarithmic—meaning that an increase of 1 on the scale implies a factor-of-10 increase in severity. INES level 0 is a *deviation* from normal but with no safety consequences; levels 1–3 are termed *incidents*; while 4–7 are *accidents* ranging from those with local consequences only (level 4) to major accidents (level 7). Figure 8.8 shows the INES levels.

Three Mile Island and Chernobyl predate the INES scale but are rated retroactively at level 5 (Accident with Wider Consequences) for TMI and level 7 (Major Accident) for Chernobyl. The only other level-7 accident to date is Fukushima—although it was initially classified as three separate level-5 events. The only level-6 event (Serious Accident) occurred in 1957 at a nuclear waste reprocessing facility in the Soviet Union, while three other events share level 5 with TMI: the 1957 Windscale fire at a graphite reactor in Sellafield, U.K.; the 1952 accident at the Chalk River experimental reactor in Ontario, Canada; and a 1987 accident in Brazil in which thieves got their hands on a medical radiation source, which they then sold as scrap metal (more on this in chapter 16). As a result, 249 people were contaminated and four died. Level-4 accidents (Accident with Local Consequences) have occurred in the U.K., the U.S.,

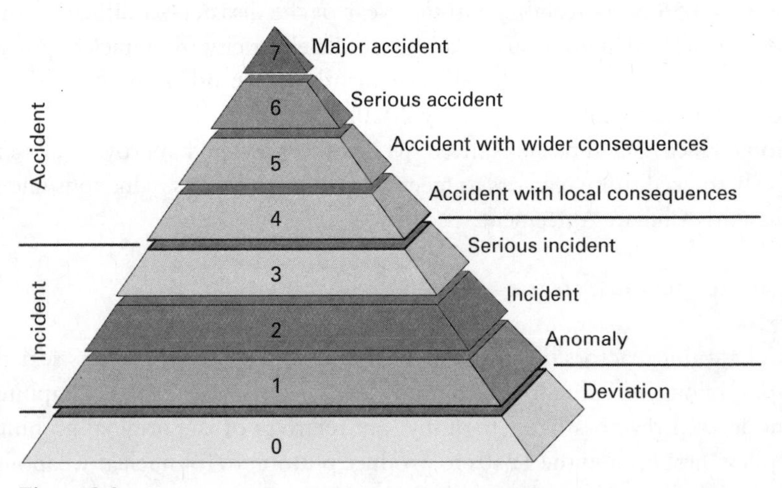

Figure 8.8
The International Nuclear and Radiological Event Scale, shown as a pyramid to emphasize increasing severity but lower probability. (Creative Commons CC BY-SA 3.0)

The International Nuclear and Radiological Event Scale (*continued*)

France, Argentina, Czechoslovakia, Japan, and India; most involved processing facilities or experimental reactors rather than commercial power reactors. Some 15 events worldwide are classified at levels 0–4. A 2019 nuclear accident at a Russian military site has not yet been classified. Because radiation monitors stopped sending information to an international monitoring network, it may be difficult to determine the radiological significance of that event.

of today's power reactors remain rooted in military designs that originated decades before safety became a paramount concern. Safety systems for aging Gen-II reactors are in this sense an afterthought, not an essential design feature.

Can a really safe reactor be made? The Gen-III+ designs outlined in the preceding chapter and being installed today are arguably safer than most of the currently operating fleet of Gen-II reactors. They're still evolutionary descendants of Gen-II designs, but Gen-III+ reactors are designed from the ground up with greater attention to safety. Their emergency cooling systems are passive, relying on gravity, rather than active pumps, to deliver cooling water (recall figure 7.8, showing the AP1000 Gen-III+ reactor). These systems typically allow 72 hours before operator intervention is required following emergency shutdown. Gen III+ reactors have fewer valves, pumps, and other failure-prone devices. Standardization of design means less confusion for operators. And Gen-III+ reactors feature more robust containment systems.

Would Gen-III+ designs have survived the Fukushima tsunami? Probably, due largely to their passive cooling. But even the 1960s-era reactors at Fukushima might have survived had the power plant's designers had the foresight to locate emergency generators and batteries on higher ground, rather than in flood-prone basements. In fact, the one generator that was higher on the Fukushima site operated throughout the crisis, providing cooling to Units 5 and 6.

Core Damage Frequencies

Probabilistic risk analysis is often used to estimate the chances of a severe reactor incident. Results are expressed as core damage frequency (CDF), or the probability per **reactor-year** (meaning one reactor operating for one year) of an incident that results in core damage. Regulatory agencies, including the U.S. Nuclear Regulatory Commission and those of many other countries, limit the CDF for older reactors to no more than 1 in 10,000. For new reactors many countries set the limit at 1 in 100,000. For today's actual reactor fleet—again, mostly Gen-II light-water reactors—calculated CDFs lie somewhere between those two limits. New reactors do better, at least on paper. For example, APR-1400 reactors being built in South Korea list a CDF of less than 1 in 100,000. A successor design, the APR+, was certified in 2014 and claims a factor below 1 in a million. Another reactor under deployment, India's Advanced Heavy Water Reactor, is an evolutionary variant of the CANDU heavy-water design that claims a CDF of less than 1 in 100 million. Generation IV reactors, still on the drawing board, are expected to have CDFs at or below those of Gen-III+ designs.

How are we to interpret these core damage frequencies, either the 1 in 10,000 that might be a high estimate for today's reactor fleet, or the 1 in a million we hope applies to Gen-III+ reactors? Today, in rough numbers, there are some 500 operating power reactors in the world, so a 1 in 10,000 chance that *one* reactor will experience core damage translates into 500 times as much chance that *some* reactor will experience core damage. Thus there's a roughly 500 in 10,000, or 5 in 100 chance of core damage in any reactor over one year—that is, an average of 0.05 events per year among the 500 reactors. Today's reactor fleet has been operating for an average of 40 years, so that implies 0.05×40 or two core damage events since the start of the nuclear age. How many have there actually been? In the incidents we've considered, we have one at Three Mile Island, one at Chernobyl, and three at Fukushima. Although too low a number to claim statistical significance, those five alone substantially exceed the two expected from our rough estimate based on a CDF of 1 in 10,000 per reactor-year. And there have been more—as many as 11—incidents involving significant fuel melting at power reactors worldwide. Indeed, a more careful study by the Natural Resources Defense Council suggests an actual CDF of 1 in 1,400 reactor-years.[13] This is almost a factor of 10 worse than the minimum calculated CDF of 1 in 10,000.

Core damage frequency isn't the only probabilistic risk assessment applied to nuclear reactors. Some reactors also cite the chance of a major radiation release, which is typically a factor of 10 lower than the CDF.

Suppose we could achieve CDFs of 1 in a million with a new generation of reactors. Would nuclear power then be truly safe? We would still have, for years to come, several hundred older reactors operating worldwide, most facing higher probabilities of serious incidents. Furthermore, the three major nuclear power accidents we've just reviewed all involved unexpected events and human errors that wouldn't have been caught in a probabilistic risk analysis. No one had thought of the hydrogen bubble at TMI, or that operators would get Chernobyl into prompt criticality, or that a tsunami could put multiple reactors in jeopardy, or that spent-fuel pools could pose a risk as serious as reactors themselves. On the other hand, the human toll even in these three worst commercial nuclear accidents pales compared with the estimated several million who die worldwide *every year* due to pollution from coal-burning power plants (more on this below).

Reactor Safety in a Warming World

Throughout this chapter, we've considered advances that could make nuclear power safer. But there's a caveat: those welcome advances are occurring in a world that's rapidly warming due to human activity, especially greenhouse gas emissions. Among the consequences are rising sea levels, increased surges, and more frequent flooding at both coasts and rivers. As you saw in chapter 5, all thermal power plants require cooling water, and for that reason most nuclear plants are built on large bodies of water. With climate change, they become more vulnerable to Fukushima-like events.

In a mandated post-Fukushima analysis of flood risks, 90 percent of U.S. nuclear plants reported facing at least one flood risk that exceeded their design capability.[14] That situation has only worsened, not only because of rapid climate change—the post-Fukushima years from 2014 onward are the warmest yet recorded—but also because of a new anti-science, anti-regulatory attitude at the top of the U.S. government. One specific example is the Turkey Point nuclear plant near Miami. It's designed to withstand a 16-foot storm surge, but new data suggest it can expect surges of more than 19 feet. Nevertheless, in 2019 Turkey Point was granted a license extension to 2053 based on its original 16-foot design specification. Nuclear plants on the Mississippi River and the Virginia coast both face surges some 10 feet above their design criteria. And the state of North Carolina, whose Brunswick nuclear plant sits just a few miles from the Atlantic Ocean, has passed a law banning coastal planners from considering scientific projections of accelerated sea level rise. That mandates ignorance, not science, in planning for future flooding!

Of course it isn't just U.S. nuclear plants that face threats from climate change. Around the world, operating plants and those under construction are supposedly built to withstand worst-case flooding scenarios. But in a warming world, those scenarios are rapidly worsening. With their 40- to 60-year lifetimes, nuclear plants built today face an uncertain future flood risk.

Summary

Is nuclear power safe? No: The huge inventory of radioactive materials within a nuclear reactor carries the possibility that an accident could have serious health consequences for thousands of people. Three Mile Island, Chernobyl, Fukushima, and dozens of less publicized incidents underscore the inevitability of nuclear accidents. And government agencies intended to protect the public have shown contemptible disregard for safety in the face of industry pressure to bring nuclear plants online and keep them operating.

Is nuclear power safe? Yes: Multiple safety systems ensure that radiation is contained even when malfunctions and operator errors occur. Major accidents are possible but so unlikely that the risk is small. You're far more likely to die in an automobile accident, a fall, or a fire than in a nuclear accident—or from the effects of other electricity-generating schemes, especially fossil fuels. Few industries can match the safety record of commercial nuclear power.

Can nuclear power be made safer? Probably: New reactors promise greater inherent safety. The nuclear industry learns from each incident and retrofits existing nuclear plants with new safety features. But even "inherently safe" reactor designs may be subject to unexpected failures that aren't predicted by computer-based risk analyses.

The question isn't whether nuclear power is safe or not. It will never be perfectly safe. The question is whether it's safe enough. In answering that question, it may help to compare nuclear risks with other risks we face, especially from electricity generation (figure 8.9).[15] Advocates for nuclear power draw attention to serious non-nuclear accidents—such as the 1984 chemical leak at Bhopal, India, which killed 2,500 people and injured 200,000, or the 1975 multiple dam failures in China's Henan Province, which killed nearly 200,000—in comparison with which, they suggest, nuclear power is relatively benign despite occasional severe accidents. Critics judge the nuclear industry on its own demerits, not in comparison with others, and find it unsafe.

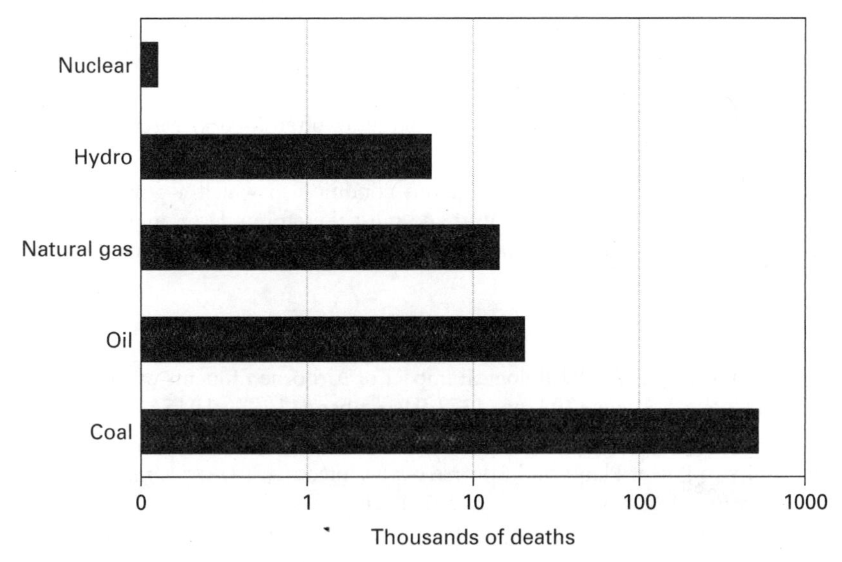

Figure 8.9
Estimated worldwide annual deaths from major electricity sources. Deaths from fossil fuels are largely due to air pollution; those from hydroelectricity and nuclear power factor in accidents as well as routine operation. Note the logarithmic scale, used because of the wide range in values, from 128 for nuclear to over half a million for coal. Other estimates are a factor of 2 or more higher than those shown. We have not included estimates for solar and wind, which vary widely but typically lie somewhere between hydro and nuclear. (Data source: statista)

Glossary

containment Structure surrounding a nuclear reactor, designed to contain radioactivity in the event of an accident. Most containments are made of heavy concrete, and are designed to withstand some pressure buildup resulting from overheating.

emergency core-cooling system (ECCS) Safety system designed to supply water to a reactor core in the event of a loss of normal coolant.

International Nuclear and Radiological Event Scale (INES) An eight-point scale (0–7) used to rate the severity of nuclear incidents and accidents. An increase of 1 on the scale is a factor-of-10 increase in severity.

loss-of-coolant accident (LOCA) A situation in which coolant is lost from the vicinity of a reactor core. Heat from the decay of radioactive fission products then builds up in the core, and may result in core damage, release of radioactive materials, and fuel melting.

meltdown Melting of a nuclear reactor's core after loss of coolant.

Nuclear Regulatory Commission (NRC) United States government agency responsible for licensing and regulation of nuclear power plants and other nuclear facilities and materials.

probabilistic risk assessment (PRA) An analysis of risk that follows possible paths leading to a serious nuclear accident, and assesses the overall probability of such an accident.

reactor-year One reactor operating for one year, often used in describing probabilities of accidents or reactor damage.

station blackout A potentially dangerous condition in which a power plant is cut off from the external power grid and must therefore rely on internal power sources to operate emergency pumps, instrumentation, and other crucial systems.

Notes

1. J. P. McBride et al., "Radiological Impact of Airborne Effluents of Coal and Nuclear Plants," *Science* 202, no. 4372 (December 8, 1978): 1045.

2. S. Ha, H. Hu, J. Roth, H. Kan, and X. Xu., "Associations between Residential Proximity to Power Plants and Adverse Birth Outcomes," *American Journal of Epidemiology* 182, no. 3 (August 1, 2015): 215–224.

3. See F. Dalnoki-Veress (with introduction by Arjun Makhijani), "What Was the Cause of the High Cl-38 Radioactivity in the Fukushima Daiichi Reactor #1?" *Asia Pacific Journal: Japan Focus* 9, no. 3 (March 28, 2011), https://apjjf.org/2011/9/14/Arjun-Makhijani/3509/article.html.

4. Full report (NUREG-1935) available at https://www.nrc.gov/docs/ML1233/ML12332A057.pdf; summary at https://www.nrc.gov/docs/ML1234/ML12347A049.pdf.

5. D. Goldberg, "Altered Molecular Profile in Thyroid Cancers from Patients Affected by the Three Mile Island Nuclear Accident," *The Laryngoscope* 127, Supp. 3 (July 2017): S1–S9.

6. L. E. De Geer, C. Persson, and H. Rodhe, "A Nuclear Jet at Chernobyl Around 21:23:45 UTC on April 25, 1986," *Nuclear Technology* 201, no. 1 (2018).

7. See S. Fesenko, "Review of Radiation Effects in Non-Human Species in Areas Affected by the Kyshtym Accident, *Journal of Radiological Protection* 39 (2019): R1–R17. The Kyshtym accident occurred in 1957 at a Soviet plutonium-processing facility.

8. See Richard L. Garwin, "Outside View: Chernobyl's Real Toll," United Press International Security & Terrorism, April 20, 2006. Garwin estimates an additional 20,000 cancer deaths over those of the "official" studies.

9. See Wladimir Wertelecki et al., "Chornobyl 30 Years Later: Radiation, Pregnancies, and Developmental Anomalies in Rivne, Ukraine," *European Journal of Medical Genetics* 60 (2017): 2–11.

10. John E. Ten Hoeve and Mark Z. Jacobson, "Worldwide Health Effects of the Fukushima Daiichi Nuclear Accident," *Energy and Environmental Science* 5, no. 9 (September 2012).

11. Jan Beyea, Edwin Lyman, and Frank N. von Hippel, "Accounting for Long-Term Doses in 'Worldwide Health Effects of the Fukushima Daiichi Nuclear Accident,'" *Energy and Environmental Science* 6, no. 3 (January 2013).

12. See Dr. Ian Fairlie's website at https://www.ianfairlie.org/news/new-unscear-report-on-fukushima-collective-doses/.

13. Thomas Cochran, "Statement on the Fukushima Nuclear Disaster and Its Implications for U.S. Nuclear Power Reactors," Natural Resources Defense Council, April 12, 2011. The statement, given as testimony to committees of the U.S. Senate, is available at https://www.nrdc.org/sites/default/files/tcochran_110412.pdf.

14. Christopher Flavelle and Jeremy C. F. Lin, "U.S. Nuclear Power Plants Weren't Built for Climate Change," *Bloomberg Businessweek*, April 18, 2019.

15. Data on deaths per TWh of energy is generated from Table 2 in Anil Markandya and Paul Wilkinson, "Electricity Generation and Health," *The Lancet* 370, no. 9591 (September 2007), except hydro, which is from Statista at https://www.statista.com/statistics/494425/death-rate-worldwide-by-energy-source/, corrected for factor-of-1,000 error in units. Data on TWh is generated from EIA International Energy Statistics for 2016, with fossil broken out according to The Shift Project at https://theshiftdataportal.org/energy/electricity.

Further Reading

Brown, Kate. *Manual for Survival: A Chernobyl Guide to the Future*. New York: Norton, 2019. Based on a decade of interviews and archival research, this book presents a darker view of Chernobyl's legacy than we do in this chapter.

Higginbotham, Adam. *Midnight in Chernobyl: The Untold Story of the World's Greatest Nuclear Disaster*. New York: Simon & Schuster, 2019. Superbly written and superbly researched account of what one reviewer calls "this breathtakingly complex accident."

International Atomic Energy Agency (IAEA). *Environmental Consequences of the Chernobyl Accident and their Remediation: Twenty Years of Experience*. Vienna: IAEA, 2006. Detailed, quantitative report on the impact of Chernobyl on the surrounding environment. Available for download at https://www-pub.iaea.org/MTCD/Publications/PDF/Pub1239_web.pdf.

International Atomic Energy Agency (IAEA). *The Fukushima Daiichi Accident, Technical Volume 1 Description and Context of the Accident*. Vienna: IAEA, 2015. Available in hardcopy or as PDF download at https://www-pub.iaea.org/MTCD/Publications/PDF/AdditionalVolumes/P1710/Pub1710-TV1-Web.pdf. A very thorough and authoritative account of what happened, and when, at Fukushima.

Kemeny, John, et al. *Report of the President's Commission on the Accident at Three Mile Island*. Washington, DC: U.S. Government Printing Office, 1979. This multivolume work details all aspects of the TMI accident. The casual reader will find more than enough in the executive summary, entitled *The Need for Change: The Legacy of TMI*.

Lochbaum, David, Edwin Lyman, Susan Q. Stranahan, and the Union of Concerned Scientists. *Fukushima: The Story of a Nuclear Disaster*. New York: New Press, 2014. Published under the auspices of the Union of Concerned Scientists, this thorough book interweaves technical and policy aspects of the Fukushima disaster.

Pinchuk, Stanislava. *Fukushima*. Paris: La Chambre Editions, 2017. Pinchuk is an artist whose work transcends both art and science. In this work, she explores the exclusion zone established after the Fukushima disaster.

Walker, J. Samuel. *Three Mile Island: A Nuclear Crisis in Historical Perspective*. Berkeley: University of California Press, 2004. A thorough account of the TMI accident by the Nuclear Regulatory Commission's official historian. Includes background on the lead-up to the accident and its subsequent consequences for the nuclear industry.

Yaroshinskaya, Alla A. *Chernobyl: Crime Without Punishment* (New Brunswick, NJ: Transaction, 2011). In this translation from the original 2006 Russian version, a journalist and former advisor to Russian president Boris Yeltsin documents the aftermath of the accident and the response, or lack thereof, of the Russian government and local bureaucracies.

9

Nuclear Waste and Fuel Cycles

You may or may not deem nuclear power plants acceptably safe, but what about nuclear waste? A nuclear reactor generates highly radioactive fission products. During normal operation—and, one hopes, even during accidents—these materials remain within the reactor. But eventually, spent fuel must be removed with its fission products still locked inside. Spent fuel straight out of the reactor will be physically hot and extremely radioactive for at least a century and will remain dangerously radioactive for far longer. Whereas the lifetime of a nuclear power plant is measured in decades, some radioisotopes in spent fuel have half-lives of 24,000 years or more. Solutions for spent-fuel storage must therefore isolate the material from the environment for thousands of years or longer.

As of 2020, some 65 years after the first commercial nuclear power plants came online, the world still has no operational long-term repository for high-level nuclear waste. The reasons are both technical and political. Today, most radioactive spent fuel is stored at individual reactor sites, in pools of water and then concrete-and-steel containers whose details we'll describe later in this chapter. But those aren't permanent solutions.

In the 1960s and 1970s, when most of today's nuclear plants were ordered, no one anticipated long-term storage of nuclear waste at plant sites. Fast forward to today, with just under a hundred nuclear reactors in operation in the United States alone and more than 70,000 metric tons (which would fill a football field to a height of 100 feet) of spent fuel in on-site storage. Yet the United States is still unable to reach a technologically and politically acceptable long-term storage plan. Other countries aren't doing much better, with the exception of France, Sweden, and Finland; Finland has a long-term storage repository set to accept nuclear waste starting in 2024.

The Nuclear Fuel Cycle

Assessing nuclear waste issues requires an understanding of the entire **nuclear fuel cycle**—the process whereby natural uranium becomes nuclear fuel, undergoes fission in a reactor, and emerges as nuclear waste. Figure 9.1 is a simplified diagram of the so-called **once-through fuel cycle**.

The first step in the fuel cycle is mining. Although uranium is widely distributed throughout Earth's crust, it's economically mineable only in ores where it occurs in concentrations of roughly 0.05 percent or more. There's abundant uranium in the western and southwestern United States. However, 77 percent of the world's uranium reserves are found in just four countries: Australia, Kazakhstan, Canada, and Russia, while 53 percent of global production comes from just 10 mines.

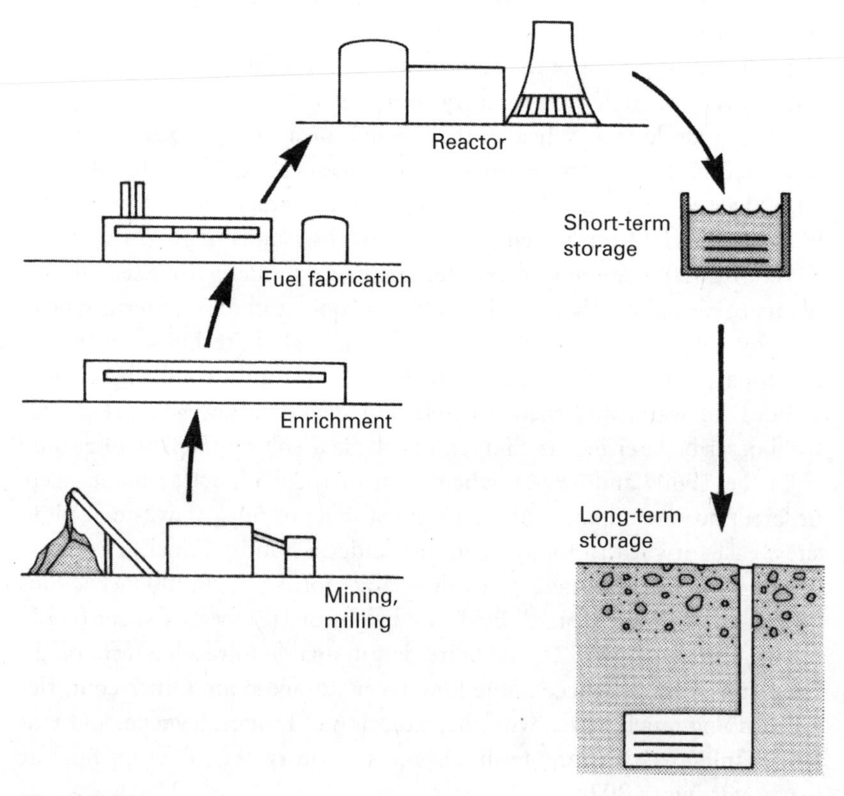

Figure 9.1
Once-through fuel cycle, taking natural uranium from mining through fuel preparation, fission in a reactor, and storage as high-level nuclear waste.

After mining, ore is pulverized and processed to remove the uranium. Left behind are vast piles of crushed rock called *tailings*. These contain radioisotopes formed in the uranium decay chain (recall figure 3.7), including radium, which decays to radon gas. Because they exude radon, tailings piles should be covered to prevent excessive radioactivity in the surrounding air. That wasn't done in the early days of uranium mining, so radon levels were high near tailings piles, and the wind dispersed slightly radioactive dust. Some tailings were even used in construction, resulting in buildings with excessive radiation levels. In addition, some currently active Canadian mines have ore exceeding 20 percent uranium, requiring dust masks and remote mining techniques to keep exposure at acceptable levels.

Removed from its ore, uranium takes the form of *yellowcake*, a chemical compound of uranium and oxygen (U_3O_8). Although yellowcake is mostly uranium, the fissile isotope U-235 occurs in yellowcake at its low natural abundance of 0.7 percent; essentially all the rest is nonfissile U-238. Heavy-water reactors, as you saw in chapter 7, can use natural uranium directly, so for these reactors the yellowcake undergoes only chemical and physical processing to become nuclear fuel. For the more common light-water reactors, however, uranium must be enriched to about 3 to 5 percent U-235.

Uranium Enrichment

Enrichment is difficult because it involves separating uranium isotopes that are chemically similar and have nearly the same mass. Therefore, we can't use chemical means to separate U-235 from U-238 but have to depend either directly or indirectly on their slight mass difference. In addition to its technical challenges, enrichment technology is politically sensitive since a nation having it can enrich uranium to the high levels needed for nuclear weapons. Here we'll review three principal uranium **enrichment** techniques, one historical, one current, and one likely to dominate in the future.

Gaseous Diffusion

Uranium enrichment in the United States was originally done by **gaseous diffusion**, a scheme that relies on the fact that lighter molecules in a gas move faster. Yellowcake itself can't be enriched in its chemical form and is first combined with fluorine to form uranium hexafluoride (UF_6) or *hex*. Hex is solid at room temperature, but when heated it skips the liquid stage and becomes a gas. Fluorine has the advantage of having only one stable isotope, so its presence as part of the UF_6 molecule doesn't complicate the separation process. The gas is introduced into a chamber containing a membrane through which hex molecules can pass but not easily. Those few molecules containing U-235 ($^{235}UF_6$) move slightly faster than their heavier but more numerous $^{238}UF_6$ counterparts, and therefore hit the membrane harder and more often, so they pass through at a higher rate. The result is a mixture on the far side that's enriched in U-235. Figure 9.2 illustrates the enrichment process.

A single chamber gives only a minuscule increase in U-235, so many chambers are connected one after another—about 1,000 for enrichment to light-water reactor fuel and more than 3,000 to make weapons-grade uranium. Gaseous-diffusion enrichment plants are huge and expensive (figure 9.3), consume enormous amounts of electrical energy, and are difficult to hide from satellites—so it's unlikely that potential proliferators would use this technology.

Gaseous diffusion became the United States' preferred process for uranium enrichment as a result of the urgent need to produce bomb-grade

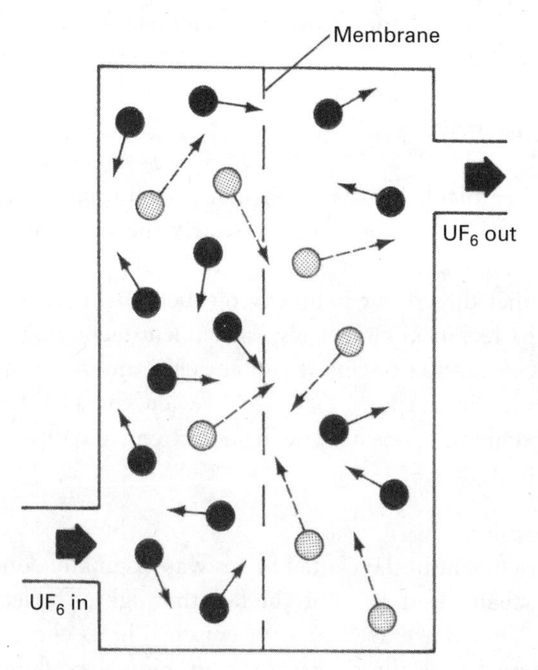

Figure 9.2
Uranium enrichment by gaseous diffusion. The lighter, faster-moving $^{235}UF_6$ molecules (light) are more likely to penetrate the membrane than $^{238}UF_6$ molecules (dark).

Figure 9.3
The uranium-enrichment complex at Oak Ridge, Tennessee, provided enriched uranium from World War II through 1985. A similar facility operated in Kentucky until 2013. (U.S. Department of Energy)

uranium during World War II. The U.S. operated three enrichment plants in Tennessee, Kentucky, and Ohio that consumed vast quantities of energy—equivalent to 0.5 percent of U.S. electrical energy production or the equivalent of six large nuclear or coal power plants. The last of these closed in 2013. France, China, and the former Soviet Union also operated gaseous diffusion plants, but these are now closed. Today we have more economical enrichment schemes that consume far less energy and that, for better or worse, are accessible to smaller nations.

Gas Centrifuge Enrichment

Gas centrifuge enrichment is currently the dominant technique globally. This works by feeding UF_6 gas into a cylindrical *rotor*, which spins inside a stationary cylinder. A vacuum between the two eliminates air resistance. The rotor spins extremely fast, some 50 times faster than an ordinary washing machine's spin cycle, and just as the washing machine separates water from clothes, the centrifuge separates $^{235}UF_6$ from $^{238}UF_6$. The heavier $^{238}UF_6$ moves toward the rotor wall, while lighter $^{235}UF_6$ stays near the axis.

It's not easy to separate the two gas layers in this configuration, so a current is established, causing gas that's richer in U-235 to flow to the top while ^{235}U-poor gas settles at the bottom. Figure 9.4 shows the workings of a gas centrifuge as well as an operating cascade of centrifuges.

Centrifuge rotors must be constructed from the strongest materials, including special alloys known as *maraging steels* and carbon fiber, to prevent them from deforming under tremendous rotational stress. Carbon fiber is light and extremely strong and is used in the aerospace industry as well as in tennis rackets and bicycles! Centrifuge separation capability improves with increasing rotational speed as well as with the strength and length of the rotor. But if the rotor length exceeds five times its diameter, it's necessary to segment the rotor to avoid bending stress that would otherwise damage it.

All commercial enrichment is currently done with gas centrifuges, which require only 2 percent of the electricity used by gaseous diffusion. And individual centrifuges are much more efficient than the gas-diffusion cell shown in figure 9.2. Centrifuges are easy to conceal underground and to fortify against potential attack. An enrichment plant two-thirds the area of a football field could produce one bomb's worth of weapons-grade uranium in less than two months. That presents a problem for monitoring clandestine enrichment activity. However, the amount of UF_6 gas that can be placed in a centrifuge is small, so while not many stages are needed in series one after the other, many centrifuges are needed at each stage to increase the quantity of enriched product. A typical example, a centrifuge facility in Iran, is a cascade with 164 centrifuges. Each cascade has 15 stages where the number of centrifuges per stage ranges from 24 down to 1. The centrifuges in each stage feed the enriched product into the next stage. However, instead of disposing of the depleted stream, called *tails*, it's recycled back into a stage upstream in order to extract as much U-235 as possible. To produce 93 percent enriched weapons-grade uranium, the quantity of natural uranium feed is some 200 times the final enriched product.

Enrichment and Proliferation

Power reactor fuel is generally enriched to 3–5 percent U-235, while research reactors have fuel enriched to 20 percent or more. Weapons grade is generally considered 90 percent and higher. It takes a great deal of work to enrich from natural uranium to weapons grade. But in the

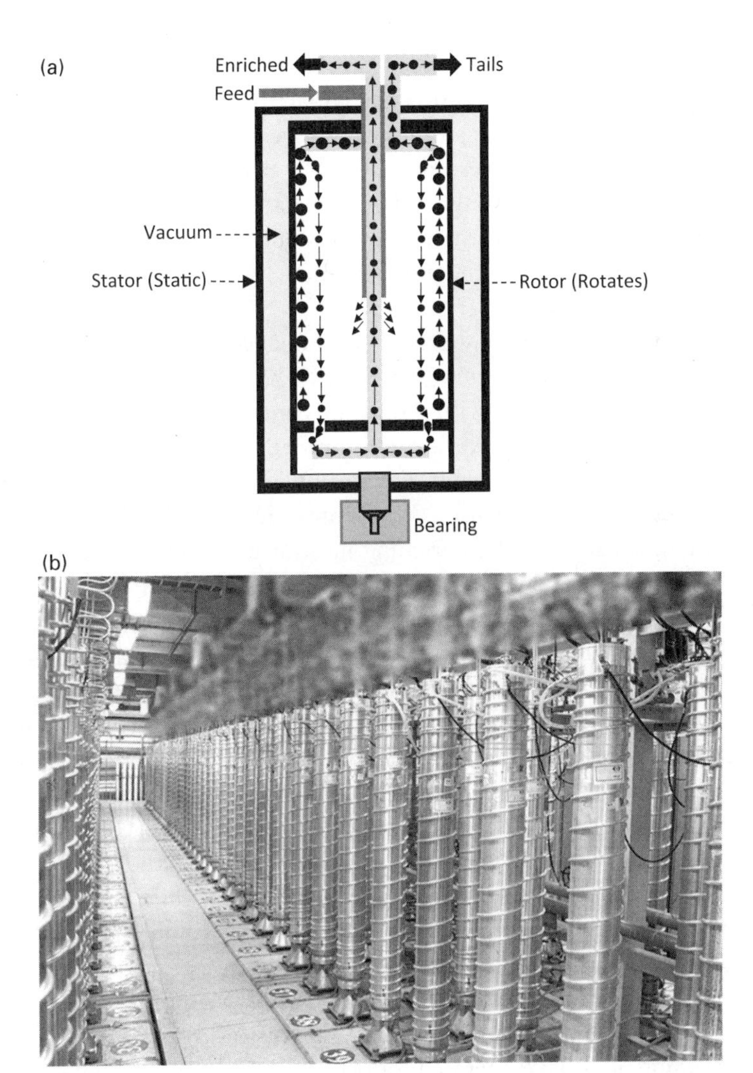

Figure 9.4
(a) Schematic diagram of a gas centrifuge. Low-enriched uranium hexafluoride is introduced toward the center of the centrifuge, and as the rotor spins at very high speed the heavier $^{238}UF_6$ molecules tend to drift toward the sides whereas $^{235}UF_6$ tends to stay closer to the axis. The two streams are usually separated at the top or bottom by heating either side to produce a convection current. (b) Centrifuges at Iran's Fordow enrichment facility. Note the blurred-out region at the top of the photo, likely done to obscure sensitive equipment obtained from outside Iran in violation of sanctions. (Photo by Atomic Energy Organization of Iran.)

box we explain that the work needed to enrich U-235 is nonlinear. That is, enriching natural uranium to 5 percent actually takes more work than to enrich it from 5 percent all the way to weapons grade. Once natural uranium is enriched to 5 percent, in fact, 70 percent of the work needed to get to weapons grade is already done! That's why even power reactor fuel is a proliferation concern. It then takes only 20 percent more work to get to 20 percent enrichment, and from there it's a mere 10 percent more work to weapons grade.

This nonlinearity has important consequences for the risk posed by countries that do their own enrichment. There are two distinct risks. The first is that countries with indigenous enrichment could chemically process regular reactor fuel into UF_6 and enrich it further, perhaps in a clandestine enrichment plant. For example, research reactors like the Tehran Research Reactor use 20 percent enriched fuel, which was produced with great effort by enriching natural uranium. But would take only 10 percent more work to get to weapons grade. That's why Iranian enrichment to 20 percent is a concern. The second risk is that countries wishing to enrich fuel for their civilian power reactors automatically gain bomb-making enrichment capability. That's because fueling just one regular 1 GW_e power reactor for a year requires 20 times more enrichment capability than needed to produce a bomb. Put another way, if you had that capability you could have enough enriched uranium each year to produce 20 bombs, or one bomb every three weeks. This is why enrichment is generally facilitated by a handful of trusted companies, with countries supplied enriched UF_6 while the centrifuge facility is essentially operated as a "black box" without the customers having access to the technology. Today the main suppliers of enriched uranium are Europe's Urenco, France's Orano (formerly AREVA), the China National Nuclear Corporation, and Russia's Rosatom. These and similar companies have operated for decades without spreading sensitive technology. However, there's been a perception that they are exclusive and discriminatory, and not all countries trust them. For example, in the past France refused to provide enriched UF_6 to Iran despite the fact that Tehran owned a substantial stake in the enrichment consortium Eurodif. That's one reason Iran pursues its own enrichment capability—something that it's entitled to as a signatory to the Non-Proliferation Treaty.

The same gas centrifuges that enrich uranium to a few percent U-235 for reactor fuel can readily be used to enrich to weapons grade. It's only necessary to reconfigure the piping between centrifuges so that there are

Proliferation Hazard! Enrichment Isn't Linear

Enrich to 5 percent U-235, at the high end of what's needed for reactor fuel, and you've already done 86 percent of the work needed to enrich all the way to 90 percent weapons-grade uranium. Why? Think of the two uranium isotopes as two different kinds of chocolates. Natural uranium is 0.7 percent U-235, meaning that if you've got 1,000 pieces of chocolate, 7 of them are tasty hazelnut chocolates and 993 are ordinary milk chocolates. Enrichment doesn't create new hazelnut chocolates, but it does change the proportion by physically removing milk chocolates from the batch. If you want to enrich to 5 percent, you need a ratio of 5 out of 100 hazelnut chocolates, or 7 out of 140. Since you started with 7 hazelnuts, you'll want $140 - 7 = 133$ milk chocolates in your 5 percent hazelnut mix. So you have to remove $993 - 133 = 860$ milk chocolates. That's a lot of work just to get just to 5 percent! In fact, it's about 86 percent of the work you would have to do to get to 90 percent in one step.

Here's why: 90 percent means 9 out of 10 hazelnuts, or about 7 out of 8. Since you started with 7 hazelnut chocolates, you'll now have $8 - 7 = 1$ milk chocolate left. So you'd need to remove 992 milk chocolates from the initial 993 milk chocolates to get right to 90 percent hazelnut chocolates. But to get to 5 percent you already had to remove 860 milks, so now you only need to remove an additional $992 - 860 = 132$ milk chocolates to have just one milk chocolate left. That's 132/992 or only about 14 percent more work. This math is a bit counterintuitive, but it emphasizes the important point that even low-enriched power reactor fuel is a proliferation hazard because it takes little additional work to enrich it to weapons grade.[1]

more stages in a cascade. In order to detect such modifications IAEA inspectors conduct so-called limited frequency unannounced access visits inside the centrifuge cascade halls.

Laser Isotope Enrichment

Uranium enrichment using gas centrifuges requires many centrifuges, because the technique exploits the minute difference in mass between the two molecules $^{235}UF_6$ and $^{238}UF_6$. This is rather like picking salt from sugar, which is difficult because they look so similar. However, if the sugar grains are red and the salt grains are white, and you've got technology to select grains by color, then separation becomes easy. Recall the visualization in chapter 2 of atoms as tiny solar systems where the electrons orbit the nucleus. In the case of uranium, the electron orbits are slightly different for the two isotopes—ultimately the result of the different nuclear masses. **Laser isotope separation** exploits this difference, because it allows a laser with precisely tuned energy to cause electrons in U-235 to jump to another orbit (we say the atom is then in an *excited state*) while leaving the U-238 electrons undisturbed in their so-called *ground state*. In this way, laser enrichment selects only particular isotopes by using the fact that each isotope is excited by a laser of a specific frequency. Left alone, the excited atom would decay back to its ground state. However, if it's stimulated once again with another laser, that laser can impart enough energy to ionize the atom—meaning the electron leaves the atom altogether like a spacecraft leaving the solar system. The remaining atom—now an *ion*—is positively charged and can be swept away using an electric field. This technique is not unique to uranium and can potentially be used to enrich tritium or plutonium. The laser technique requires considerably fewer separation units for high-level enrichment, and that in turn means that laser enrichment facilities will be smaller and harder to detect.

Several approaches to laser isotope enrichment are being explored. In *atomic vapor laser isotope separation* the feedstock is vaporized uranium metal, and individual atoms are selectively ionized. In *molecular laser isotope separation*, lasers excite not individual atoms but entire molecules such as UF_6. One molecular technique, called SILEX (separation of isotopes by laser excitation), has been developed in Australia and is close to commercialization. The process involves excitation of molecular vibrations in $^{235}UF_6$, but little else is known publicly because SILEX has been classified in both Australia and the United States.

Why Is Laser Isotope Separation a Proliferation Concern?

Laser enrichment requires advanced technology. However, it demands far less power than other techniques and achieves higher enrichment in a single stage. These factors give laser enrichment facilities a much smaller physical footprint and contribute to the danger that laser enrichment could be used clandestinely to produce weapons-grade uranium.

In addition, atomic vapor laser isotope separation could separate plutonium-239 from other plutonium in spent reactor fuel, whose Pu-240 content makes it less suitable for weapons. The United States had a program to process reactor-grade fuel using the Special Isotope Separations plant, which "was expected to turn reactor-grade plutonium into weapons-grade plutonium in a single pass."[2] The program was eventually cut not because of technological problems but because of a change in priorities. It has been reported that the program successfully separated kilogram quantities of plutonium-239.

Fuel Fabrication

After enrichment, uranium is processed chemically to make uranium dioxide (UO_2), then formed into **fuel pellets** about one-half inch long and three-eighths inch in diameter (roughly a centimeter in each dimension; see figure 9.5). Each pellet is heated (a process called *sintering*) to further harden it. Thanks to the nuclear difference (chapter 2), each pellet contains the energy equivalent of 150 gallons of gasoline or a ton of coal. The pellets are loaded into 13-foot (4-meter) long cylindrical zirconium-alloy **fuel rods** (figure 9.6). Two hundred or more fuel rods are bundled into a square structure to make a **fuel assembly**, also called a **fuel bundle** (figure 9.7). Several dozen fuel assemblies comprise the reactor core.

In the Reactor

What happens to uranium once it's in the reactor? Uranium-235 undergoes fission, of course, producing energy that ultimately leaves the power plant as electricity and waste heat. Fission also results in the highly radioactive fission products discussed in chapter 5. Something else happens in the reactor fuel too: some of the uranium-238 absorbs neutrons, becoming plutonium-239. Pu-239 is fissile, so it too can undergo fission to produce energy and additional fission products. Or a plutonium nucleus may absorb a neutron without fission, resulting in a still heavier nucleus. After a while, the original U-235 and U-238 have become a complicated mix of uranium isotopes, fission products, plutonium, and heavier transuranic isotopes. The details of that isotopic mix have implications for the disposal of nuclear waste, the proliferation of nuclear weapons, and even the economics of nuclear power.

In a typical light-water reactor, a given fuel bundle remains in the core for about three years. At that time the decrease in U-235 and the buildup of fission products mean it's no longer economical to keep the fuel in the reactor. Figure 9.8 shows the mix of substances at the end of those three years. Every 1,000 pounds of 3.3 percent enriched uranium contains 967 pounds of U-238 and 33 pounds of U-235. Three years later the U-235 is down to 8 pounds, the other 25 pounds having fissioned. Some of the U-238 is gone, too, 24 pounds of it having transformed into plutonium and other transuranics. Much of the plutonium-239 thus formed has fissioned, contributing with the U-235 to make a total of 35 pounds of fission products. In fuel near the end of its

Figure 9.5
Uranium dioxide fuel pellets. (Courtesy of GE Hitachi Nuclear Energy Americas LLC.)

Figure 9.6
Cutaway of a zirconium-clad fuel rod, showing individual fuel pellets. (Courtesy of GE Hitachi Nuclear Energy Americas LLC.)

Figure 9.7
A fuel assembly being lowered into a reactor core during refueling. (Courtesy of GE Hitachi Nuclear Energy Americas LLC.)

useful life, in fact, most of the energy comes from fissioning plutonium rather than uranium.

Before proceeding to the nuclear waste dump, let's look more closely at plutonium. The fissile isotope Pu-239 is only one neutron away from U-238, and it builds up quickly. Heavier plutonium isotopes require additional neutron capture, so they increase more slowly. Fuel that's been in the reactor for a short time thus has a significant amount of Pu-239 and not much in the way of other plutonium isotopes. But Pu-239 is fissile, and as it builds up it begins to undergo significant fission. After three years, the rate at which Pu-239 is produced and the rate at which it's destroyed by fission are nearly equal, and Pu-239 buildup essentially stops. But other plutonium isotopes continue to increase, giving older

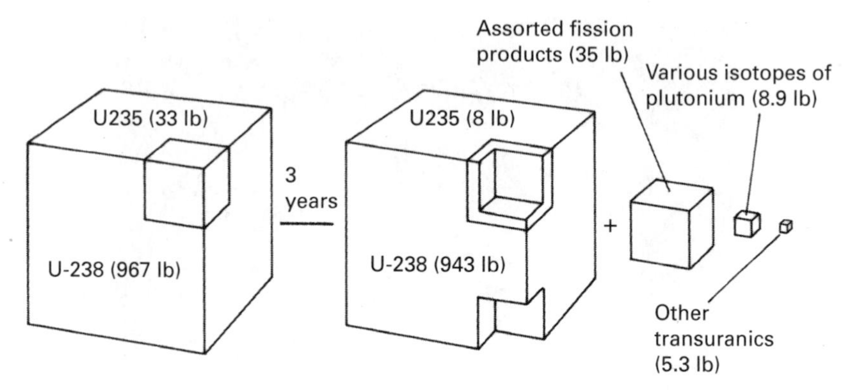

Figure 9.8
Evolution of 3.3 percent enriched uranium over three years in a nuclear reactor. (Adapted with permission from Bernard L. Cohen, "The Disposal of Radioactive Wastes from Fission Reactors," *Scientific American* 236, no. 6, 1977; all rights reserved.)

reactor fuel a higher proportion of these heavier plutonium isotopes in relation to Pu-239.

So what? As you'll see in subsequent chapters, the presence of plutonium-240 makes the construction of plutonium weapons more difficult and renders those weapons somewhat less predictable and less effective than bombs made with pure Pu-239. As a result, it's often argued that spent reactor fuel is safe from a weapons-proliferation standpoint, and that the connection between nuclear power and nuclear weapons is therefore tenuous. It's true that a weapon contaminated with Pu-240 would have a smaller explosive yield, but it would still be far more powerful than any conventional explosive. However, for fusion-boosted designs (see chapter 12) this is not a serious obstacle. In fact, in 1962 the United States conducted a nuclear test where reactor-grade plutonium was substituted for weapons-grade; it produced an explosive yield equivalent to 20,000 tons of TNT, more than the bomb that destroyed Hiroshima. The former head of the Theoretical Division of Los Alamos National Laboratory, Dr. Carson Mark, stated, "The difficulties of developing an effective design of the most straightforward type are not appreciably greater with reactor-grade plutonium than those that have to be met for the use of weapons-grade plutonium."[3] Complications of making a bomb using reactor-grade plutonium include a critical mass 30 percent larger along with heat and radiation exposure about five times that of weapons-grade plutonium. These problems need to be managed but aren't insurmountable. Fuel that's spent a long time in a reactor is certainly not the best choice for weapons—but it would do the trick, bringing a nation or a terrorist group into the nuclear weapons "club." That's why chapter 7 stressed the weapons-proliferation potential of continuously refueled reactors: with those designs it's easy to remove fuel after any desired time in the reactor. "Cooking" nuclear fuel for a shorter time is an inefficient way to make electricity, but it results in much better bomb material.

Out of the Reactor: Spent Fuel

Fresh from the reactor, spent fuel is a nasty radioactive mix of several hundred isotopes. It's so intensely radioactive that a few minutes' exposure to a spent-fuel bundle would be fatal. Energy released in radioactive decay makes the fuel physically hot, so it requires active cooling. The heat and the radioactivity impose severe constraints on facilities for transportation and storage of spent fuel.

The most intense radiation, and therefore also the most intense heating, comes from short-lived fission products. Figure 9.9 shows that radiation and heat drop rapidly as these fission products decay. Holding spent fuel at the reactor site for several years reduces the problems associated with transportation and long-term storage. But even years-old spent fuel is far from benign. Longer-lived fission products—especially strontium-90 and cesium-137, both with half-lives around 30 years, retain significant radioactivity for centuries. Beyond that time, as figure 9.9 shows, fission products have essentially decayed. But the waste remains radioactive for many tens of thousands of years, thanks to longer-lived plutonium and other transuranic isotopes and their decay products. Removal of those heavier isotopes from spent fuel would, as figure 9.9 suggests, result in waste that was no more radioactive than natural uranium ore after 1,000 years. But it would take more than 100,000 years to reach that level if transuranics remain in the spent fuel.

How dangerous is spent fuel? The world's nuclear power reactors produce enough waste each year to kill the planet's human population many times over if the wastes were to be ingested or inhaled. On the other hand, radioactive decay takes its inexorable toll on the wastes' lethality. Nuclear advocates point out that 1,000-year-old nuclear waste may be less toxic than cleaning products you probably store under your kitchen sink. (This presumes removal of plutonium from the waste, something we'll consider shortly.) Critics counter that nuclear waste is an entirely new class of hazardous materials that have never before existed on Earth, and that

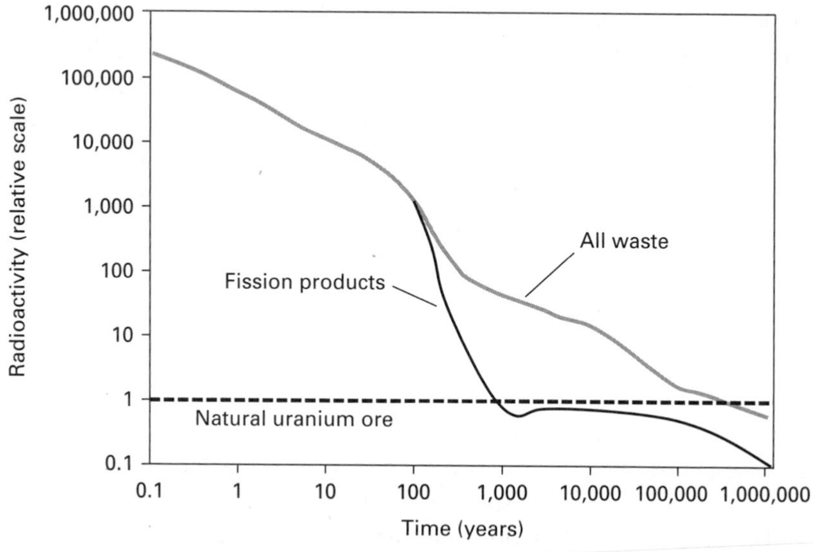

Figure 9.9
Radioactivity of spent fuel from a typical light-water reactor versus the time it's been out of the reactor. Note the logarithmic scale, with divisions on both axes corresponding to factors of 10 increases in the plotted quantities. Fission products (and their daughter nuclides) dominate the first century, after which plutonium, other transuranics, and their decay products become more important. Dashed horizontal line represents the radioactivity of natural uranium ore. (Adapted, with permission, from figure 3.5 of Allan Hedin, "Spent Nuclear Fuel—How Dangerous Is It?" Swedish Nuclear Fuel and Waste Management Technical Report 97–13, 1997.)

1,000-year-old nuclear waste decays far more slowly than the rapid chemical breakdown of most household poisons. In any event, nuclear advocates and critics agree that spent fuel is dangerous, and that it requires long-term isolation from human beings—indeed, from the entire biosphere.

How much nuclear waste do we produce? In volume, not much. Today, the entire global nuclear power industry produces about 1,400 cubic yards of high-level waste annually. That sounds like a lot, but it would fit in 100 dump trucks if they didn't need shielding. Contrast that with the sprawling mountains of landfill that handle our more mundane domestic and industrial waste, or the huge volumes of gaseous waste that spew every second from coal-burning power plants. The small volume of radioactive waste is the reason on-site storage is a viable temporary option, and why an entire country can consider a single waste-storage facility occupying at most a few square miles. On the other hand, its virulent radioactivity and its long life make nuclear garbage unlike any other waste.

Nuclear Waste: Short-Term Solutions

Cooling Pools

Once removed from the reactor core, spent-fuel bundles are moved through an underwater channel to a water-filled pool (called a **cooling pool** or **spent-fuel pool**) typically about 30 feet (10 meters) deep and adjacent to the reactor (figure 9.10). The fuel at this point is still highly radioactive and physically hot. The water shields against radiation and also cools the fuel, although active water circulation is needed to prevent boiling and subsequent exposure of radioactive fuel to the atmosphere. In pressurized-water reactors the cooling pools are embedded in the ground whereas in boiling-water reactors they sit several stories above

Figure 9.10
Spent-fuel pool at a boiling-water reactor. Deep in the circular pit at rear is the reactor vessel with its cover removed. A narrow channel connects the reactor compartment with the spent-fuel pool in the foreground. The spent-fuel pool is normally flooded with water for radiation shielding. (Courtesy of GE Hitachi Nuclear Energy Americas LLC.)

the reactor itself. In neither case do they enjoy the same level of protection as does fuel in the reactor core. Spent fuel needs to remain in the pool for typically 5 to 10 years until it no longer needs water for cooling. Unfortunately, fuel assemblies often remain in the vulnerable cooling pools for decades. As pools fill up, rather than transferring the spent fuel into dry storage, fuel assemblies are sometimes placed closer together. This process, known as *re-racking*, carries a risk of criticality. Criticality in a spent-fuel pool doesn't mean it will explode like a bomb. Rather, large intermittent bursts of neutrons may occur causing a dangerous dose of radiation. To decrease the criticality risk, neutron absorbers are added to the racks themselves. Besides re-racking, another method to house more spent fuel in a cooling pool is fuel consolidation, where individual fuel rods are placed in submerged storage canisters where the fuel rods are close together. Critics have long sounded the alarm about the risks posed by re-racking and high-density consolidation of fuel assemblies and have recommended moving spent fuel to hardened dry casks, but as of 2020 the NRC still permits these techniques in U.S. nuclear plants.

Dry Casks

Once spent fuel has cooled, it can be transferred to storage containers called **dry casks** (figure 9.11). Dry-cask storage employs a defense-in-depth approach with sealed, leak-tight stainless-steel containers holding closely spaced spent fuel in a compartmentalized basket. The stainless-steel containers go into larger containers and then into concrete casks providing a final level of protection and shielding. Shielding materials include reinforced concrete, lead, and steel. Dry casks weigh as much as 100 tons when fully loaded and can hold 10 to 15 tons of spent fuel—about 32 fuel assemblies. A 2006 study[4] of the comparative risks of dry casks and spent-fuel pools found that dry casks are less susceptible to risk of sabotage or accident since pools hold an order of magnitude more spent fuel than individual casks, while a 2014 NRC report suggests that dry casks are far safer in the event of an earthquake.[5] In addition, fuel rods in cooling ponds hold younger fuel than do dry casks, and radioactive material released as a result of a zirconium cladding fire would produce deadly radioactive aerosols. In any scenario, after being removed from the reactor, the fuel will need to be cooled for 5 to 10 years, at which point it can be either reprocessed or stored for decades in dry casks. Dry casks are scattered over a relatively large area, further reducing their vulnerability as possible terrorist targets.

The Fukushima disaster was a wake-up call about the dangers of cooling pools. The pool of Fukushima's Unit 4 reactor contained fuel assemblies

Figure 9.11
Workers checking dry casks storing spent fuel at a nuclear power plant. (Nuclear Regulatory Commission)

recently unloaded from the reactor. After a hydrogen explosion the pool was entirely exposed to the environment. The water didn't completely evaporate, but that was only because of an accidental and fortuitous water leak from the reactor well into the spent-fuel pool. Japan's prime minister at the time, Naoto Kan, who managed the Fukushima crisis, noted that if the spent-fuel pools of 10 reactors at Fukushima Daiini and Fukushima Daiichi had evaporated the entire Tokyo metropolitan area would have had to be evacuated, a nightmare scenario involving 50 million people. However, at Fukushima Daiichi there were also nine dry casks containing 408 fuel assemblies, and while the damage to the reactor and spent-fuel pools was serious, the dry casks survived unscathed. This emphasizes the importance of transferring spent fuel into dry casks as soon as possible. Currently, there are more than 244,000 spent-fuel assemblies in storage in the United States alone, with the majority (70 percent) in pools regardless of their age; the rest are in dry casks.

Storage of spent fuel in dry casks appears to be safe and secure for decades more than originally thought and is used at numerous sites around the world. In the 1980s, the U.S. Nuclear Regulatory Commission (NRC) estimated that spent fuel "could be stored safely for at least 30 years after a reactor's operating license expired." That estimate was pushed further out in 1990, when the NRC stated that it was safe "30 years beyond a 40-year initial license and a 30-year license renewal period, for a total of at least 100 years."[6]

Even dry casks, though, are not without issues. For example, casks aren't necessarily designed to withstand terrorist attacks or impacts from large aircraft. Given their immense weight and construction, it would be difficult to repair a cask that develops cracks through mishandling. And there are potential conflicts of interest; in the U.S., a leading dry-cask manufacturer, Holtec, is also buying up closed nuclear plants to turn a profit on their decommissioning—a profit that's augmented when they then provide waste storage casks. An authoritative study undertaken by the U.S. National Academies explores dry-cask vulnerabilities and concludes that radioactive releases are possible in cases of extreme insults to the casks but are unlikely to be catastrophic.[7]

Nuclear Waste: Long-Term Solutions

Can we find a way to ensure that radioactive spent-fuel waste remains safely isolated for thousands of years? The problem is daunting, in part because no human institutions have lasted that long. How can we know enough about future conditions, both social and physical, to be sure that nuclear waste doesn't put impossible burdens on our descendants?

Many who have worked on nuclear waste issues claim that satisfactory means of disposal are at hand now, and that the reason they haven't been implemented is lack of political will. Others lack confidence in long-term predictions of geological, climatic, and sociological factors involved in waste isolation. Some argue for further study; others claim that no solution will be forthcoming. Meanwhile, nuclear waste accumulates. Even if nuclear power plants ceased operating immediately, we would still need to dispose of more than half a century of spent fuel now stored at reactor sites. Here we discuss some of the options for disposing of spent-fuel waste.

Transmutation

The big problem with nuclear waste is that it remains radioactive for hundreds, thousands, even tens of thousands of years and longer. Most radioisotopes decay quickly, so spent fuel rapidly becomes less radioactive. But the remaining long-lived isotopes give the waste its long-term radioactivity. As figure 9.9 suggests, after 1,000 years all the fission products except a handful of long-lived ones will have decayed, so the remaining radioactivity is dominated by transuranic isotopes. Of particular concern are those of plutonium, neptunium, and americium. **Transmutation** can convert these into shorter-lived isotopes that will quickly decay, thus reducing the long-term burden of radioactive waste management.

Transmutation occurs when long-lived isotopes either absorb neutrons or undergo fission, in either case resulting in shorter-lived species that quickly decay. For transuranics, transmutation would occur through fission. This can be done in reactors where transuranic isotopes are mixed into fuel elements or incorporated into dedicated targets. Fast-neutron reactors (discussed in chapter 7) would be preferable because fission probabilities for transuranics beyond Pu-239 are greater with fast neutrons. However, large quantities of transuranic waste can't go into power reactors because its presence makes the chain reaction harder to control. That's one motivation for the accelerator-driven subcritical reactor (ADSR) discussed in chapter 7.

Transmutation could also eliminate the few long-lived fission products; for example, iodine-129 has a half-life of 16 million years, but when it absorbs a neutron it becomes iodine-130, which decays with a mere 12-hour half-life to stable xenon-130. Iodine-129 and the more radioactive technecium-99 (211,000-year half-life) are candidates for transmutation.

Astute readers will notice a problem: fissioning transuranics does indeed destroy them, but it also introduces more fission products to be dealt with in the short term, creating more immediate danger and challenges for storage repositories. And a small fraction of those fission products, like I-129, are very long lived and might then themselves need transmutation.

Geologic Repositories

The simplest way of managing nuclear waste is to dispose of it underground far from the biosphere. This is the concept of the **geologic repository**, a space constructed inside a stable rock or salt formation that provides a barrier between radionuclides and the environment. The repository requires no attention over many generations, but the material should still be retrievable if that proves desirable. Further isolation is provided by the chemical form of the fuel itself, since uranium dioxide is very hard and has a high melting point. If spent fuel is reprocessed, with uranium and plutonium extracted, the resulting liquid waste is chemically converted into a boron-loaded glass (a process known as *vitrification*) which is chemically inert, highly insoluble, and compact. The purpose of the boron is to absorb neutrons that might result from spontaneous fission of transuranic isotopes. In these forms the spent fuel may be disposed in artificially constructed underground mines or deep boreholes.

Artificial Mines as Geologic Repositories

Sweden, Finland, and France are close to finishing underground mines for spent fuel storage. Finland will be the first country licensed to dispose of spent fuel from civilian reactors in a mined repository that's expected to start accepting nuclear waste in 2024. Fuel assemblies will be placed into iron canisters which themselves will be inside 15-foot-long copper-lined canisters. The canisters will be emplaced in mined tunnels some 1,500 feet underground. Clay padded around the canisters will serve a dual purpose: being fully saturated with water, the clay will prevent water laden with radionuclides from migrating through it, and over time the clay will expand and seal the tunnel to protect the canisters from rock movement. The largest barrier to the eventual release of radionuclides is the stability of the geologic environment in which mined repositories are excavated. Hard crystalline rock is often chosen, because it's chemically and seismically stable and thus can ensure isolation from the surface for upwards of 100,000 years. The Achilles heel of the Scandinavian system is the ability of the copper lining to withstand corrosion. The company that manufactures Sweden's KBS-3 canisters claims the canisters will maintain their integrity for as much as a million years. However, critics say the copper lining may fail in only 1,000 years. There's an ongoing debate about who's correct. We hope and expect that humankind will be around in 1,000 years, but what about 10,000 or a million years? This question isn't purely philosophical, and settling it represents an important nuclear choice.

How should the waste-disposal site be commemorated to prevent future generations from digging into it? Some argue that it's important to warn future generations even if language as we know it may have evolved. Some suggest placing a plaque like the one proposed in 1993 by the U.S. Department of Energy and shown in figure 9.12. Others suggest Edvard Munch's painting *The Scream* as a visual warning! The opposing view holds that no announcement of the repository's existence should be made, to prevent future people from thinking there may be valuable materials underground.

You might think that what determines the amount of waste that can be placed in a geologic repository is the volume of the tunnels. However, studies show that the crucial parameter is the temperature of the tunnel walls, and whether adjacent tunnels will affect one another over thousands of years. The studies find that if 99.9 percent of the original cesium,

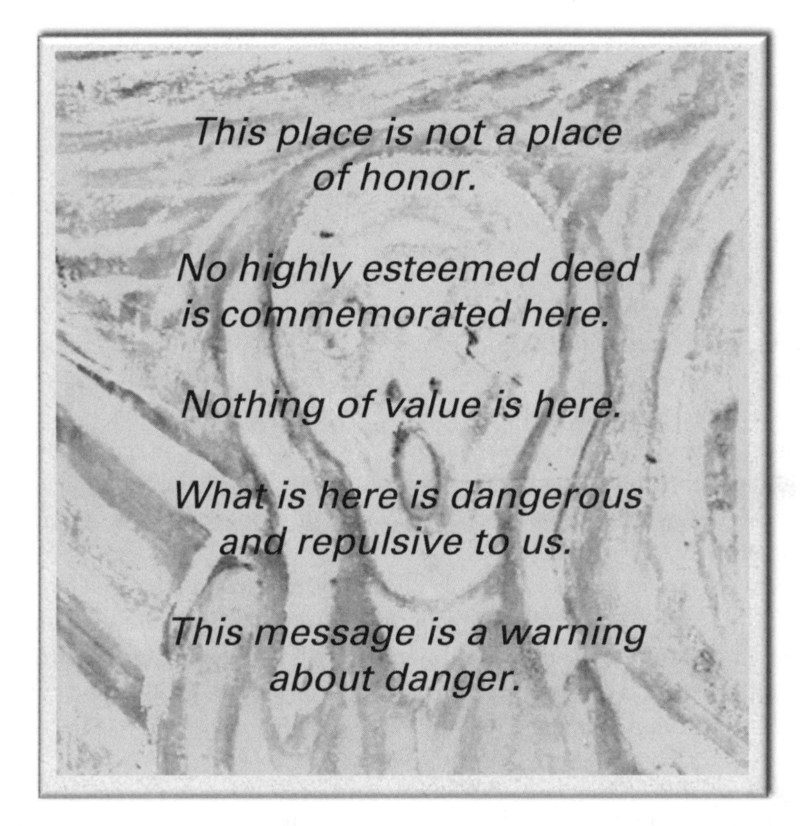

This place is not a place
of honor.
No highly esteemed deed
is commemorated here.
Nothing of value is here.
What is here is dangerous
and repulsive to us.
This message is a warning
about danger.

Figure 9.12
Wording proposed by the U.S. Department of Energy for plaque suggested to mark geologic repositories.

strontium, and transuranics are eliminated via repeated recycling and reprocessing, then the tunnels can accommodate 225 times more light-water reactor waste than without reprocessing. This is part of the motivation for reprocessing, which we'll discuss in the next major section.

Deep Borehole Disposal

Deep borehole disposal is a simple concept, involving placement of waste in vertical shafts three miles deep or more. The lower halves of the shafts could accommodate some 400 canisters with clay between them, then backfilled to the surface with clay, concrete, or sand. The canister designs anticipated would be about 15 feet (5 meters) long and roughly the diameter of a car tire. Borehole disposal has been proposed since the 1950s, but it's garnered more interest recently because truly deep boreholes are becoming economically feasible. Deep boreholes are also being considered for disposal of plutonium from weapons dismantlement, where safeguarding the material is important. The top section of the borehole can even be obliterated, making it difficult for nefarious actors or future generations to access the nuclear materials. Proponents argue that the sheer depth—miles below the water table—will prevent radioactive material from reaching the surface.

One promising idea by the California startup Deep Isolation is to drill horizontally after reaching sufficient depth. This takes advantage of advances in the oil and gas industry where drilling is routinely done over long-radius curves, so canisters could navigate the curves without getting stuck. The horizontal tunnels would be up to several miles long and miles underground. Such horizontal holes have the same advantages as vertical deep boreholes. In addition, by fanning out tunnels horizontally, like spokes of a wheel, the heat produced in neighboring tunnels is no longer a concern.

An advantage of deep borehole disposal is that it could be used in seismically active regions where mined repositories may be more difficult to site. In addition, boreholes could potentially be drilled offshore in a host of different types of rock or where a thick layer of clay would prevent the canisters from moving.

What's Happening in the United States?

What's happening with nuclear waste in the United States? In two words: not much. The plan to dispose of spent fuel in a geologic repository hinged on a mined repository in Yucca Mountain near the Nevada Nuclear Test Site (figure 9.13). The repository itself would be about 1,000 feet (300 meters) below the surface, with waste placed in double-shelled metal canisters with a corrosion resistant outer shell and an inner container housing spent fuel assemblies. The U.S. design also includes a massive titanium drip shield on top to prevent water or rocks from damaging the canister. The shield would be installed 100 years after the waste is placed in the repository. This is to address the possibility of more water dripping into the repository and onto the canisters than was initially expected.

The history of Yucca Mountain is one of ups and downs. In 1987, the Nuclear Waste Policy Act was modified to designate Yucca Mountain as the only repository in the United States. Eleven years later, the Department of Energy submitted an application for construction. In 2009, the Obama administration cancelled the project, but in 2013 a federal appeals court ordered the NRC to continue its review of the facility. This culminated in

Figure 9.13
Entrance to the Yucca Mountain waste repository. (Nuclear Regulatory Commission)

two documents: an environmental impact assessment and a technical safety review where the repository was considered safe for more than a million years. However, critics point out that there's more water at Yucca Mountain than originally expected, hence requiring the drip-shield fix a century out. Critics claim that canisters would corrode in less than 1,000 years without the drip shield and that drip shields to be installed after 100 years require future financing that's far from guaranteed and that places an undue burden on future generations. Most recently, in 2020 the Trump administration reversed its previous support for Yucca, saying it would abandon the project in favor of "innovative approaches" to waste management.

Some Good News

Despite problems with the Yucca Mountain repository for civilian waste, the United States has established the first licensed facility for military nuclear waste at the Waste Isolation Pilot Plant (WIPP) in Carlsbad, New Mexico. The facility was built in a salt bed 2,000 feet thick. Since salt dissolves in water, its very presence indicates that no significant water has been present for a very long time. Salt has the additional benefit of sealing itself in the event that cracks develop. At WIPP, waste is in large steel drums placed in excavated underground areas. Military waste shipped to the facility amounts to almost 100,000 cubic yards of transuranic waste. In 2014, 15 years after receiving the first waste, an accident occurred that injured 13 people and contaminated one-third of the repository's underground area. The problem was due to a typographical error where an organic kitty litter was used as a sorbent material instead of clay-based kitty litter, which is a commonly used sorbent. This mix-up resulted in an explosive chemical reaction! This unfortunate incident resulted in more than three-year suspension of the facility and 2 billion dollars wasted. It's the second most expensive nuclear incident in the United States after Three Mile Island.

Despite these setbacks, many industry and government scientists are confident that underground waste repositories will ultimately prove safe and effective. They point to studies of the natural reactor at Oklo in West Africa (see box "A Natural Reactor" in chapter 7), which show negligible migration of fission products and transuranics over 2 billion years. Even if waste containers fail—as critics expect after 1,000 years or so—and if water seeps into a repository, the Oklo studies suggest it could be hundreds of thousands of years before waste would make its way to the surface. By that time, it would be essentially harmless. Critics respond with skepticism that realistic predictions can't be made about events thousands of years hence, and point to a host of unknown factors—from intentional sabotage to earthquakes—that could compromise a waste-storage facility.

Other Disposal Schemes

Other locations that have been proposed for nuclear waste disposal include outer space, ice sheets, and on or under the ocean floor. All have major flaws and none are particularly realistic. Launching waste into space, possibly to the Sun, would be prohibitively expensive and expose Earth to dangers of a launch accident. Letting waste melt its way into miles-thick Greenland or Antarctic ice sounds attractive, but hot waste might change the behavior of ice in unexpected ways, and in any event climate change is making ice sheets vulnerable to melting and collapse. Sea-bottom or undersea burial could run afoul of international treaties and interfere with plans for mining and other seabed development. So these ideas are probably out, at least for the foreseeable future.

Waste Transport and Preparation

It's one thing to store dangerously radioactive waste in a secure facility below Earth's surface; it's quite another to move waste to a repository from commercial power reactors scattered over a whole country. Many nuclear critics regard transportation as a weak link in the nuclear fuel cycle. However, the recognition that on-site storage in dry casks is viable for up to a century reduces the pressure to move waste into a repository and means that by the time waste is moved its radioactivity will be greatly reduced.

Nuclear critics worry about using highways and railroads to transport nuclear waste. Nuclear advocates counter with a record of over 3 million shipments of radioactive materials yearly in the United States, with no documented loss of life due to radiation. Only a minute fraction of these shipments involve the high radioactivity typical of spent reactor fuel. This high-level waste is transported in special casks that provide for radiation shielding, cooling, and containment in the event of an accident. Among other criteria, casks must survive falling 30 feet onto a hard surface, exposure to a temperature of 1,475°F for 30 minutes, and immersion in water for 8 hours without leaking any radioactive material. More dramatic tests have confirmed the integrity of some shipping-cask designs (figure 9.14). Critics fault these seemingly impressive test results by noting that the casks used were obsolete designs that might prove more crashworthy than their contemporary counterparts. Critics also claim that widespread publicity failed to note that cask leakage occurred in several tests, and that cask temperature ratings might be exceeded in fires involving commonly transported fuels and chemicals. Finally, critics see transportation as especially vulnerable to theft and sabotage, perhaps

Figure 9.14
In a test, a locomotive moving at 82 mph slams into a flatbed truck carrying a cask designed for transport of highly radioactive waste. (U.S. Department of Energy)

by nations or terrorist groups hoping to build nuclear explosives or simpler "dirty bombs." Is transportation of radioactive waste a serious danger to society? That nuclear question bears on our overall assessment of the safety of nuclear power.

Before it's entombed in a repository, waste must be prepared for its final disposal. The simplest preparation is to load intact fuel bundles into the canisters we described earlier. More efficient use of repository space can be achieved by compacting, perhaps by vitrification—incorporating ground-up nuclear waste into molten glass. The solidified glass rods would be encased in stainless steel for emplacement in the repository. Even after the steel corrodes, the glass should permit only very slow release of radioactivity.

Reprocessing

As figure 9.8 showed, spent fuel still contains significant amounts of uranium-235 and plutonium-239. Removing these substances would greatly reduce the long-lived radioactivity of the waste and would allow further use of these fissile materials as nuclear fuels—either for power generation or for weapons. **Reprocessing** is the separation of fissile uranium and plutonium from spent fuel.

Conventional reprocessing involves chopping up the spent fuel rods and dissolving the fuel pellets in acid. Chemical separation then removes uranium and plutonium. The remaining fission products are ready for vitrification and subsequent disposal as nuclear waste. The uranium, containing both U-235 and U-238, can be centrifuged to enrich the portion of U-235. The plutonium is mixed with enriched uranium to make **mixed oxide fuel** (MOX)—a mixture of uranium and plutonium oxides that can substitute for uranium as a reactor fuel. Alternatively, plutonium can be diverted for nuclear weapons production. Figure 9.15 shows a nuclear fuel cycle involving reprocessing.

An alternative to chemical reprocessing is **pyroprocessing**, a technique that separates transuranics (including plutonium) from fission products using high-temperature electrorefining. The bulk of the radioactivity in spent fuel is due to short-lived fission products that decay to manageable levels for pyroprocessing after only 100 days of cooling. Proponents of pyroprocessing claim it's more proliferation resistant because plutonium is not extracted separately from other transuranics. Opponents counter that pyroprocessing facilities are smaller than conventional reprocessing plants and therefore easier to conceal, and that

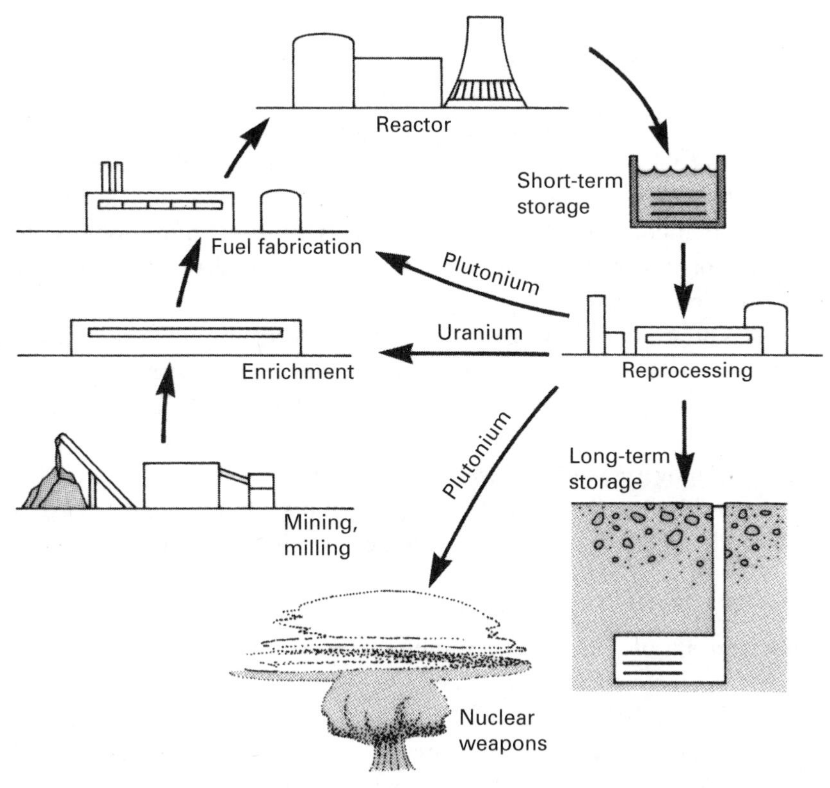

Figure 9.15
Nuclear fuel cycle with reprocessing. Uranium-235 and plutonium-239 are removed from spent fuel and recycled into new reactor fuel. Alternately, Pu-239 can be used in weapons. Compare with the once-through cycle of figure 9.1.

crude weapons can be made even when plutonium isn't separated from other transuranic waste.

Most nuclear critics—and even some proponents of nuclear power—consider reprocessing undesirable. Highly fissile plutonium could be widely available in a world economy encouraging reprocessing. Through theft, terrorism, or legal purchase, a nation or group seeking nuclear weapons capability might obtain the necessary plutonium. And although plutonium from spent reactor fuel isn't ideal bomb material, it is, as we noted earlier, sufficient to construct a crude fission bomb. And a nation with reprocessing facilities could easily produce bomb-grade plutonium from specialized reactors.

In the 1970s, the United States chose to forgo reprocessing. Other countries pursued reprocessing technology, with the goal of extracting as much energy as possible from their nuclear fuels. In actual fact, while this recycling (as nuclear proponents call it) sounds attractive, it turns out to be expensive and fraught with problems. As a result, both France and the United Kingdom have concluded that reprocessing isn't economical. One problem is that plutonium quality worsens each time the fuel is cycled through the reactor because of the buildup of neutron absorbing isotopes that compete for fission neutrons. This is less of a problem for fast reactors where there are high-energy neutrons that can fission the undesirable isotopes.

It was hoped that the large-scale use of MOX fuel would incinerate the world's plutonium stock as commercial breeder reactors were developed. This didn't happen, and in 1989 Électricité de France (EDF) deemed that MOX would not be competitive with the once-through fuel cycle. In fact, the French government has investigated MOX in a reprocessing cycle for 28 nuclear reactors and found an additional cost of $35 billion. A 2003 MIT study found that if the United States adopted MOX, the cost could be as much as four times that of the once-through cycle. The message here is that development of advanced fuel cycles must be done carefully and cognizant of underlying assumptions such as the expectation of fast reactor commercialization.

Would a global reprocessing ban help limit the spread of nuclear weapons? Possibly: almost all states possessing nuclear weapons used plutonium in their first nuclear explosives. (Only China and Pakistan chose the enriched-uranium route. Plutonium and uranium weapons evolved simultaneously in the United States.) How to handle the international development of nuclear power and its plutonium by-product, and manage peaceful use of enriched uranium, are among the most sensitive nuclear issues.

Waste Issues with Advanced Reactor Designs

Chapter 7 introduced several advanced reactor designs that have their own unique implications for nuclear waste management.

Fast Reactors and Integral Waste Management

One of the Gen-IV designs, the sodium-cooled fast reactor (SFR), has been touted as a waste-management device for burning transuranics in spent fuel from light-water reactors. The *integral fast reactor* would include an SFR, a fuel fabrication plant, and facilities for waste handling all in the same reactor park. By isolating the full cycle to one complex, fuel transport is simplified and the probability of unlawful access to nuclear materials is decreased.

Proponents of sodium-cooled fast reactors point to the availability of many more neutrons than in light-water reactors, with the excess neutrons useful for fissioning transuranic waste. In practice, though, waste would have to be cycled through a fast reactor many times to achieve a high level of burnup. And the SFR's sodium coolant is sufficiently radioactive that it must be treated as high-level waste (HLW), presenting challenges for disposal of this chemically reactive substance. Furthermore, the waste from fast reactors carries a higher initial heat load than does light-water waste, meaning that considerably less fast-reactor waste can be stored in a given geologic repository.

What about Thorium?

Another advanced reactor concept discussed in chapter 7 is the molten salt reactor, which burns uranium-233 bred from thorium-232. Thorium-232, with a half-life of 14 billion years, is more common than uranium, and large deposits are found around the world, especially in India, where considerable research on the **thorium fuel cycle** is ongoing. The breeding of U-233 from Th-232 is directly analogous to how U-238 absorbs a neutron and becomes Pu-239. Thorium reactors need another fissile material (known as the *driver*) to produce neutrons to jump-start the reaction until enough U-233 is available to sustain a chain reaction. One possibility is to mix thorium with plutonium, producing a mixed fuel in which plutonium acts as the driver.

The products of thorium reactors have unique implications for proliferation and waste management. The critical mass of the U-233 fissile fuel is only 10 percent greater than for plutonium-239. Moreover, U-233 doesn't produce many spontaneous fission neutrons, so, unlike Pu-239, it can be used in a simple gun-type bomb. The disadvantage for nuclear weapons production is that neutron bombardment of Th-232 also produces the isotope U-232, whose decay chain includes the intense gamma emitter thallium-208. Proponents of the thorium fuel cycle suggest that the intense radiation from Tl-208 makes it difficult to fashion a nuclear weapon. However, the high radiation can be overcome by limiting exposure to personnel or by handling the material remotely—which is becoming increasingly easy with robots. Furthermore, continuous refueling can limit the time that Th-232 spends in the reactor and therefore limits the buildup of U-232. It's also possible to extract intermediate isotopes in the U-232 decay chain, thus preventing the formation of Tl-208. So, although it may present challenges to a potential bomb maker, the thorium fuel cycle isn't entirely proliferation proof.

Other Nuclear Waste

So far we've concentrated exclusively on the intensely radioactive **high-level waste** (HLW) from spent reactor fuel. But there are many other sources of nuclear waste. In the United States, the nuclear weapons industry has produced considerably more high-level waste than all the nation's power reactors. Early on that waste was handled very poorly, and we're still paying the price—as in the $320 billion cleanup ongoing at the Hanford Nuclear Reservation in Washington State. A great many activities produce **low-level waste** (LLW), whose radioactivity is low enough and/or sufficiently short lived to permit less stringent handling and disposal requirements. Low-level waste is generated in the nuclear fuel cycle during refinement, enrichment, and fabrication of nuclear fuel. Contaminated clothing and tools, filters used to clean primary-loop cooling water, piping, and hardware are among the low-level wastes generated in nuclear power plants. Radioisotope tracing techniques used in medicine, research, and industry provide additional low-level waste; in states lacking nuclear power plants or weapons facilities, hospitals and universities are generally the largest producers of low-level radioactive waste. A third officially designated waste category is **transuranic waste** (TRU), consisting of mildly radioactive but very long-lived transuranic radioisotopes from the production of weapons or the reprocessing of nuclear fuel.

High-level waste from the weapons program and spent fuel from commercial reactors are ultimately destined for permanent underground repositories. In the absence of such repositories, on-site storage of weapons wastes is currently practiced under far less safe conditions than prevail at commercial nuclear power plants. Low-level waste from all sources is handled differently, typically by burial in shallow trenches at designated dump sites. Still another source of mixed nuclear waste is the nuclear power plants themselves, as they reach the ends of their roughly 40-year life spans. Current practice has them decommissioned and their sites returned to uncontaminated conditions. **Decommissioning** of commercial reactors in the United States is allowed to take up to 60 years, although enterprising companies are buying up reactors as they close and promising a much shorter decommissioning time. In return they get to spend the decommissioning trust funds that have accumulated over the plants' decades of operation.

Summary

Nuclear waste is intimately connected with the nuclear fuel cycle, the types of reactors in operation, and political/technical decisions about reprocessing.

The fuel cycle comprises mining uranium, processing it into fuel, fissioning in a reactor, and management of the resulting nuclear waste. For most power reactors, processing uranium into reactor fuel requires enrichment to increase the proportion of fissile uranium-235. You've seen how the enrichment process is not linear, in that a great deal of work is required to enrich to just 5 percent U-235, but only a fraction more work takes uranium all the way to weapons grade. That makes enrichment for power reactor fuel a proliferation-sensitive technology.

Eventually spent fuel must be removed from the reactor. The spent fuel is both hot and radioactive, and will remain radioactive for hundreds to thousands of years or more. Spent reactor fuel contains shorter-lived but highly radioactive fission products and longer-lived transuranic isotopes, including plutonium. Disposal of these materials with safety ensured for thousands of years poses technical and political challenges that have prevented the opening of a waste repository anywhere in the world more than 65 years since the first commercial nuclear reactor went critical. You've seen many waste management techniques, including transmutation, reprocessing, pyroprocessing, and disposal in geologic repositories and deep boreholes. Although many would argue that technical solutions are at hand, others point to continuing scientific uncertainty and lack of confidence in government agencies as evidence that the waste-disposal problem may not be solved soon.

Reprocessing of spent fuel lessens the burden of long-lived radioactive materials and reduces the need for fresh uranium, but reprocessed plutonium and the technology to prepare it may facilitate the spread of nuclear weapons. Transmutation may seem like an ideal solution but at this point is neither practical nor economical. Geologic repositories are the waste-disposal option that's closest to fruition, but repositories are difficult to site for technical as well as political reasons. Deep borehole disposal may offer a glimmer of hope, but like other disposal techniques, it suffers from the fact that solving the nuclear waste problem is seldom a priority. Nuclear waste remains among the thorniest problems of the nuclear age.

Glossary

cooling pool A water-filled pool located adjacent to a reactor and used for temporary storage of spent fuel as it cools and becomes less radioactive. Also called a **spent-fuel pool.**

decommissioning The process of disassembling or entombing a nuclear reactor at the end of its useful life and safeguarding the associated radioactive material.

dry cask A large concrete cylinder with an inner stainless-steel cylinder holding spent fuel; designed for storage at reactor sites for up to 100 years.

enrichment The process of increasing the proportion of fissile U-235 in uranium, to 3–5 percent for light-water reactor fuel and to 90 percent or more for weapons.

fuel bundle An assembly of fuel rods (also called a **fuel assembly**).

fuel pellet A small (~ ½ inch) cylinder of uranium dioxide (UO_2) that is the basic fuel unit in light-water reactors.

fuel rod A zirconium-alloy tube containing uranium fuel pellets.

gas centrifuge enrichment A uranium enrichment process in which uranium hexafluoride (UF_6) gas is injected into high-speed centrifuges that separate lighter $^{235}UF_6$ from heavier $^{238}UF_6$.

gaseous diffusion A uranium enrichment process in which uranium hexafluoride molecules containing U-235 pass more readily through a membrane than do slower-moving molecules containing U-238.

geologic repository An underground storage facility for nuclear waste.

high-level waste (HLW) Nuclear waste derived from spent fuel in power or weapons-production reactors, and typically characterized by a mix of intensely radioactive fission products and longer-lived but less radioactive transuranic isotopes.

laser isotope separation A uranium enrichment process in which a precisely tuned laser selectively excites U-235 atoms, which are then ionized and separated from unexcited, unionized U-238 by an electric field.

low-level waste (LLW) Any radioactive waste not classified as high-level or transuranic, and characterized by low radioactivity and relatively short half-lives.

mixed-oxide fuel (MOX) Nuclear reactor fuel made from a mix of uranium and plutonium in oxide form.

nuclear fuel cycle The sequence by which uranium progresses from its natural state through fuel fabrication, fission in a reactor, processing or reprocessing, and eventual long-term storage of nuclear waste.

once-through fuel cycle Simple nuclear fuel cycle in which spent reactor fuel is stored as radioactive waste, without any attempt to remove and reuse fissile U-235 or Pu-239.

pyroprocessing A process for separating transuranic isotopes from fission products in nuclear waste, using high temperature electrorefining.

reprocessing Removal of fissile uranium-235 and plutonium-239 from spent reactor fuel, making these materials available for use as reactor or weapons fuel.

thorium fuel cycle The fuel cycle for reactors that use uranium-233 as their fissile fuel. The U-233 is bred by neutron bombardment of thorium-232.

transmutation Conversion of radioactive nuclei to other, shorter-lived species, usually by neutron bombardment in a reactor and subsequent neutron absorption or fission.

transuranic waste (TRU) Radioactive waste consisting of transuranic isotopes (heavier than uranium) and characterized by modest radioactivity and half-lives that range to tens of thousands of years or more.

Notes

1. This analogy assumes that there is no waste stream at all, so, in our chocolate analogy, all the hazelnut chocolates end up in the enriched batch of chocolate. In reality, the enrichment process also produces a waste stream of depleted uranium containing from 0.2 percent to 0.4 percent U-235. In that case, enriching to 5 percent U-235 requires about 70 percent of the work needed to get to 90 percent enrichment (that's the figure given in the text). And if you enrich to 20 percent, then you only need to do 10 percent more work to get to weapons grade.

2. S. Fetter et al., "Fissile Materials and Weapon Models, Appendix 11.A," in *Reversing the Arms Race: How to Achieve and Verify Deep Reductions in the Nuclear Arsenals*, ed. F. Von Hippel and R. Sagdeev (New York: Gordon and Breach Science, 1990).

3. J. Carson Mark, "Explosive Properties of Reactor-Grade Plutonium," *Science and Global Security* 4, no. 1 (1993): 111–128.

4. National Academy of Sciences, *Safety and Security of Commercial Spent Nuclear Fuel Storage* (Washington DC: National Academies Press, 2006), available at https://www.nap.edu/catalog/11263/safety-and-security-of-commercial-spent-nuclear-fuel-storage-public.

5. U.S. Nuclear Regulatory Commission, "Consequence Study of a Beyond-Design-Basis Earthquake Affecting the Spent Fuel Pool for a U.S. Mark I Boiling Water Reactor," NUREG-2161, 2014, available at https://www.nrc.gov/docs/ML1425/ML14255A365.pdf.

6. See NRC, "Continued Storage of Spent Nuclear Fuel, NRC-201200246," available at https://www.nrc.gov/docs/ML1417/ML14177A477.pdf.

7. "Safety and Security of Commercial Spent Fuel Storage: Public Report" (Washington, DC: National Academies Press, 2006). Available at https://www.nap.edu/read/11263/chapter/35. Some findings of the study are classified and do not appear in this public report.

Further Reading

Cohen, Bernard L. "The Disposal of Radioactive Wastes from Fission Reactors." *Scientific American* 236 (June 1977): 21–31. A well-written but now dated article detailing many of the issues associated with underground nuclear waste disposal. Excellent graphs and diagrams accompany the text. The author was an independent scientist who, until his death in 2012, was among the United States' most eloquent and persistent advocates of nuclear power.

MacFarlane, Alison, and Rodney Ewing, eds. *Uncertainty Underground: Yucca Mountain and the Nation's High-Level Nuclear Waste*. Cambridge, MA: MIT Press, 2006. A compilation of authoritative articles on the political and technical issues surrounding Yucca Mountain.

Massachusetts Institute of Technology. *The Future of the Nuclear Fuel Cycle: An Interdisciplinary MIT Study*, 2011. Available at https://energy.mit.edu/wp-content/uploads/2011/04/MITEI-The-Future-of-the-Nuclear-Fuel-Cycle.pdf. An authoritative survey of all aspects of nuclear fuel cycles, including waste disposal, economics, nonproliferation, and fuel cycles for advanced reactors.

Muller, Richard A., et al. "Disposal of High-Level Nuclear Waste in Deep Horizontal Drillholes." *Energies* 12, no. 2052 (May 29, 2019). Available at https://www.mdpi.com/1996-1073/12/11/2052/pdf. A thorough and highly technical analysis of the potential for nuclear waste storage in deep horizontal boreholes.

National Research Council et al. *Going the Distance? The Safe Transport of Spent Nuclear Fuel and High-Level Radioactive Waste in the United States*. Washington, DC: National Academies Press, 2006. A report by the NRC's Nuclear and Radiation Studies Board and Transportation Research Board describes the risks associated with transport of nuclear waste. Somewhat dated given the report's expectation that Yucca Mountain would soon be open, but nevertheless a valuable look at nuclear waste transport.

Walker, J. Samuel. *The Road to Yucca Mountain: The Development of Radioactive Waste Policy in the United States*. Berkeley: University of California Press, 2009. Going beyond Yucca Mountain itself, the former historian of the Nuclear Regulatory Commission describes the full political and technical history of nuclear waste policy in the United States.

U.S. Department of Energy. "Blue Ribbon Commission on America's Nuclear Future: Report to the Secretary of Energy (2012)." Available at https://www.energy.gov/sites/prod/files/2013/04/f0/brc_finalreport_jan2012.pdf. Commissioned during the Obama administration, this report emphasizes nuclear waste management, which the authors recognize as being at an impasse. The lengthy report details the technical background, lessons learned from Fukushima and other nuclear incidents, and makes concrete proposals for moving forward.

10

A Future for Nuclear Fission?

Nuclear power plants accounted for some 10 percent of global electricity generation in 2020, down from a high of nearly 18 percent in 1996 (figure 10.1). Although actual nuclear generation has risen slightly in recent years, the world's total electricity generation has risen more rapidly, thus reducing the nuclear share. That trend is likely to continue, given the increasing electrification of our energy supply. If it does, nuclear power will have a hard time maintaining its current share of world electricity. Yet a healthy nuclear power enterprise has its benefits, especially given the increasing urgency of human-caused climate change.

A Brief History of Nuclear Power

"Too cheap to meter" is a 1950s-era phrase evoking the optimistic prediction of virtually limitless, inexpensive electricity from nuclear sources. Today it's used by nuclear critics as an example of just how wrong early nuclear power advocates were.

Actually, it's not clear that "too cheap to meter" was intended to apply to nuclear fission. The phrase was first used by Atomic Energy Commission (AEC) chair Lewis Strauss in a 1954 speech to the National Association of Science Writers. Strauss was a strong advocate of nuclear fusion, whose prospects we'll explore in the next chapter, but at the time the United States' fusion program was classified. So it's entirely possible that he was referring to fusion. Or maybe he was being intentionally ambiguous. In any event, Strauss's optimism wasn't universal; neither the AEC nor the budding nuclear industry shared his view. Yet "too cheap to meter" has found its way into the lore of nuclear power and isn't likely to be dislodged.

The year 1954—the year of "too cheap to meter"—also saw the world's first nuclear-generated electricity, when a modified plutonium-production

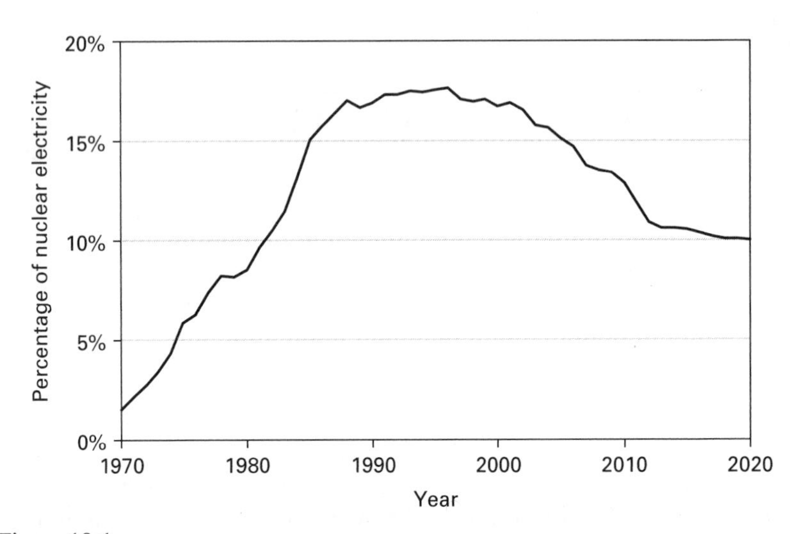

Figure 10.1
Percentage of the world's electricity produced by nuclear fission peaked in the 1990s and has been declining ever since. (Data sources: U.S. Energy Information Administration, International Atomic Energy Agency, and International Energy Agency)

reactor in the Soviet Union's closed city of Obninsk began supplying electrical energy at the rate of 5 megawatts. The AM-1, whose letters abbreviate "peaceful atom" in Russian, was a water-cooled, graphite-moderated reactor that eventually evolved into the RBMK design used at Chernobyl and elsewhere in the Soviet bloc. In 1956, the United Kingdom brought Unit 1 at the Calder Hall nuclear plant online. This was a graphite-moderated gas-cooled reactor producing 50 megawatts of electric power as well as making plutonium for the U.K.'s nuclear weapons program. It was soon joined by three identical units. Calder Hall later ceased plutonium production but continued to supply electric power through 2003. Calder Hall inspired most of the U.K.'s current fleet of gas-cooled reactors. France brought a similar dual-use graphite/gas reactor online in 1956 but later switched to light-water reactors. The first electric power reactor in the United States was a 60-MW demonstration plant at Shippingport, Pennsylvania, which operated from 1957 through 1982. Canada, following its own heavy-water path to nuclear power, started with a 20-MW CANDU reactor that went online in 1962. Japan's first power reactor was a 12.5-MW boiling-water prototype in 1963, followed by a 160-MW U.K. graphite/water design, followed by a return to light-water reactors.

The late 1960s and early 1970s saw a surge in reactor orders and construction, primarily in North America, western Europe, the Soviet Union, and Japan, and most of these units came online in the 1970s through the mid-1980s (figure 10.2). But by the mid-1970s it became clear that demand for electricity wasn't growing as fast as anticipated, and that made nuclear power less economical. The result was cancellation of many outstanding reactor orders. Some projects were canceled even after construction had begun. Antinuclear protests often delayed plant construction, and, globally, average construction times doubled from about 6 years in the late 1970s to more than 12 years in the 1990s. Costs were escalating, too, with many reactors coming in at more than twice their initial cost estimates and some as much as five to eight times greater. The 1979 Three Mile Island accident revived nuclear fears and, with new post-TMI safety requirements, increased nuclear costs still further. The result of all this—not just TMI—was a precipitous drop in worldwide reactor orders, followed, after completion of projects still under construction, by a plunge in new reactors coming online—obvious in figure 10.2 after the mid-1980s. Nowhere was the nuclear decline as great as in the United States, where 1978 saw the start of a 34-year period with no reactor construction reactor permits issued.

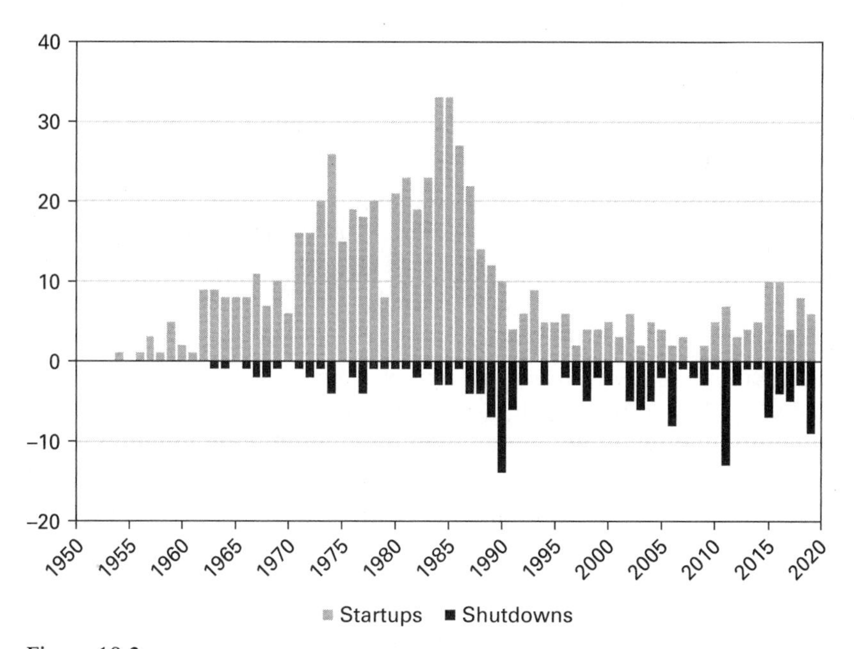

Figure 10.2
Commercial reactor startups and shutdowns worldwide from 1954 through 2019. "Startup" here means the reactor began supplying electric power to the grid. (Data source: World Nuclear Association)

Nuclear power is expensive, but aren't all power plants? Yes—but nuclear has its own unique economic challenges. Most of a nuclear plant's investment is in the plant itself, which has to be completed before the owner and investors can start making money. Long permitting and construction times mean that financial planners have to look far into an uncertain future, and ratepayers, investors (which often include municipal and state bondholders—i.e., the public), or owners must pay interest on loans to finance the project during the many-years-long construction period. In contrast, a modern gas-fired power plant can be built for roughly one-fifth the cost of a nuclear plant and in well under half the time. And you may have seen a few acres of land go from undeveloped to a fully operating solar photovoltaic power facility in a matter of weeks—at a cost per kilowatt of capacity that's about a third that of a comparable nuclear plant. Finally, nuclear plants run almost full time and are therefore forced to take whatever the market rate is for the energy they produce—so they lose money if that rate drops too low.

Most of today's reactors were built for decades of operation; in the United States, a standard initial operating license was for 40 years. Some reactors shut down before that time, for either economic or technical reasons or, as with TMI, accidents severe enough to render the reactor inoperable. But many reactor owners applied for, and were often granted, 20-year extensions of the original license; in 2019, a Florida reactor was granted a 40-year extension. Nearly half of the U.S. reactors operating today are beyond their original 40 years, with the oldest—New York's Nine Mile Point Unit 1—having been in operation since 1969. The average age of the U.S. reactor fleet is 37 years. Yet U.S. reactors are now closing at an accelerating rate, many of them years before their licenses expire. The reason isn't so much safety concerns, or fallout from Fukushima, or antinuclear protesters; rather, it's economics. You've just seen why economics favor lower-cost, nimbler technologies like natural gas and solar. Add to that the dramatic drop in the price of natural gas that accompanied the fracking revolution of the early 2000s, and it became cheaper for utilities to abandon perfectly functional nuclear plants in favor of natural gas generation, or to shutter plants when costs of repairs or routine maintenance seemed economically discouraging. So it's primarily cheap natural gas that's driving the latest shutdown of nuclear power in the United States.

As recently as the early 2000s—before the fracking revolution fully took hold—it looked like nuclear power in the United States might enjoy a turnaround. The U.S. Department of Energy's "Nuclear Power 2010" initiative was to be a government-industry partnership with a streamlined licensing process intended to get at least one new nuclear plant, using a Gen-III+ design, operating in the U.S. by 2010. Although that deadline came and went, interest in new nuclear plants grew with the 2005 Energy Policy Act that included a tax credit for nuclear power, comparable to subsidies for other low-carbon energy sources like solar and wind, and a government guarantee to cover cost overruns due to regulatory delays. The nuclear industry responded with applications to the Nuclear Regulatory Commission for some 25 new reactors. But competition from natural gas and the chilling effect of Fukushima took their toll, and nearly all those applications were eventually withdrawn or suspended. Construction began in 2013 on four newly licensed reactors, but in 2017 two of those were canceled well into the construction phase, leaving South Carolina electric customers to pay billions of dollars on the unfinished project. As of 2020, twin reactors were still under construction at Georgia's Vogtle nuclear plant, where they'll join two reactor units that have been operating since the late 1980s. The new Vogtle reactors are the AP1000 Gen-III+ design (recall figure 7.8) and are scheduled to begin supplying power by 2022—although delays and cost overruns make that a tentative date. In 2019, the U.S. Department dedicated an additional $3.6 billion in loan guarantees to Vogtle—on top of $5 billion already pledged.

Prior to the Vogtle startups anticipated for 2022, the newest U.S. reactors were the two units at Tennessee's Watts Bar nuclear plant, which came online in 1996 and 2016. But they're not all that new. Both are pressurized-water reactors on which construction started in 1973—giving Unit 2 the dubious honor of taking 43 years from start to commercial operation (although the project was on hold for 22 years, from 1985 to 2007). But during the nearly quarter century between 1996 and 2019, 17 U.S. reactors were shut down, mostly for economic reasons. Another seven are scheduled for shutdown before 2022, when the two new Vogtle reactors may come online.

The ongoing closure of U.S. nuclear power plants drives a downward spiral that's rippling through the nuclear industry. Companies that service nuclear plants or produce nuclear fuel see declining revenues. Enrollments in universities' nuclear engineering programs drop—with national security implications because that drop reduces the pool of qualified personnel not only for power reactors but also for the U.S. Navy's fleet of nuclear submarines and nuclear surface ships (although the navy also has its own nuclear training schools). There's a corresponding decline in U.S. companies that manufacture reactors for sale abroad—to the extent that today the United States is a minor player in a nuclear market dominated by Russia, China, and France. This could eventually diminish the U.S. role in setting reactor safety standards or weapons proliferation safeguards. One bright light in this otherwise grim picture of the United States' nuclear industry is the development of small modular reactors (SMRs) by several U.S. companies. These are among the most mature of the next-generation power reactors and, if certified and then deployed in the 2020s, might help revive the U.S. nuclear industry.

Although reactor closings have been particularly notable in the United States, figure 10.2 shows that shutdowns became significant globally in the late 1980s, then continued to fluctuate with an average of about four shutdowns per year through the present. The prominent negative spike in 2011 includes the reactors at Japan's Fukushima Daiichi and eight German reactors that were permanently shuttered immediately after Fukushima. Japan temporarily closed all its nuclear plants within two years of Fukushima but has been gradually reopening them and anticipates that more will be back online in the 2020s. Meanwhile, Germany continues a phaseout of its remaining nuclear plants and is scheduled to be nuclear-free by 2022.

Figure 10.2 also shows a fluctuating number of new reactor startups worldwide since the precipitous drop of the late 1980s, averaging about six per year. Given the average of four shutdowns per year in this period, that means the world has averaged a net addition of some two reactors per year—even as figure 10.1 shows the proportion of nuclear-generated electricity worldwide continuing to drop. But a closer look at figure 10.2 shows an uptick in reactor starts since 2010. Figure 10.3, which shows reactors currently under construction by location and reactor type, confirms that this uptick is driven largely by projects in Asia, especially China, India, and Korea, and to a lesser extent in eastern Europe. Western Europe and North America, which drove the reactor boom in the 1970s and 1980s, are all facing net declines in their reactor fleets despite the handful of construction starts shown in figure 10.3.

Energy Generation and Capacity Factors

What we want out of nuclear reactors is electrical energy, so the total energy generation is a more important measure than the number of reactors. That energy generation depends, first and foremost, on the rated power output of a nuclear power plant, termed **capacity**. Capacity is usually expressed in megawatts or gigawatts, and is the power generated when plants are running at their maximum rated output. The 450

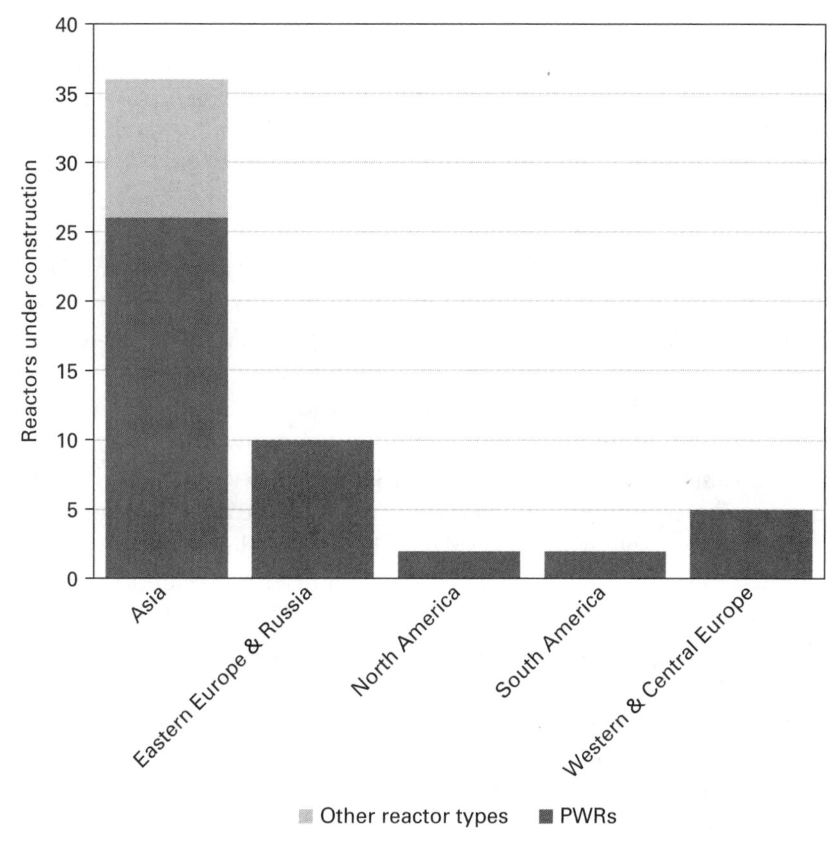

Figure 10.3
Reactors under construction as of 2020, by location and reactor type. Nearly all are pressurized-water reactors. The 10 reactors of other types, all in Asia, include four BWRs, one each fast-neutron and high-temperature reactor, and four heavy-water reactors, the latter all in India. (Data source: World Nuclear Association)

power reactors in operation worldwide today have a total capacity of just about 400 gigawatts (GW) of electric power. Nuclear power plants generally do run at full power when they're operating, but they experience downtime for refueling, routine maintenance, and unexpected incidents. A better measure of nuclear power's contribution to the world's energy supply accounts for such downtime. The **capacity factor** for any power plant, nuclear or otherwise, is the ratio of the energy it actually produces in a given time (usually a year) to the energy it would produce were it generating electricity continuously at full power. Taking account of capacity factors is especially important in comparing different methods of generating electricity, as we'll do later in this chapter. But it's also important in evaluating a single power source, such as nuclear, because capacity factors vary over time depending on such things as experience with the technology, age of power plants, electricity demand, and fuel-cycle issues.

Nuclear power plants are **baseload** plants, meaning they're best run continuously to help supply the minimum power that's always needed on the grid they service. Varying the power output of a nuclear plant, although technically feasible, isn't an efficient mode of operation. Turning a nuclear plant on and off based on power demand is even worse. So most nuclear plants run at full power except during maintenance and refueling.

In the 1970s and 1980s, worldwide nuclear capacity factors were in the range from 60 to 70 percent, increasing to just over 80 percent as the industry matured. Today, however, the global capacity factor is back to around 75 percent because of Japanese reactors idled, but not slated for formal closing, following Fukushima. The annual global nuclear electrical energy generation that results is about 2,500 terawatt-hours (TWh—a trillion watt-hours or a billion kilowatt-hours). That number constitutes the 10 percent of the world's electricity that's generated by nuclear power plants. The capacity factor for the United States' nuclear plants today is significantly higher than the global average, at around 92 percent.

Challenges for Fission's Future

So where is nuclear power headed? If you're not happy about nuclear power, you can take solace in nuclear fission's declining share of world electricity and in the accelerating shutdowns of commercial power reactors, especially in the West. If you're a nuclear advocate, you might cheer what could look like a budding nuclear renaissance in Asia.

Whether you advocate for more or less nuclear power, the reality is this: There are some 450 power reactors operating worldwide today and, as you saw in chapter 7, two-thirds of them are more than 30 years old. Even if many continue beyond their typical 40-year design lifetimes, the world is still looking at replacing hundreds of reactors in the next decade or two just to keep nuclear fission producing as much electricity as it does today. And with replacement alone, the nuclear portion of world electricity will continue to decline as the global energy supply becomes more electrified. A world in which nuclear power plays an increasingly important role would require massive new reactor construction. Some 55 reactors under construction today are a start, but alone they won't be enough to replace just the reactors that will be retiring soon. Even in a world that was fully supportive of nuclear power—not our world—building hundreds of reactors would be a daunting challenge. But not an impossible one—the heyday of reactor construction from 1970 to 1985, as shown in figure 10.2, amounted to some 20 reactor startups each year—about one every 18 days. Were that rate possible today—which it probably isn't for a host of logistical, economic, political, and technical reasons—we could replace the world's 300 oldest reactors in 15 years.

At present, replacement and expansion of the world's reactor fleet is being done largely with Generation III+ designs, which, as you saw in chapter 7, are supposed to offer both safety and economic advantages over older reactors. Enhanced safety is evident in some aspects of the new designs and is backed by probabilistic risk assessment, although we haven't yet accumulated enough operating experience with these new reactors to confirm safety projections. On the economic front, however, the picture for Gen-III+ reactors is decidedly mixed. The two new reactors in South Carolina were Westinghouse AP1000s, canceled in 2017 when they were 40 percent complete, largely because of soaring costs. That debacle led Westinghouse, then a subsidiary of Japan's Toshiba, to file for bankruptcy; the company has since been acquired by a Canadian hedge fund. The reactors still under construction in Georgia, also twin AP1000s, face delays and cost overruns as well as economic fallout from the Westinghouse bankruptcy. Elsewhere in the world, projects involving the European Pressurized Reactor (EPR), a Gen-III+ design from the French nuclear company Orano (formerly AREVA), are experiencing delays and cost overruns. Finland, which had planned for two EPRs at its Olkiluoto nuclear plant, canceled the second EPR in 2015 after the first reactor's cost nearly tripled in the midst of a nine-year delay. A similar EPR project in France has been plagued by delays, cost overruns, and a defect in the reactor pressure vessel. The Hinkley Point C nuclear plant in the U.K., scheduled for operation in 2025, features two EPRs, but costs continue to mount and delays threaten. Two EPRs went online in China in 2018 and 2019, following nine years under construction. In South Korea, four planned APR+ reactors—an advanced Korean design—were canceled in 2018 after a new government pledged to phase out Korean nuclear power. Korea has been a leading exporter of power reactors, so that phaseout may have international ramifications.

Chapter 7 also listed a diverse range of Generation IV designs, most going beyond light-water technology and with many offering greater fuel utilization and less nuclear waste. Gen-IV reactors have been touted as spearheading a true nuclear renaissance and as crucial technologies for mitigating climate change. But Gen-IV designs are mostly still on the drawing board. Given the delays, cost overruns, and technical glitches with the current Gen-III+, it's simply far too soon to know what to expect, and when, from Gen-IV nuclear power.

Public Opinion

Except in the most authoritarian countries, a nuclear renaissance is probably impossible without substantial public support. Unfortunately for the nuclear industry, that support has been declining. A survey of citizens in 24 countries, conducted just after the Fukushima disaster, found only 38 percent supporting nuclear power (figure 10.4). Even coal did better, at 48 percent. But support varied by country; even post-Fukushima, majorities in the United States, Poland, and India still supported nuclear power. U.S. support has subsequently slipped, however; according to the annual Gallup Environment Poll, nuclear support had dropped to 45 percent by 2018, after peaking at 62 percent in 2010.

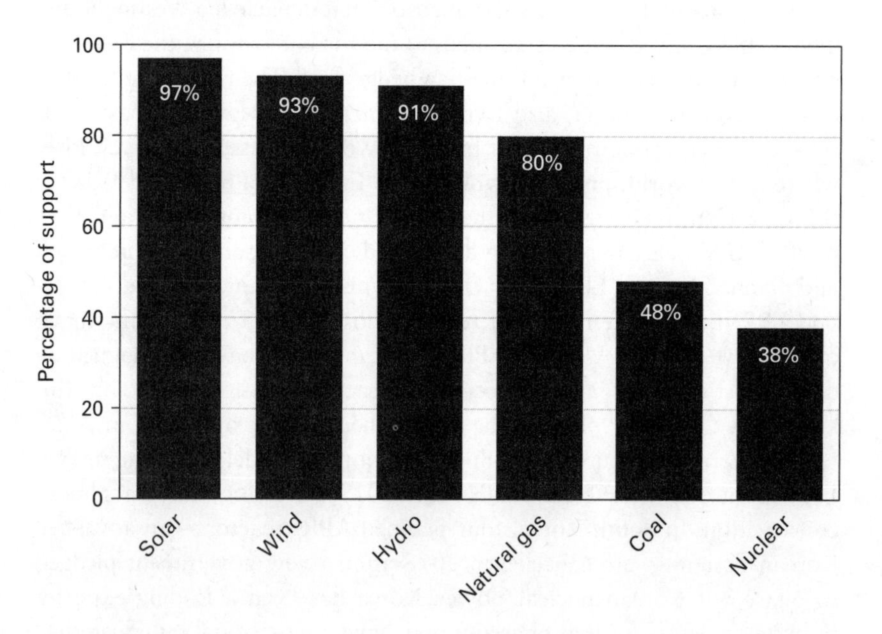

Figure 10.4
A poll of citizens in 24 countries just after the Fukushima disaster shows support for nuclear power at only 38 percent—an abrupt drop from 54 percent before Fukushima. (Data source: Ipsos)

Nuclear Power and Climate Change

Environmentalists—a group that was once solidly antinuclear—are now divided on nuclear power. The reason is global climate change. Whatever your opinion of nuclear power, the fact remains that nuclear fission today represents one of only of two technically proven sources of baseload electricity that don't result in substantial climate-changing carbon emissions and that are already deployed on a large scale (the other is hydroelectric power). Indeed, some of the world's most prominent environmentalists and climate activists are urging growth in nuclear power, including the American climatologist James Hansen; the British Gaia-hypothesis originator James Lovelock; Californian Elizabeth Muller, founder of the climate science organization Berkeley Earth; two former U.S. Secretaries of Energy; and a host of Nobel laureates and climate scientists worldwide.

Today, the fossil fuels—coal, oil, and natural gas—supply some 86 percent of the world's total energy and 63 percent of our electrical energy. Because fossil fuels contain carbon, burning them produces carbon dioxide (CO_2). Carbon dioxide is a **greenhouse gas,** meaning it absorbs infrared radiation emitted by Earth's surface. Radiated energy needs to escape to space so as to balance the energy absorbed from incoming sunlight and thus keep the planet at a constant temperature. Adding greenhouse gases to the atmosphere—by burning fossil fuels, for example—therefore increases Earth's temperature. That, in an oversimplified nutshell, describes the **greenhouse effect,** the primary cause of anthropogenic climate change. Recent research shows that existing and proposed fossil-fueled energy systems, especially for generating electricity, will emit enough CO_2 to take Earth past the temperature goals set in the Paris climate agreement.[1] So the most direct solution to the problem of global warming is to switch to low-carbon energy sources—of which nuclear fission is one. We'll discuss other low-carbon alternatives in the next section.

Except that in the case of nuclear fission, *switching* may be too ambitious. If we want nuclear as part of a low-carbon energy mix, our first task would be to staunch the ongoing closure of viable operating reactors and to plan for replacement of older reactors that must inevitably be retired. Recognizing that need has inspired the surprising sight of some environmentalists cheering for nuclear power.

A 2018 MIT study looks at prospects for nuclear power in the context of climate change and recognizes that nuclear could play a significant role in mitigating our climate problem. However, the study concludes that, under current economic and regulatory conditions, nuclear power

Saving Nukes with Zero Emissions Credits

As nuclear plant closures accelerated in the United States through the decade of the 2010s, some states responded with measures intended keep their existing nuclear plants viable. Motivations include minimizing carbon emissions from electricity generation, as well as maintaining the economic and employment benefits of nuclear plants to the surrounding communities.

In 2016, New York and Illinois adopted *zero emissions credits* (ZECs) that treated nuclear plants somewhat like the renewable energy sources solar, wind, and hydro. New York State's Clean Energy Standard calls for half the state's electricity to come from renewable sources by 2030. The plan favors solar and wind for new renewable energy facilities, although biomass and small hydro (without damming rivers) also qualify. Generators of electricity are required to purchase renewable energy credits to support new renewable energy. They're also required to purchase ZECs, revenue from which goes to subsidize existing nuclear plants. The Illinois ZEC plan is similar. In both states the price utilities pay for ZECs is tied to the *social cost of carbon*, an estimate of the costs to society of damage associated with climate-changing carbon emissions. The New York and Illinois ZECs are currently tied to a carbon price of $42 per ton. Utilities ultimately pass the cost of ZECs on to their customers; for New York customers, the increase in individual utility bills is estimated at less than $2 per month. Adoption of ZECs in the two states has resulted in the continued operation of five nuclear plants that were either economically marginal or had been scheduled for closure.

New York and Illinois faced immediate court challenges to their requirements that utilities purchase zero emissions credits to support nuclear power. But the ZECs were upheld in federal appeals courts, and in 2019 the U.S. Supreme Court declined to overturn the lower court rulings. Although some environmental groups had opposed the ZECs' support of nuclear power, others celebrated the survival of low-carbon electricity generation and also the courts' implications that states were free to adopt climate-friendly policies involving electrical energy generation. The court rulings on ZECs in New York and Illinois will likely clear the way for similar programs in other states. Indeed, just days after the Supreme Court decision, New Jersey granted zero emission credits to its three operating nuclear reactors, two of which had been about to close.

will struggle to compete with fossil fuels and renewables, especially in the West.[2] But, the study suggests, that could change with better power-plant construction management, carbon-reduction incentives and regulations, accelerated development of advanced reactor technologies and, surprisingly, greater reliance on renewable energy sources like wind and solar. This last factor helps nuclear power because the incremental cost of adding renewables to the power grid grows as the renewable portion of electricity generation increases—a result of the intermittency of these renewable sources. That, in turn, makes baseload sources like nuclear relatively more economical.

So far we've considered nuclear fission as an alternative to fossil fuels—mostly coal and natural gas—for generating electricity. But advanced Gen-IV high-temperature reactors could also produce hydrogen gas (H_2) directly, and that could serve as a transportation fuel, replacing oil. In the longer term, then, nuclear fission has the potential to displace nearly all our fossil fuel consumption.

So, can nuclear power help save us from impending climate change? Technically, yes—but it would require a prompt and massive effort to replace fossil-generated electricity with nuclear power. How likely is that?

Projections for the Future

As you've seen, today's 450 operating reactors generate some 2,500 TWh of electricity annually, comprising 10 percent of global electrical energy. Where are those numbers likely to go in the future?

Each year the International Atomic Energy Agency issues a report with projections for the future of nuclear power in the context of global energy consumption and electrification (see Further Reading at the end of this chapter). The IAEA report includes both high and low estimates for nuclear power's contribution through the year 2050. These aren't the most extreme possibilities imaginable. Rather, they represent contrasting nuclear futures based on realistic assumptions of what's technologically feasible and where political will might take nuclear power. They also reflect economic conditions, including projected costs of competing power sources, especially natural gas and renewables. Finally, the projections are made in the context of assumptions about growth in energy consumption and increasing electrification of our energy mix.

The IAEA's 2019 report estimates a growth in total world energy consumption of 1 percent annually, compounding to a 16 percent increase over 2019 by 2030 and 38 percent by 2050. Over the same period the growth in global electrical energy consumption is estimated at 2.2 percent annually through 2030 and 2 percent thereafter. As a result, electricity's share of total energy consumption is projected to increase from about 19 percent today to some 26 percent by 2050. It's that extra-strong rise in electricity consumption that makes it a challenge for nuclear power to maintain, let alone grow, its share of the world's electricity supply.

In the context of these energy and electricity assumptions, the IAEA's low estimate projects that the nuclear share of world electricity will decline to some 6 percent in 2050, while the high estimate has it increasing by only 2 percent from today to 12 percent by 2050. That's despite a projected high-estimate increase of 220 percent—to 2.2 times today's value—in nuclear-generated electricity. Figure 10.5 summarizes these IAEA projections. Other authoritative bodies make similar projections; for example, the U.S. Energy Information Administration projects only a modest 1.6-percent annual growth in global nuclear capacity, from just under 400 gigawatts today to just over 500 GW by 2040—unlikely to be sufficient to maintain even today's 10 percent share of global electricity.

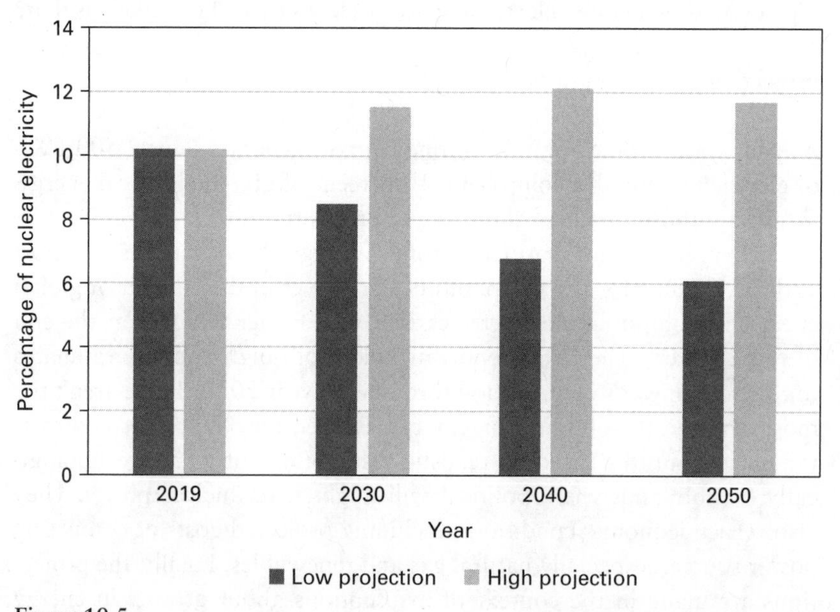

Figure 10.5
IAEA projections for percentage of electricity generated by nuclear fission through 2050. Even the high projection barely increases the nuclear share and falls far short of the 1990s values shown in figure 10.1.

Alternatives to Nuclear Fission

Maybe you aren't happy with nuclear power and would like to see a nuclear-free world. Or maybe you recognize nuclear fission's potential for mitigating climate change, but you don't think that's economically or politically realistic. What are the alternatives?

If you aren't concerned about climate change—although you should be—then the fossil fuels themselves constitute alternatives to nuclear fission. The fossil fuels are literally that: fossil energy that ultimately came from the Sun and was captured by photosynthetic plants hundreds of millions of years ago. A minuscule fraction of those plants escaped prompt decay and were buried to become the fossil fuels.

Because today's nuclear facilities supply baseload power, they're most suitable for replacement with large power plants fueled by coal or natural gas. Coal, in particular, operates best as baseload power—just as do nuclear plants. Gas can be more flexible, although many gas plants are baseload suppliers. (These considerations mean that, going the other way, nuclear plants fit most directly into today's energy infrastructure as replacements for coal or natural gas plants.) In addition to their carbon emissions, though, fossil fuels have a host of downsides for the environment and for human health. The World Health Organization estimates that each year at least a million people worldwide die prematurely from pollution emitted in coal burning. Coal combustion also results in mercury emissions that contaminate fish and other organisms high on aquatic food chains. Coal mining levels entire mountaintops, contaminates streams with acid runoff, and threatens human communities with land subsidence and underground fires. Oil extraction takes its own environmental toll, especially when oil spills into marine environments. Natural gas, although cleaner burning than either coal or oil, can be dangerously explosive when it leaks into confined spaces. Unburned natural gas contributes some 30 times the greenhouse warming of an equivalent amount of carbon dioxide, meaning that any natural gas leakage reduces the climate benefits of gas over coal.[3] All three fossil fuels are limited in quantity; we have at most several centuries of coal left, probably less of oil and gas—and if we were to burn all that, we would do unacceptable damage to Earth's climate. All in all, our species would do well to wean ourselves soon from fossil fuels, with their climate-changing carbon emissions and other deleterious environmental and health impacts.

Contrary to popular opinion, however, there is no carbon-free energy source. There are, however, low-carbon sources whose associated carbon emissions are far below those of the fossil fuels. Nuclear power is among them, as shown in figure 10.6.[4] The figure shows high and low estimates for CO_2 emissions per unit of electrical energy produced. For the fossil fuels, CO_2 comes almost entirely from combustion, and the high and low figures represent traditional versus more advanced combustion technologies. Note that natural gas has the lowest emissions of the fossil fuels. That's because it's largely methane (CH_4), which is mostly hydrogen, so when methane burns (that is, combines with oxygen) it produces primarily water (H_2O), along with considerably less CO_2. Coal, in contrast, is mostly carbon, and oil's chemical makeup is in between. Emissions for other energy sources reflect both operation and manufacturing. For example, cement production emits CO_2, and large amounts of cement go into nuclear-plant containment structures, hydroelectric dams, and offshore wind facilities. Traditional approaches to manufacturing solar photovoltaics are energy-intensive and may incur significant carbon emissions especially if coal-fired electricity is used. More advanced solar cells, including so-called thin-film photovoltaics, take a lot less energy to manufacture. And either way, if photovoltaics are made using renewably generated electricity, then the associated carbon emissions go way down. All this explains the large variation in carbon emissions from photovoltaics, as well as the surprising fact that, at the high estimate, PV appears only about twice as good as natural gas.

Carbon emissions associated with nuclear power have been a subject of lively debate, and the numbers in figure 10.6 span a realistic range that doesn't, however, include the most extreme claims. These are life cycle emissions numbers, accounting for both construction and operation of nuclear plants. Their massive containment structures and multiple safety systems mean that much more material goes into a nuclear plant than a fossil plant, and that means more carbon emissions associated with construction. Although the quantity of nuclear fuel needed is minuscule compared with the coal or gas that go into a fossil plant (thanks to the nuclear difference; recall figure 2.1), there's nevertheless significant energy expenditure in mining, processing, and enriching uranium to make nuclear fuel. The carbon footprint of nuclear fuel manufacture was higher when gaseous diffusion was the dominant enrichment technique (recall figure 9.3). Today, though, as figure 10.6 shows, carbon emissions associated with nuclear power are small, and comparable to those of other nonfossil energy sources.

If we want to phase out fossil fuels, and we either don't want nuclear or don't think it can realistically fill the fossil fuel void, then we're left

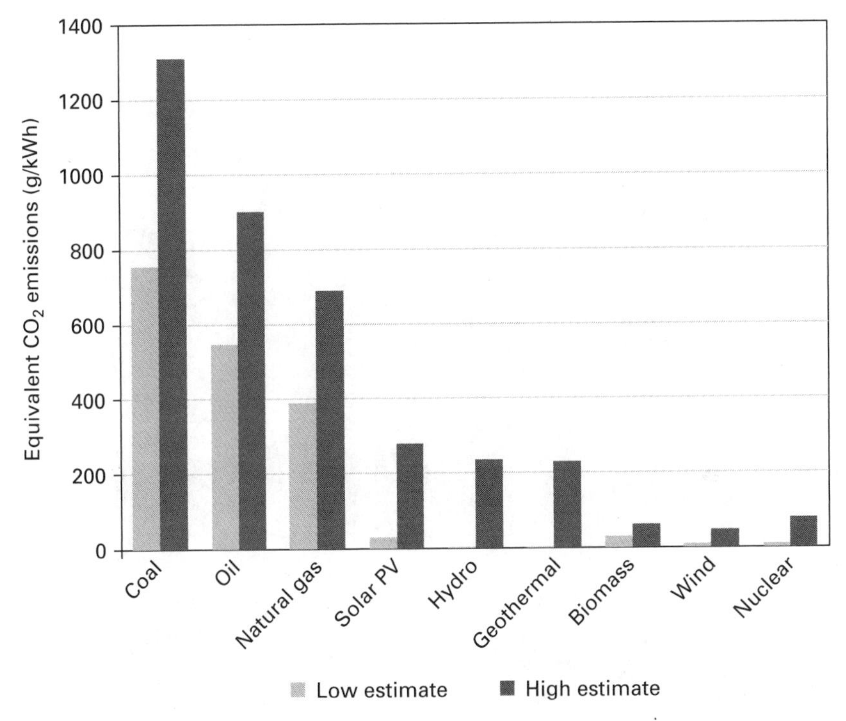

Figure 10.6
Comparative greenhouse gas emissions per unit of electrical energy produced (grams CO_2 equivalent per kilowatt-hour). Greenhouse emissions include CO_2 and other greenhouse gases, notably methane (CH_4), which are converted to CO_2 equivalents based on their global warming effect. Data are from a variety of studies.

with other sources shown in figure 10.6. Here we consider each of these in turn.

Hydroelectricity

Earlier in this chapter, we noted that nuclear power is one of only two low-carbon energy sources currently deployed at large scales (figure 10.7). The other is **hydroelectric power.** This is one of our oldest ways of producing electrical energy and utilizes the same power of flowing or falling water that drove mechanical systems in nineteenth-century factories. Hydro is, in principle, one of the cleanest and most environmentally friendly energy sources. It's ultimately solar energy that went into evaporating water, which then fell as rain and found its way into rivers. It's renewable, because it depends on the continuous flow of solar energy rather a limited fuel resource. When sited at natural waterfalls, hydro doesn't require major

Figure 10.7
Grand Coulee Dam in Washington State produces nearly 6,500 MW, the equivalent of six large nuclear or coal plants. (Bureau of Reclamation)

disruption of river systems. With hydro there's no combustion, so no emissions of combustion products. However, hydro facilities built in tropical locations may result in methane emissions from decay of vegetation flooded by hydro dams, and in the worst case those emissions can make a tropical hydro plant as bad as a coal plant from a climate perspective. Natural waterfalls are limited, so most hydro plants use dams to block river flows. That causes a host of environmental issues, including changes in riverine ecology and problems for migratory fish, evaporation from lakes that form behind dams and thus cause a loss of freshwater that's needed for natural and human communities downstream, and the mass relocation of human populations. China's Three Gorges hydro plant, at 22.5 gigawatts the world's largest power plant of any kind, required relocation of more than a million people. And dams can break, spreading death and havoc from sudden flooding downstream. Collapse of a large hydro dam upstream from even a modest city could be a disaster greater than the worst imaginable nuclear reactor accident. When the Banqiao dam in China failed in 1975, some 26,000 people died in the resulting flooding.

Today, hydropower accounts for some 16 percent of world electricity, second after the fossil fuels and substantially greater than nuclear power's 10 percent. Hydro provides somewhat over half the electricity that's from renewable sources—a fraction that's dropping rapidly with the rise of solar and wind generation. Hydro's capacity factor, averaging some 45 percent globally, means that hydro plants don't run as much of the time as do nuclear plants, but this isn't a major problem in regions with multiple hydro plants. Some countries and regions, including Norway, Canada, the Pacific Northwest of the United States, and many developing countries get most of their electrical energy from hydropower. There's some 10,000 GW of untapped hydro resources in the world, nearly all of it divided among Africa, South America, and Asia. The developed countries of North America and Europe, in contrast, have been utilizing most of their hydro resources for decades. Hydro, then, offers the potential to replace nuclear power in much of the world but not in the long-industrialized countries of the West. There's one caveat, though: climate change will alter patterns of rainfall and could reduce hydropower's potential in some regions.

Solar Energy

There's plenty of solar energy. The Sun supplies Earth with energy at the rate of 174,000 terawatts—nearly 10,000 times humankind's 20-TW energy-consumption rate. Solar, long a favorite of environmentalists, has in the past decade become economically competitive with other energy sources. Solar technologies include passive (no moving parts) and active (pumps circulate heat-transfer fluids) designs for heating buildings and providing hot water. Solar-heated water can even drive so-called *absorption chillers* to cool buildings. Concentrating sunlight with mirrors develops high enough temperatures to drive steam turbines that generate electricity as in conventional power plants (figure 10.8). But the most elegant and rapidly spreading solar technology involves **photovoltaic** (PV) **cells** that turn sunlight directly into electricity with no moving parts. PV, which used to be so costly that it was used predominantly in spacecraft, is now inexpensive enough that PV panels are sprouting on rooftops worldwide (figure 10.9) and being deployed in solar power plants (figure 10.10), the largest of which approach the capacities of large nuclear plants. By 2020, world photovoltaic capacity had reached some 700 GW and was growing at more than 20 percent annually.

Figure 10.8
A pair of solar-thermal power plants in Andalusia, Spain. Fields of Sun-tracking mirrors reflect sunlight to the towers at left. (Wikimedia Commons CC BY 3.0)

Could solar replace nuclear? Not directly at present, because solar energy is intermittent. Today's power grids have difficulty when intermittent sources like solar and wind exceed 20 to 40 percent of a grid's power production. A large-scale PV installation, in a sunny desert location, might have a capacity factor of 30 percent—meaning it's producing electricity 30 percent of the time. Rooftop systems in non-ideal locations average much less. Some of the downtime is predictable—nighttime—but the impact of clouds is uncertain. That 30 percent capacity factor for solar PV contrasts with more than 90 percent for a well-run nuclear plant. However, development of large-scale energy storage technologies and deployment of continent-spanning power transmission lines could smooth out the intermittencies associated with solar generation and make this almost-limitless energy source a true replacement for nuclear power.

Wind

A small fraction of the solar energy incident on Earth drives the air movement we call wind. Wind turbines tap into that flow, turning generators to produce electricity (figure 10.11; see also figure 6.6b). Wind, like solar, is expanding rapidly, with some 700 GW installed by 2020 and producing

Figure 10.9
Photovoltaic panels on author RW's home generate more energy than the home uses—despite being in Vermont, one of the cloudiest parts of the United States. Arrows highlight two separate photovoltaic arrays, with total peak power of 4.05 kW. (Drone photo by Brett Simison.)

about 6 percent of the world's electricity. Wind's capacity factor averages about 35 percent but can exceed 60 percent for offshore wind farms.

The global wind resource amounts to about 2,000 TW—some 100 times humankind's energy consumption rate. However, it will never be possible to harness all that wind energy—and if we tried, we would likely alter weather and climate. But wind can make a significant contribution to world electricity; already, Denmark gets over 40 percent of its electricity from wind and several states in the midwestern United States exceed 30 percent wind-generated electricity.

Like solar, wind's intermittency means that it can't directly replace baseload nuclear power today. But, again, improvements in energy storage and long-distance transmission lines to link distant wind farms could help overcome wind's intermittency. Furthermore, wind and solar together complement each other, since wind often blows strongly at night while the Sun, of course, shines in the daytime.

Figure 10.10
The largest photovoltaic power plant in the Western Hemisphere is Mexico's 828-MW Villanueva facility. (Alfredo Estrella/Getty Images)

Biomass

Biomass, like the fossil fuels, is energy-containing biological material that originated when photosynthetic plants stored solar energy in their chemical makeup. Unlike fossil fuels, biomass is new; it's from plants that grew recently or at most a few centuries ago. Biomass can be burned directly, as is done with wood, or processed into liquid or gaseous *biofuels* such as ethanol, biodiesel, or biomethane derived from manure or waste. The carbon in biomass was drawn out of the atmosphere only a short time ago, and when the biomass is burned as fuel that carbon goes back to the atmosphere as carbon dioxide. So overall the process is nearly carbon neutral; unlike fossil fuel combustion, biomass in principal puts no net carbon into the atmosphere. However, agricultural and forestry practices affect carbon storage in soils and therefore result in some carbon emissions, and transportation and processing of biomass may entail carbon emissions from fossil fuel usage. In most cases, though, carbon emissions from biomass are, as figure 10.6 shows, far below those from fossil fuels.

Each year, photosynthetic plants pull some 120 billion tons of carbon out of Earth's atmosphere. We humans release about 10 billion tons

Figure 10.11
A Danish wind farm. (Environmental Images/UIG/Science Source)

of carbon to the atmosphere annually through fossil fuel combustion. Comparing those two numbers tells us that there's plenty more biomass being produced than we would need to replace fossil fuels. However, plants' photosynthetic effort supplies energy, as food, for nearly all living things—including ourselves. In fact, we already divert some 40 percent of global photosynthetic productivity for human uses. Tapping more than a small fraction of that for energy production would be majorly disruptive of natural ecosystems and the human food supply. So could biomass replace nuclear? Probably, given today's modest nuclear contribution to world energy. But biomass alone won't solve our climate problem.

Incidentally, biomass offers a possible way to remove carbon from the atmosphere permanently—something that's looking increasingly necessary if we're to limit global warming to safe levels. **Bioenergy with carbon capture and storage** (BECCS) would have us burn biomass and then capture the resulting CO_2 for storage underground—thus creating a net flow of carbon out of the atmosphere. But it's unclear whether BECCS can be technically or economically feasible on the large scales needed.

Geothermal and Ocean Energy

Geothermal energy is energy flowing from Earth's interior, where it originates both from decay of naturally occurring radioactive materials and from primordial heat left over from Earth's formation. The total geothermal flow amounts to some 40 terawatts, only about 0.025 percent of the rate at which solar energy reaches Earth. That's roughly twice humankind's energy consumption rate. Given that it would be impractical to harness more than a tiny percentage of this geothermal flow, it's clear that geothermal energy will never supply a significant portion of humankind's energy needs, nor is it likely to replace even nuclear power's modest contribution. However, that doesn't mean we can't tap geothermal energy in those places where substantial geothermal resources lie close to Earth's surface. Nigeria gets nearly half its electricity from geothermal sources, while Iceland gets about a third—and more than half of its total energy because it uses geothermal heating. California's electricity is 6 percent geothermal, but for the United States as a whole it's only 0.4 percent—a figure that describes the global average as well.

The ocean tides contain energy driven by mechanical and gravitational effects associated with Moon and Sun and with Earth's rotation. The idea of harnessing this energy sounds attractive, and it's been done in a handful of places with exceptionally large tidal ranges. But the tidal energy flow amounts to less than one-tenth of the geothermal flow—making it insignificant in the context of humankind's global energy consumption and not a substitute for nuclear power.

Finally, ocean waves and currents are, like atmospheric wind, a minor manifestation of the solar energy flow reaching Earth. Technologies to harness these energy flows are minimally developed, although estimates suggest that we might someday tap as much as 450 GW (about half a terawatt) from ocean currents—just about the total output of today's nuclear power plants. But "someday" is a long way off.

Using Less Energy

If we want or need to replace nuclear power, there's another way, and it's obvious: use less energy, either by doing without or, more intelligently, using energy more efficiently. The world has made modest gains in energy efficiency in recent decades, and there's the potential to go much farther. Just consider Europe, a developed region with a high standard of living whose per-capita energy consumption is half that of North America. Efficiency improvements are often the most economical and technically simplest way to displace undesirable energy consumption. In the case of nuclear, though, we're talking about baseload electricity in a world whose energy system is becoming increasingly electrified. That move toward electrification is in part a response to climate change and the associated need to phase out fossil fuels. As we convert our transportation system to electric vehicles and move to electric heat pumps for heating and cooling, we open the option to use renewably generated electricity for those formerly fossil-fueled tasks. In that context, achieving greater efficiency in the electric sector is crucial if the goal is to displace nuclear power.

Prospects for Fission Alternatives

Here we've briefly described the main alternatives that we might consider to replace nuclear fission, were that desirable or necessary. Have we covered every possibility? Just about. With one exception, there simply aren't any hidden or undiscovered sources of energy available to us earthlings. You may hear of new technologies for harnessing those known sources, but they can't be fundamentally new because we've covered the fundamentals of all but one of Earth's available energy sources. Four of the sources we've mentioned here—hydro, solar, wind, and biomass—do indeed have the potential to replace nuclear fission. However, hydropower's potential is largely tapped in the developed world, and solar and wind require advances in energy storage and power grids before they can fully replace the continuous baseload power from nuclear plants. As a fallback, that leaves fossil fuels—the very sources we need to avoid if we're to avert a climate crisis. In fact, power from the nuclear plants currently being shut down in United States is mostly being replaced by power from natural gas—not surprising given that the low price of gas is what's driving nuclear out of business. Solar, wind, and hydro are playing more modest roles in replacing nuclear power. Their role could grow, especially given the exponential increase in solar and wind installations, but it will take a greater commitment to these renewables if nuclear shutdowns aren't to mean more climate-changing fossil fuel consumption. Your authors' views are that renewable energy sources must play a major role in the transition away from fossil fuels, and also away from nuclear power if we humans want to take that route—although it's clear that a robust nuclear power enterprise could help in the transition from fossil fuels.

The one exception to this chapter's catalog of energy sources, alluded to above, is nuclear fusion. We'll cover fusion's prospects in the next chapter. But suffice it to say for now that fusion power is at least decades away, so—unlike the renewables solar, wind, and hydro—it's not available today to replace either nuclear fission or fossil fuels.

Summary

Whatever your view of nuclear power, it's clear that for the foreseeable future nuclear fission's share of the global energy supply is at best going to hold about steady and may well diminish. Alternatives, especially renewable solar and wind, are now more economical than nuclear, as well as nimbler in their potential for rapid deployment. So the answer to this chapter's title question, "A future for nuclear fission?" is a future that's at best minimal at least for the next few decades—and those are the decades that matter in confronting our climate crisis.

Glossary

baseload Refers to an electric power plant that runs continuously, helping to supply the minimum power needed on the grid.

bioenergy with carbon capture and storage (BECCS) A technique for removing carbon dioxide from the atmosphere, wherein CO_2 from biomass combustion is captured and stored underground.

biomass Material of biological origin containing energy stored in chemical compounds through the process of photosynthesis.

capacity The maximum rated electric power output of a power plant or plants, usually expressed in megawatts (MW) or gigawatts (GW).

capacity factor The ratio of energy actually produced by a power plant or other power source over a given time (usually a year) to the energy it would produce if it ran continuously at its maximum rated power.

geothermal energy Energy derived from the thermal energy in Earth's interior.

greenhouse effect Absorption of infrared radiation by greenhouse gases in a planetary atmosphere, resulting in surface warming.

greenhouse gas A gaseous constituent of a planetary atmosphere that absorbs outgoing infrared radiation.

hydroelectric power Electric power generated by falling or flowing water.

photovoltaic cells Semiconductor devices that turn sunlight energy directly into electricity with no moving parts. Usually assembled into photovoltaic panels, also called *solar panels*.

Notes

1. Dan Tong et al., "Committed Emissions from Existing Energy Infrastructure Jeopardize 1.5°C Climate Target," *Nature* 572 (August 13, 2019): 373.

2. See *The Future of Nuclear Energy in a Carbon Constrained World: An Interdisciplinary MIT Study*, listed in Further Reading.

3. This 30-year figure takes into account the much shorter atmospheric lifetime of methane versus carbon dioxide.

4. World Nuclear Association, *Comparison of Lifecycle Greenhouse Gas Emissions of Various Electricity Generation Sources* (WNA, 2011), available at http://www.world-nuclear.org/uploadedFiles/org/WNA/Publications/Working_Group_Reports/comparison_of_lifecycle.pdf.

Further Reading

International Atomic Energy Agency. *Country Nuclear Profiles*. Vienna: IAEA. Published annually, this report updates the status of nuclear power by country. Available at https://www.iaea.org/publications/13448/country-nuclear-power-profiles.

International Atomic Energy Agency. *Energy, Electricity and Nuclear Power Estimates for the Period up to 2050*. Vienna: IAEA. Published annually as the IAEA's Reference Data Series No. 1, this publication is available for download at https://www-pub.iaea.org/books/IAEABooks/13412/Energy-Electricity-and-Nuclear-Power-Estimates-for-the-Period-up-to-2050. An authoritative discussion of prospects for nuclear power in the coming decades, including analyses for major world regions. The report gives high and low estimates for future nuclear capacity, both worldwide and regionally.

Kramer, David. "US Nuclear Industry Fights for Survival." *Physics Today* 71, no. 12 (December 2018): 26–30. Paints a grim picture of the U.S. nuclear industry, in contrast to what's happening in Russia and Asia.

Lovins, A. B. "How Big Is the Energy Efficiency Resource?" *Environmental Research Letters* 13, no. 9 (September 2018). A champion of energy efficiency argues that we can dramatically reduce our energy consumption by rethinking the way we design systems from houses to industries.

Makhijani, Arjun. *Carbon-Free and Nuclear-Free: A Roadmap for U.S. Energy Policy*. Muskegon, MI, and Takoma Park, MD: RDR Books, IEER Press, 2007. A joint project of the Institute for Energy and Environmental Research and the Nuclear Policy Research Institute (the latter founded by antinuclear activist Helen Caldicott), this study argues that we don't need nuclear power to achieve a carbon-free, climate-friendly future. Available at https://carnegieendowment.org/files/CarbonFreeNuclearFree.pdf.

MIT Energy Initiative. *The Future of Nuclear Energy in a Carbon Constrained World: An Interdisciplinary MIT Study*. Cambridge: Massachusetts Institute of Technology, 2018. PDF available at http://energy.mit.edu/wp-content/uploads/2018/09/The-Future-of-Nuclear-Energy-in-a-Carbon-Constrained-World.pdf. Pessimistic about the economics of nuclear power today, this report paints a more optimistic nuclear future in a world that takes the cost of carbon seriously.

Oreskes, Naomi. "There's a New Form of Climate Denialism to Look Out For—So Don't Celebrate Yet." *Guardian*, December 16, 2015. Oreskes is a highly respected historian of science and champion of scientific truth. Here, in the wake of the Paris agreement, she takes issue with environmentalists who support nuclear power, in effect calling them renewable energy deniers who don't believe that renewables are capable of displacing fossil fuels.

United States Energy Information Administration. *Annual Energy Outlook*, https://www.eia.gov/outlooks/aeo/pdf/aeo2019.pdf. Projections for future energy generation by different sources through 2050.

11

Nuclear Fusion

We've just spent three chapters dealing with nuclear fission reactors—the many different types, how they work, their safety issues, reactor accidents, fuel cycles, and nuclear waste. More than 70 years into the nuclear age, with hundreds of reactors operating, we have a lot of experience with the technology to produce energy from nuclear fission—the process of splitting heavy nuclei to produce lighter ones. But the curve of binding energy, introduced in chapter 5, shows that there's another way to extract energy from the nucleus: fusion of lighter nuclei to produce heavier ones.

Fusion offers enormous potential for energy production without most dangers associated with nuclear fission and with greatly diminished production of radioactive waste. We've been working on fusion power since the 1950s—almost as long as we've been doing fission—yet we haven't produced a single kilowatt-hour of fusion-generated electricity. This chapter explores the potential and the challenge of harnessing nuclear fusion for energy generation. Although we're still far from controlling fusion, we have considerable experience with uncontrolled fusion as it occurs in nuclear weapons. More on this in chapter 12.

Fusion Fuel Resources

If anything is an unlimited source of energy, it's nuclear fusion. The first fusion reactors will use **deuterium-tritium fusion** (D-T fusion), their fuel a mix of the hydrogen isotopes deuterium and tritium. Recall from chapter 2 that deuterium is the heavy isotope ${}^{2}_{1}H$. It's stable but rare, occurring once in about 6,500 hydrogens. But there's a lot of water in the world's oceans, and even at that 1-in-6,500 ratio, seawater contains enough deuterium to satisfy humankind's current energy consumption rate for some 25 billion years—about five times as long as the Sun will continue to shine. And its deuterium content makes a gallon of ordinary water the energy equivalent of 350 gallons of gasoline.

The first reactors will also use tritium ($^{3}_{1}H$), the heaviest hydrogen isotope. Tritium is radioactive with a 12.4-year half-life, so it doesn't persist in nature. But it can be bred by neutron bombardment of lithium-6, using a reaction we'll describe later. Lithium itself is abundant in Earth's crust and in the oceans, and lithium-6 constitutes 7.4 percent of it. There's enough lithium to fuel D-T fusion reactors supplying energy at today's consumption rate for tens of millions of years. If we do succeed in developing fusion reactors, though, we'll be onto better fuel cycles within at most a few centuries, so lithium abundance won't be an issue. And millions of years still outlives any of our other fuels—although not streams of renewable energy based on sunlight.

The Fusion Challenge

What makes fusion difficult is the electric force. Atomic nuclei contain positive protons and neutral neutrons, so overall they're positively charged. Therefore nuclei experience a repulsive electric force, which must be overcome if they're to get close enough to fuse. Recall from chapter 2 that the electric force has a relatively long range, so nuclei experience electric repulsion when they're far apart. The nuclear force, although very strong, has a much shorter range. So the goal in any fusion scheme is to get nuclei so close that the attractive nuclear force is stronger than the repulsive electric force. Then they can fuse and release nuclear energy.

Getting one nucleus to fuse with another is like dropping a ball into a deep hole surrounded by a barrier. A lot of energy is released when the ball drops into the hole, but to get it there you first have to supply enough energy to get it over the barrier (figure 11.1).

If the ball in figure 11.1 moves rapidly toward the hole, it may have enough energy to overcome the barrier. Similarly, if two nuclei approach each other fast enough, they can surmount the barrier of their mutual electrical repulsion. In most fusion schemes, heating of the fusing material brings the nuclei to the required speeds. But that takes very high temperatures. For deuterium-tritium fusion, the D-T mixture must at about 100 million degrees![1] How do you get something that hot? And what do you keep it in? Those two questions—heating and confinement—are the essential challenges facing fusion scientists.

The stars conveniently meet both fusion challenges with their immense gravity. A star forms when a cloud of gas collapses under its own gravity,

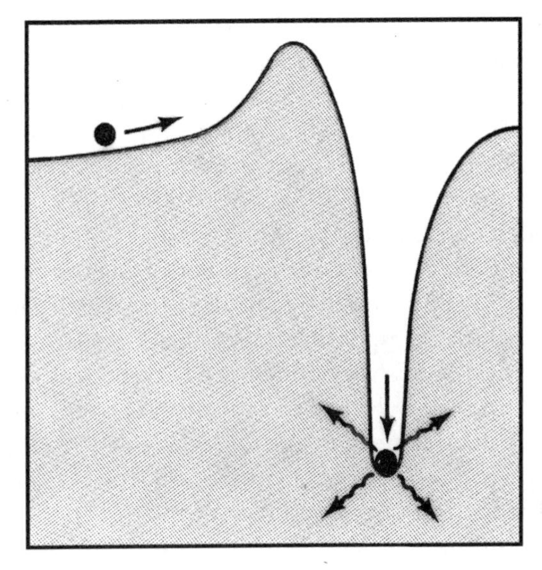

Figure 11.1
A gravitational analogy for fusion. The energy released in fusion is like the energy released when the ball falls into the deep hole. But to get to the hole, the ball first has to overcome a high barrier, analogous to electrical repulsion between nuclei. Only if the ball approaches the hole very rapidly will it have enough energy to overcome the barrier.

compressing and heating as it does so. A sufficiently large cloud gets hot enough to initiate fusion, and a star is born. Once fusion starts, the energy generated from fusion reactions keeps the gas hot enough to sustain ongoing fusion, and the star's immense gravity confines the hot gas.

On Earth, we don't have access to a star's strong gravity. Terrestrial scientists have had to devise other ways to heat and confine fusing material. Here you'll see two main approaches to heating and confinement in schemes aimed at controlling nuclear fusion for peaceful energy production. Later you'll see how the same challenges are met in thermonuclear weapons.

Magnetic Confinement

Anything that's hot enough to fuse is necessarily in a gaseous state. And at fusion temperatures there's so much energy around that atoms of lighter elements, including those used as fusion fuels, are stripped entirely of their electrons. The resulting gas therefore consists of electrically charged particles—negative electrons and positive nuclei. Such a gas of charged particles is called a **plasma**, and exhibits behaviors so unique that plasma is often called "the fourth state of matter," adding to the usual solids, liquids, and gases. The behavior of plasmas under fusion conditions remains a subject of active study, and the quest for fusion energy is intimately tied with our understanding of plasma physics. The confinement challenge in fusion energy becomes a search for ways of taming the often-unruly behavior of hot plasmas.

In **magnetic confinement**, the fusion plasma is contained in what's basically a *magnetic bottle* constructed not of matter but of magnetic fields. You've probably seen diagrams depicting Earth's magnetic field (figure 11.2), which is responsible for compass needles pointing north. The **magnetic field lines** shown in figure 11.2 represent the direction in which a small magnet—which is what a compass needle is—would align itself. The lines show something else too: where they're close, the field is strong, and where they're farther apart the field is weak. Thus, figure 11.2 shows that Earth's field is strongest near the poles.

So how can magnetic fields confine hot plasma? To answer this question, we need to understand the origin of magnetism as well as how magnetism exerts forces on matter. Those two understandings come from a fundamental fact of nature: Magnetism involves *moving electric charges*. Moving electric charges are the source of magnetism. Moving charges, in the form of currents of electrons flowing in wires, enable electromagnets like the one you might have made in school by winding wire around a nail and connecting a battery. The role of moving electric charges is less obvious in the magnet that holds a note to your refrigerator, but moving charges are still the essence of the refrigerator magnet—this time in the

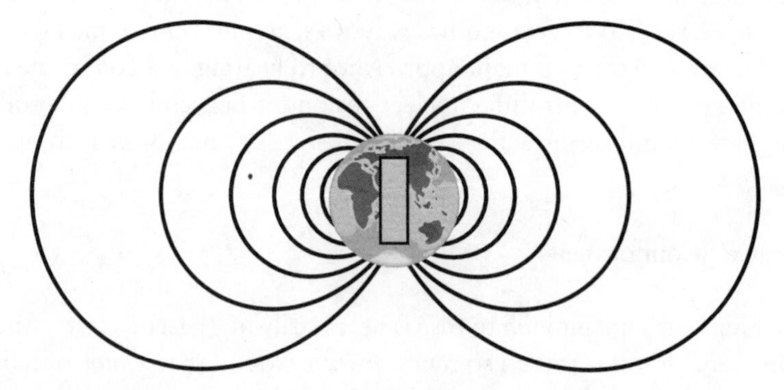

Figure 11.2
A depiction of Earth's magnetic field as represented by field lines similar to those of a bar magnet. This diagram is highly simplified and doesn't show distortions associated with a flow of plasma from the Sun.

form of individual atomic electrons spinning and orbiting within atoms of the magnet.

Similarly, magnetism most fundamentally affects *moving electric charges*. Again, you don't see that with the refrigerator magnet, but it's there in the response of atomic electrons in the steel of the refrigerator. It's easier to see in considering a single electrically charged particle that finds itself in a magnetic field. The charged particle's response to the magnetic field is complicated. It experiences a force that depends on the particle's charge, the strength of the magnetic field, and, most interestingly, on the particle's speed and the direction of its motion. If the particle happens to be moving in the direction of the field—along a magnetic field line, that is—then it experiences no force whatsoever no matter how strong the field. But if it moves at right angles to the field, then it experiences a force that causes it to circle a magnetic field line. The size of the circle depends on the field strength and the particle's speed; which direction it circles depends on whether it's positively or negatively charged. Of course, a particle need not be moving strictly along or perpendicular to the magnetic field. Any other direction, and the part of its motion that's along the magnetic field isn't affected by the field, and the part that's perpendicular becomes a circle. The result is that charged particles in magnetic fields move in spiral paths, as shown in figure 11.3a.

Try making the particle in figure 11.3a move faster. If you push it *along* the field, then the field has no effect and it moves easily. But if you push it sideways to the field, it just goes into a bigger circle. So it's easy for charged particles to move along the field but difficult for them to move across the field. As a result, they behave somewhat like beads strung on a wire (figure 11.3b), trapped on the magnetic field lines and able to move

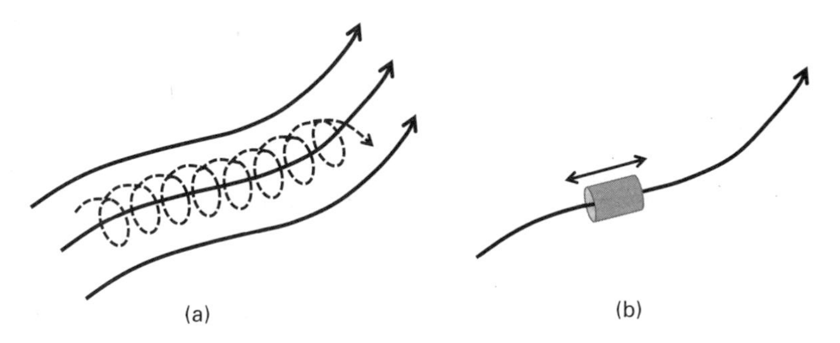

Figure 11.3
(a) A charged particle in a magnetic field moves in a spiral path about the magnetic field. Solid curves represent magnetic field lines. (b) The particle behaves like a bead strung on a wire, able to follow the wire but unable to move sideways to the wire.

easily along them but not sideways to the field. This trapping of charged particles on magnetic field lines is what enables magnetic confinement of the charged particles comprising a fusion plasma.

One caveat that's important in harnessing fusion: the discussion summarized in figure 11.3 is strictly correct only if the magnetic field is perfectly uniform, meaning it doesn't change with position in either strength or direction. If the field changes very slowly with position—and by slowly we mean with much less curvature than the spiral motion of the particles around the field—then the "beads on a wire" analogy is pretty accurate. But even so, particles *drift* slightly at right angles to the field. Furthermore, positive and negative charges drift in opposite directions. This sets up electric fields that further complicate attempts to trap the charged particles with the magnetic field.

There's yet another challenge to magnetic confinement: The collective motion of the charged particles comprising a plasma results in waves, which can grow to the extent that they disrupt the plasma so it's no longer confined. Such *instabilities* have long frustrated fusion physicists, and they're partly responsible for slowing progress toward fusion energy. Clever design of the magnetic field configuration can minimize instabilities, as can making the field and plasma region larger. A larger field region exhibits decreased field-line curvature, and that also helps minimize drifting of plasma across the field. Both instability and curvature considerations motivate the drive toward ever-larger fusion experiments.

So how do we confine fusion plasmas with magnetic fields? Because charged particles move easily *along* the magnetic field, it won't do to have a field whose lines penetrate the walls of the fusion chamber. Such a field would give hot plasma an easy route to the walls, where it would cool and quench the fusion reactions. An obvious alternative is to construct a magnetic field whose lines form closed loops that don't end anywhere. That's fairly easy to do. Wind wire around a cylinder and pass electric current through it and you get a straight, uniform magnetic field inside the cylinder. Bend the cylindrical coil into a circle and you've got a **toroid**—a doughnut-shaped structure whose magnetic field lines are closed circles (figure 11.4). Plasma trapped on those field lines can't reach the chamber walls because no field lines intersect the wall. This endless-field property is the reason that most experimental fusion devices are toroidal. But a problem arises once we bend the coil into a toroid: as you can see in figure 11.4, the coils are closer on the inside of the toroid. That makes the magnetic field stronger on the inside—also evident because the

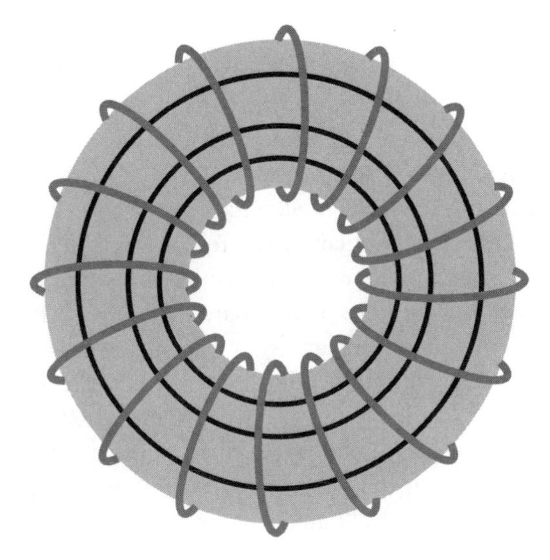

Figure 11.4
A toroidal coil of wire produces a magnetic field whose field lines form closed circles. Wire coils are the gray loops wrapping around the toroid; magnetic field lines are the black circles inside the toroid.

field lines are closer near the inside. So the field isn't uniform, and therefore plasma particles drift across the field lines, as we described a few paragraphs ago, and may eventually hit the walls of the plasma chamber.

Tokamaks

Clever techniques can help stabilize the plasma in a toroidal magnetic field. One technique is to give the field a twist, so it spirals around the toroid rather than running straight around the toroid's long dimension. A second is to give it *shear*, so the field direction—equivalently, the degree of twist—changes as you move across the short dimension of the toroid. One way to produce these effects would be with coils that don't wrap strictly around the toroid, as they do in figure 11.4, but instead spiral as they wrap around. But a much more effective approach is to let the plasma itself carry an electric current. After all, a plasma is made of charged particles, and it's moving charged particles—here in the form of an electric current—that produces magnetic fields. A device that implements this plasma-current approach is the **tokamak**, a Russian invention dating from the 1950s that today offers what is generally considered the most promising path to fusion energy.

Plasma Heating

With today's tokamaks, we're close to solving the plasma confinement problem. But we've still got to heat the plasma to 100 million degrees or more. Today's experimental fusion devices often use a combination of the three heating techniques described below.

Ohmic heating works essentially like an electric stove burner, an incandescent lightbulb's filament, a toaster, or a hair dryer: electric current flowing in a wire or in a plasma results in heating as charged particles collide with one another or, in a wire, with atoms. But plasmas are excellent electrical conductors, and they become even better conductors as they get hotter. So ohmic heating isn't all that effective in a fusion plasma, although it can bring the plasma to some 20–30 million degrees.

Neutral beam injection takes over where ohmic heating falters, or it can stand on its own. In this scheme, neutral atoms are first stripped of their electrons and then accelerated to high energies using electric fields, which work on charged particles but not neutrals. They're then recombined with electrons to make neutral atoms, atoms that are now moving rapidly due to their high energy. Beams of these high-energy atoms shoot into the plasma in the tokamak. Since they're neutral, the atoms aren't affected by the magnetic field, so they cross the magnetic field lines and penetrate deep into the plasma. In the high-temperature plasma environment they're once again stripped of their electrons, and they collide with plasma particles to spread their energy and thus heat the plasma. Neutral beam injection typically uses deuterium and thus serves the dual purpose of heating the plasma and replenishing the deuterium fuel.

A third approach is *radiofrequency* heating. This works somewhat like a microwave oven, with powerful radio waves exciting plasma particles into circular or wave-like motions that ultimately spread energy as heat. Today's experimental fusion devices often use a combination of these three heating techniques.

Progress toward Magnetic Confinement Fusion

Some 70 years of research with dozens of experimental fusion machines worldwide have brought us ever closer to an energy-producing fusion device. To get there, we need both high temperature and confinement. Confinement doesn't have to be perfect, but it does need to hold the hot plasma long enough that there's a good chance of fuel particles—at first, deuterium and tritium—colliding and fusing. The higher the density of plasma particles, the more likely are such collisions. (In magnetic-confinement fusion, however, even the most advanced experiments exhibit densities only 10 millionths that of Earth's atmosphere.) An important measure of fusion progress is the product of plasma density, confinement time, and plasma temperature. Think of this product in terms of what it takes to start and sustain a fire: You need dense enough fuel (the plasma density), a high enough temperature to ignite the fuel and keep it burning (the plasma temperature), and you want it to burn for a significant time (the confinement time). Figure 11.5 is a plot of this *triple product* over time, showing achievements of different fusion experiments. Figure 11.5 also shows **breakeven**, which occurs when power released in fusion equals the power that was supplied to heat the plasma. A more remote goal is **ignition**, which occurs when fusion reactions become self-sustaining. Ignition occurs in the region of figure 11.5 marked "Power Plant." For a practical plant using D-T fusion, that would mean some 10 times as much fusion power as is needed to heat and confine the plasma.

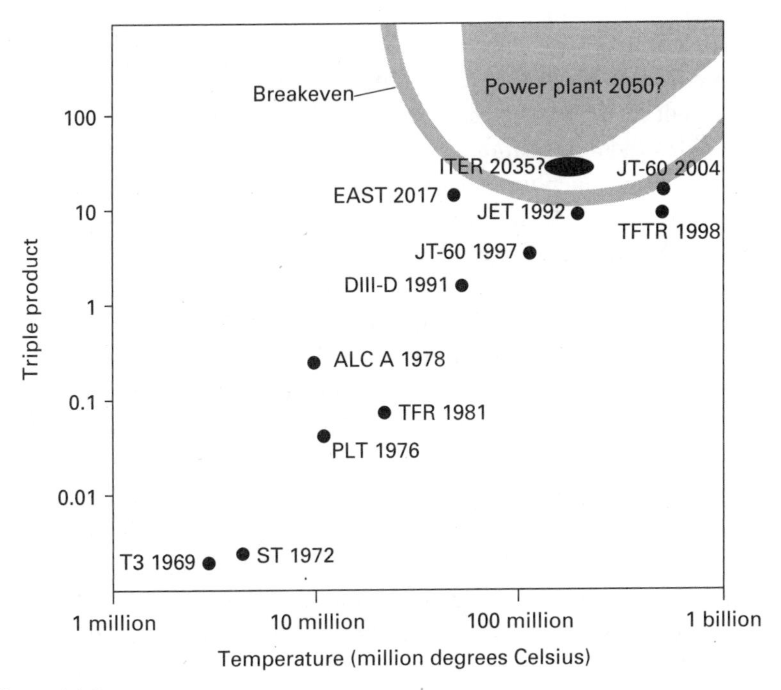

Figure 11.5
Progress in nuclear fusion, as measured by the triple product of plasma density, confinement time, and temperature plotted against temperature (at these high temperatures degrees Celsius and kelvins are indistinguishable). Labels name individual fusion experiments and dates they achieved the values shown. Note the logarithmic scales, in which a fixed distance on either axis corresponds to a 10-fold increase. Units of triple product are 10^{20} s·keV/m^3.

ITER

The region marked ITER on figure 11.5, with the tentative date of 2035, represents humankind's most ambitious fusion experiment. ITER—originally an acronym for International Tokamak Experimental Reactor but now just called ITER—was at first a collaboration involving the European Union, Japan, the Soviet Union, and the United States. Design of the ITER tokamak began in the late 1980s and was more or less finalized in 2001. China, South Korea, and India subsequently joined the ITER collaboration. The United States withdrew in 1998 but rejoined in 2003.

Following contentious debates over where to build ITER, a site in southern France was chosen in 2005, and construction began in 2010. By 2017, the project was halfway to the point where ITER would produce its first plasma, scheduled for 2025. The first experiments involving fusion with deuterium-tritium plasma are scheduled for 2035. These dates have been pushed back several times as logistical difficulties of coordinating and funding a huge multination scientific project increased.

ITER will be a large tokamak, with a toroidal vacuum vessel measuring some 60 feet across and containing a plasma volume 10 times greater than current fusion machines. Figure 11.6 is a cutaway of ITER, showing the toroidal vacuum chamber with D-shaped cross section to minimize curvature on the inner and outer edges. ITER is designed to operate at 150 million degrees Celsius and produce 500 megawatts of fusion power—10 times the 50 MW needed to establish and heat the plasma. ITER is expected to be the first fusion experiment to achieve a true burning plasma—one in which heat from fusion is sufficient to sustain the high plasma temperature.

ITER should achieve many of the milestones needed for a fusion power plant, including a sustained burning plasma and a substantial net energy from fusion. However, ITER is still an experimental device, not a power plant. It won't produce electricity. But it will enable experiments with tritium breeding, with heat removal, and with the behavior of materials in the harsh environment of the tokamak—all of which should help guide the design of a fusion power plant. Success with ITER should establish the technological and scientific feasibility of fusion as an energy source.

Successors to ITER are already on the drawing board, under the general name DEMO. It's not clear whether DEMO will be an international collaboration like ITER or separate efforts by individual countries. Plans call for construction of one or more DEMO projects in the 2030s, with operation in the 2040s. The most ambitious DEMO ideas would result in what's essentially a prototype of a fusion-powered electric power plant. The goal is to demonstrate industrial-scale generation of electricity by 2050.

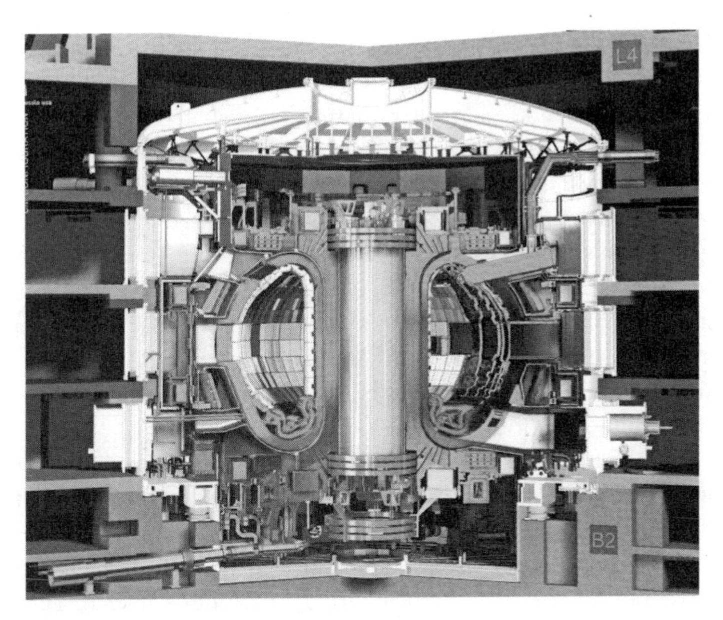

Figure 11.6
Cutaway diagram of the ITER fusion experiment, which is intended to produce 500 MW of fusion power. Note the toroidal vacuum chamber with D-shaped cross section. (Copyright ITER Organization, https://www.iter.org/)

Fusion: Safety and Radiation

Fusion reactors are inherently safer than fission reactors. There's no chain reaction to go out of control, no inventory of radioactive spent fuel, and essentially no residual heat if the reaction stops. There's only enough fuel in the reactor for a few seconds of energy production. The challenges of heating and confinement mean that if anything goes wrong the plasma will escape its magnetic bottle and immediately cool below fusion temperatures. Recall that the density in magnetic-confinement plasma is orders of magnitude below normal atmospheric density, so there isn't all that much material in the plasma of a fusion reactor.

What about radiation? The easiest fusion reaction to achieve is the deuterium-tritium (D-T) reaction that we illustrated in figure 5.2:

$$ {}^{2}_{1}\mathrm{H} + {}^{3}_{1}\mathrm{H} \rightarrow {}^{4}_{2}\mathrm{He} + {}^{1}_{0}\mathrm{n} + \text{energy} \qquad (1)$$

This reaction requires the lowest temperature and is more likely than other possible fusion reactions, so it's almost certainly what would be used in the first commercial fusion reactors. But its copious neutron production is problematic. Because they're neutral, neutrons easily penetrate matter, and they themselves constitute potentially lethal radiation. So an operating fusion reactor would be a high-radiation zone. Stop the fusion reaction, though, and neutron radiation immediately stops. But there are more subtle issues involving neutrons. First, when they interact with nuclei in the solid structure of the fusion reactor, they create new isotopes through the process of *neutron activation*, introduced in chapter 4. That makes the walls of the reactor's vacuum chamber mildly radioactive. The particular radioactive isotopes produced depend on the materials used in construction of the reactor, but in any event the radioactivity on a per-mass basis would be much lower than that of fission products in a fission reactor, and most half-lives would be short enough that significant radioactivity would last for only about a century.

In the D-T fusion reaction, 80 percent of the energy produced is in the motion of the neutrons. Because they're neutral, the neutrons don't interact significantly with the fusion fuel, so they can't provide energy to sustain the plasma temperature. The remaining 20 percent of the fusion energy is in the motion of the alpha particles (${}^{4}_{2}He$, or helium-4 nuclei). Those are charged particles, so they can help heat the plasma. But because the alpha particles carry only one-fifth of the fusion energy, a D-T fusion reactor needs a large ratio of fusion energy production to heating energy in order to sustain the plasma temperature. The fact that most of the D-T fusion energy comes out in the neutrons also presents an engineering challenge of capturing that energy as heat that's then used to run a steam turbine cycle. Because charged particles are readily stopped in matter while neutrons aren't, it would be easier to generate heat from a reaction that produced most of its energy in charged particles. And with charged particles of high enough energy, it's possible to generate electricity directly without the intermediary of a steam cycle.

Finally, the relentless neutron bombardment of a fusion reactor's vacuum-chamber walls results in structural damage to wall materials, as the original atoms get replaced by neutron-activated isotopes of other species. Much of the practical engineering work needed before we have a fusion power plant will be in determining the best materials to use, especially for the innermost surface of the vacuum chamber.

A final issue neutron issue involves weapons proliferation. Placing uranium-238 in a fusion reactor anywhere it's exposed to fusion neutrons would result in production of plutonium-239 via the reactions we illustrated in figure 5.8. The plutonium could be separated by means described in chapter 9, then used for nuclear weapons. Tritium is also a weapons-related substance, added to fission weapons to boost their explosive yield. A typical fusion reactor might contain several kilograms of tritium, which could be diverted to weapons use. However, a Princeton University study of fusion's proliferation potential concluded that diversion of weapons materials from the fission fuel cycle is much more likely—although the study suggests that it might be wise to consider international safeguards for fusion reactors similar to those that apply to peaceful uses of nuclear fission.[2]

There's one advantage to that copious flow of high-energy neutrons from the D-T reaction: they can breed tritium fuel ($^{3}_{1}H$) by neutron bombardment of lithium-6:

$$^{6}_{3}Li + ^{1}_{0}n \rightarrow ^{3}_{1}H + ^{4}_{2}He \quad (2)$$

The other product is harmless, inert helium-4. Research continues on how best to take advantage of this reaction, with most schemes envisioning a lithium *blanket* on the inside wall of the vacuum chamber. The ITER experiment won't depend on breeding for its tritium supply, but ITER experiments with lithium blankets should pave the way for DEMO and future fusion power plants that will breed their own tritium.

Inertial Confinement

Magnetic confinement fusion schemes use relatively low density plasmas confined for relatively long times. In ITER, for example, the confinement time will be some six minutes and the density ten millionths of normal atmospheric density. Inertial confinement fusion (ICF) takes the opposite approach: very high density but very short confinement time. The term **inertial confinement** implies a time so short that particles' inertia prevents them from moving significant distances during the time they interact and fuse. Hence inertial confinement doesn't require magnetic fields to confine the fusion reaction. What it does require is a colossal compression of fusion fuel to densities far greater than normally encountered on Earth.

Most experiments with inertial confinement use lasers to compress a tiny fuel pellet of frozen deuterium and tritium only a millimeter or so in diameter. Enormous laser power is required, with the lasers precisely aimed so they compress the pellet uniformly from all directions. The most advanced inertial confinement experiment to date is the National Ignition Facility (NIF) at Lawrence Livermore National Laboratory in California. This facility, the size of a football stadium, focuses 192 laser beams on its target, delivering energy at the rate of 500 terawatts—some 30 times humankind's total energy consumption rate. The energy is stored gradually in huge banks of *capacitors*, devices that store energy in electric fields. The capacitor banks are then discharged almost instantaneously, delivering a laser pulse that lasts only 20 nanoseconds. You saw in chapter 6 that power is energy per time, and it's this very short time that results in such a huge power. Figure 11.7 shows the NIF target chamber.

Some inertial confinement experiments focus their lasers directly on a D-T fuel pellet, but NIF uses so-called *indirect drive*, with the lasers focused into a small gold cylinder surrounding the fuel pellet. The cylinder absorbs the laser energy and heats to some 3 million degrees. As a result, it emits intense X-rays, which compress the fuel pellet more symmetrically than the laser beams themselves could do. At NIF, the fuel pellet reaches densities some 100 times that of lead, a pressure 100 billion

Figure 11.7
The NIF target chamber being prepared for installation. The chamber weighs 130 tons and is some 30 feet in diameter. At its center sits a millimeter-size D-T pellet. Holes in the chamber are for the 192 laser beams that converge on the pellet, as well as for diagnostic equipment. (Lawrence Livermore National Laboratory)

times Earth's atmosphere, and a temperature of 100 million degrees. At that point it explodes like a miniature fusion bomb.

NIF's National Ignition Campaign (NIC) had the goal of reaching fusion ignition conditions by 2012 but failed to meet that goal. NIF did reach a minor milestone when a fusing pellet released more energy from fusion than the laser beams had delivered to the pellet. However, the fusion energy was still far below what was needed to drive the lasers. In 2012, NIF's focus was directed away from the goal of fusion ignition. Today the facility is used for what was always its primary purpose, namely to simulate nuclear explosions as a means of maintaining confidence in the United States' weapons stockpile without testing actual weapons (more on this in chapter 19). It's also used for research on materials under conditions of extreme density and temperature, and a modest program continues to research inertial confinement fusion. That program passed another milestone in 2018, producing twice as much fusion energy as was delivered to compress the fuel pellet. But that's still a factor of 30 below what's needed, and NIF is unlikely to live up to its name as the National *Ignition* Facility.

Research on inertial confinement fusion continues at other sites around the world, some using laser-driven fusion while others experiment with magnetic compression schemes. However, it's not clear how to go from exploding fuel pellets to a power plant producing steady electric power. So, despite impressive advances in inertial fusion, most experts believe that magnetic fusion will be the way to fusion-generated electricity.

Alternative Approaches to Fusion

Even the most advanced magnetic fusion experiments are still just that—experiments. Some may point the way to practical fusion energy plants, but they're by no means proven. Fusion researchers are therefore exploring myriad other schemes, even though they're less developed than tokamak-based magnetic fusion. As the cost and timeframe for big experiments like ITER increase, some are becoming convinced that smaller, less expensive schemes just might achieve sustained fusion earlier. Here we explore alternatives to mainstream D-T magnetic fusion.

Other Fusion Reactions

We've emphasized deuterium-tritium fusion because that's the reaction most likely to be used in the first commercial fusion reactors. Figure 11.8 shows why. The graph plots reaction rates for several different fusion reactions versus temperature.[3] D-T fusion stands as the easiest to achieve,

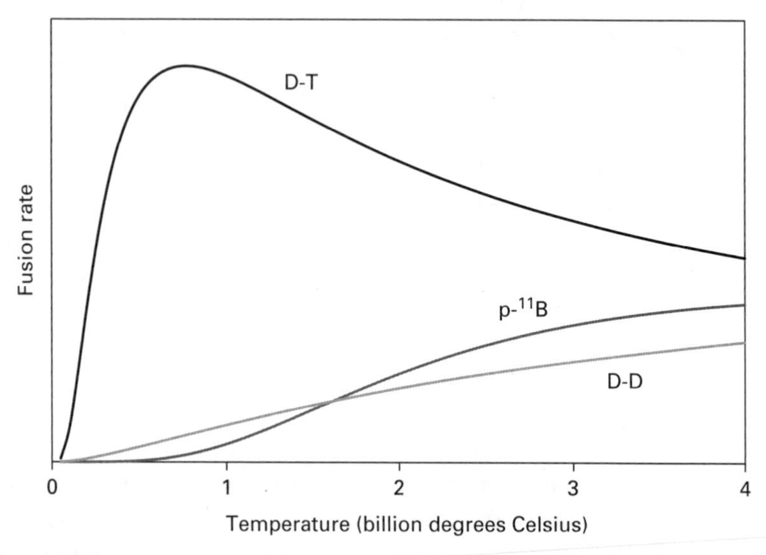

Figure 11.8
Reaction rates versus temperature for several fusion reactions. The graph makes clear why the D-T reaction will be used in the first fusion reactors. The vertical scale is linear and describes the number of fusion reactions per second occurring in a given volume.

both in terms of its much higher rate and the fact that the rate peaks at a much lower temperature than the other reactions. But we've also seen that D-T has disadvantages, most associated with the copious high-energy neutrons that carry most of the fusion energy. The other reactions shown in figure 11.8 have lower rates and require higher temperature, but they either reduce or essentially eliminate neutron issues.

Deuterium-deuterium fusion (D-D fusion) could supply truly unlimited energy because, as you saw at the beginning of this chapter, there's enough deuterium in the world's oceans to keep us in energy for five times as long as the Sun will continue to shine. There are two possible outcomes of fusion between two deuterons, which occur with nearly equal probability:

$$ {}^{2}_{1}\mathrm{H} + {}^{2}_{1}\mathrm{H} \rightarrow {}^{3}_{1}\mathrm{H} + {}^{1}_{1}\mathrm{H} + \text{energy} \qquad (4) $$

$$ {}^{2}_{1}\mathrm{H} + {}^{2}_{1}\mathrm{H} \rightarrow {}^{3}_{2}\mathrm{He} + {}^{1}_{0}\mathrm{n} + \text{energy} \qquad (5) $$

The first of these reactions produces tritium (${}^{3}_{1}\mathrm{H}$) along with ordinary hydrogen (${}^{1}_{1}\mathrm{H}$, which is just a proton). The second reaction produces the rare but harmless isotope helium-3 (${}^{3}_{2}\mathrm{He}$) and a neutron. The neutron energy here is much lower than in D-T fusion, greatly reducing radioactivity from neutron activation and the associated damage to vacuum chamber walls. Unfortunately, though, there's no way to stop the tritium produced in the first reaction from reacting with a deuteron in the D-T reaction that produces high-energy neutrons. So although D-D fusion is considerably "cleaner" than D-T fusion alone, it still produces some radioactivity and neutron damage—albeit much lower than with D-T fusion.

The reaction labeled p-^{11}B is truly clean, producing no neutrons and therefore no radioactivity:

$$^{1}_{1}\text{H} + {}^{11}_{5}\text{B} \rightarrow 3^{4}_{2}\text{He} + \text{energy} \tag{6}$$

Here a proton ($^{1}_{1}$H) fuses with a boron-11 nucleus to yield three alpha particles (ordinary helium-4 nuclei, $^{4}_{2}$He). Superficially, this **proton-boron fusion** might sound more like *fission*, since a particle (here a proton rather than a neutron) hits the relatively heavy boron nucleus that then splits into three lighter helium nuclei. But the energy release isn't in the breakup into the three alpha particles, a process that actually requires energy. Rather, energy is released in the fusion of the proton with the boron to make a highly energetic carbon-12 nucleus. That almost immediately decays into the three alpha particles, using some of its excess energy to free the alpha particles while the rest appears as energy of the alphas' motion. Since the alphas are charged particles, their energy could ultimately be converted directly into electricity without the need for a steam cycle.

Although the proton-boron reaction produces no neutrons whatsoever, a rare variant yields carbon-11 and a low-energy neutron. C-11 is radioactive but with a 20-minute half-life, so it's not a significant concern, while the neutron's energy is low enough that activation-induced radioactivity is negligible. Alpha particles from the fusion reaction can also react with boron-11 to produce ordinary nitrogen-14 and a neutron. A final but rare reaction path has the carbon-12 formed in proton-boron fusion remaining intact but emitting gamma radiation to shed its excess energy. These reaction variants mean that a proton-boron reactor would need some radiation shielding, but it wouldn't produce significant amounts of radioactive materials.

Boron occurs in Earth's crust with about half the abundance of lithium, and boron-11 comprises 80 percent of it, so we're again talking about a fuel resource that could last for tens of millions of years. There's also plenty of boron in the oceans. We'll describe current work on p-^{11}B fusion schemes shortly.

Alternative Fusion Devices

Scientists around the world are experimenting with a plethora of alternative fusion devices. Some are modifications of traditional magnetic or inertial fusion, while others are hybrids of the two or entirely new schemes. Some have government backing, others are private ventures. None are close to achieving breakeven or ignition, but supporters of alternative fusion schemes claim that some may offer cheaper, faster ways to a fusion power plant. Here we describe just a few of these alternative schemes.

Stellerators are magnetic confinement devices that date to the 1950s. They use a complicated, twisted magnetic field and plasma structure (figure 11.9). They fell out of favor when tokamaks came on the fusion scene but are now enjoying something of a revival. Although far behind tokamaks in the race to achieve fusion conditions, stellerators offer the promise of continuous operation as opposed to the pulsed mode of today's tokamaks.

Laser-boron fusion experiments utilize the proton-boron reaction described above. They achieve fusion not with high temperatures but by using high-power lasers to accelerate a blob of plasma and then confine it briefly with intense magnetic fields provided by a second high-power laser. Today a group based in Australia, but including collaborators around the globe, is experimenting with this technique and envisioning a laser-boron fusion power plant that could produce electricity without radioactive waste or climate-changing carbon emissions.[4]

TAE Technologies, a California startup formerly known as Tri Alpha Energy, has a different approach to proton-boron fusion. Its machine accelerates two beams of plasma and delivers them from opposite directions into a central chamber where the plasma spins in a strong magnetic field.

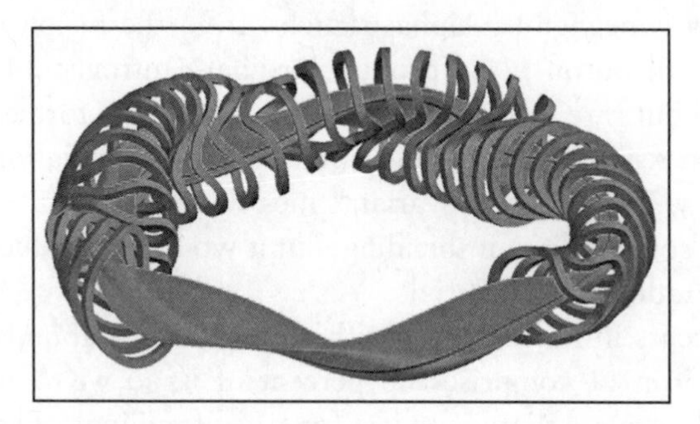

Figure 11.9
Magnetic coils of Germany's Wendelstein 7-X stellerator wrap around the twisted plasma. (Max-Planck Institut für Plasmaphysik)

Carefully controlling the spin helps mitigate instabilities, and high-energy plasma injection heats the merged plasma toward fusion temperatures.

General Fusion, a Canadian company based in British Columbia, has what may sound like an improbable approach to fusion. Their reactor contains a sphere of molten lead, which is set spinning to form a whirlpool-like vortex of liquid metal. A magnetized plasma is injected into the vortex, and mechanical pistons then compress the liquid metal and the plasma trapped within it, creating fusion conditions. Lithium added to the lead can breed tritium for D-T fusion. The first wall surrounding the plasma is liquid, enabling easier heat transfer and minimizing the radiation damage that would occur in a solid. General Fusion's steam-powered pistons are a low-tech, commercially available alternative to high-power lasers or superconducting magnets.

Helion Energy, of Redmond, Washington, is developing what it calls a *fusion engine*, a kind of hybrid of magnetic and inertial fusion. This truck-sized device uses pulsed magnetic fields to accelerate two blobs of magnetized plasma to some 300 miles per second. The blobs collide in a narrow region where magnetic fields compress the mixed plasma to fusion conditions. The engine operates on a cycle that repeats this process once a second. Helion claims they will be able to build 50-megawatt fusion power plants that will fit in shipping containers. The company's name, Helion, refers to the helium-3 nucleus—and thus to the fuel mix of helium-3 and deuterium used in the company's reactor.

Prospects for Nuclear Fusion

When fusion research started in the 1950s, the claim was that fusion power plants might be operating within 20 years. But no matter what year you choose since then, fusion has always seemed several decades away. No wonder many people—including many scientists—regard fusion as an impractical "pie in the sky" solution to our energy problems. And even if we solve the technical problems of fusion energy, will the cost of fusion power plants be prohibitive—especially in a world where the cost of renewable energy keeps falling? For others, though, fusion's virtually unlimited fuel resource and the growing urgency of a changing climate that can't tolerate fossil-fueled energy justify continued fusion research. And we are making progress: a successful ITER, for example, should generate 10 times much fusion energy as it consumes and point the way toward a true fusion power plant. And it's just possible that

Cold Fusion?

In 1989, chemists Martin Fleischmann and Stanley Pons reported on a simple tabletop-scale experiment done at room temperature, which they claimed resulted in energy production from deuterium-deuterium fusion. Fleischmann and Pons passed electric current through a solution of heavy water (D_2O or ${}^2_1H_2O$) that contained an electrode made of the element palladium. The idea was that D_2O molecules would split into deuterium and oxygen by the well-known process of electrolysis, and that deuterium atoms would nestle onto the palladium surface—also a process that had been known, at least for ordinary hydrogen, since the nineteenth century. Fleischmann and Pons claimed that deuterium atoms on the palladium would get close enough that their nuclei could undergo fusion. Their claimed evidence for fusion was not only excess heat but also the appearance of neutrons and tritium—both expected products of D-D fusion.

Fleischmann and Pons chose an unorthodox way of first presenting their results: a press release and a subsequent press conference. The world took note, and news articles touted the possibility of inexhaustible and inexpensive energy. Scientists around the world rushed to performed their own experiments with cold fusion. Most failed, but a few groups noted excess heat and some claimed to see expected products of D-D fusion. All were hampered by having only vague descriptions of the Fleischmann-Pons experiment. The few claims of neutron and tritium production were soon retracted, with explanations ranging from nuclear instrumentation being affected by heat from chemical reactions to contamination by external sources of tritium. Within months the U.S. Department of Energy (DOE) published a report downplaying any possibility of obtaining useful energy from the Fleischmann-Pons scheme. The DOE revisited cold fusion in a 2004 report, and again found no convincing evidence for cold fusion. In 2019, a group of scientists, spread across many institutions, reported the results of a careful two-year study of conditions that might lead to cold fusion.[5] Again, they found no evidence for excess heat or for neutron production—although they did suggest that additional research might be warranted. Although a handful of researchers continue to pursue cold fusion, the scientific community now looks on the whole episode as an example of shoddy science, undocumented procedures, exaggerated claims, and outright fraud.

There remains one possibility for cold fusion unrelated to the Fleischmann-Pons chemical approach. It's based on a 1968 idea by Nobel laureate Luis Alvarez, also known for helping formulate the now widely accepted hypothesis that dinosaur extinction was caused by an asteroid impact. Alvarez suggested replacing the electron in hydrogen isotopes with a *muon*, an unstable cousin of the electron with 207 times the electron's mass. That would result in atoms 207 times smaller than those containing ordinary electrons. When those atoms joined to form molecules, their nuclei would be much closer than usual. A water molecule formed this way

Cold Fusion? (*continued*)

with deuterium and tritium in place of the ordinary H in H_2O would have D and T nuclei in close proximity—close enough that they would, on occasion, undergo fusion. And if fusion occurred the muons—which last some two microseconds before decaying—could go on to catalyze hundreds of additional fusion events. Thus the process is termed **muon-catalyzed fusion.**

Muon-catalyzed fusion has been observed, but it occurs only at a very slow rate. And the energy balance isn't good: it takes a lot more energy to create the muon than can be obtained from fusion, even if one muon catalyzes 100 fusion events. Nevertheless, muon-catalyzed fusion may be worth pursuing as a long-shot approach to fusion. But don't hold your breath!

one of the new, smaller-scale fusion schemes will achieve plasma ignition before ITER and lead more quickly to fusion-generated electricity.

If we do achieve fusion, we'll have an energy source with virtually unlimited fuel, whose environmental and health impacts will be far less than those of fossil fuels or even nuclear fission. One final caveat, though: were we to develop *inexpensive* fusion technology, then electricity might become, in the 1954 words of Atomic Energy Commission chair Lewis Strauss, "too cheap to meter." Absent financial or resource limitations, our energy consumption could grow exponentially. Now, all the energy humankind uses ultimately ends up as heat that Earth must radiate away to space. Unlike sunlight or geothermal energy, the energy released in fusion is energy that wouldn't otherwise have been delivered to Earth. So we risk adding substantially to Earth's heat burden. That could provoke a climate crisis more direct and serious than what we now face at the hands of fossil fuels and their heat-blocking greenhouse emissions. But that's a long way off—and fusion that gets deployed by the mid-twenty-first century could help stave off our more immediate fossil-fueled climate crisis.

Glossary

breakeven A goal for fusion experiments, in which the energy produced from fusion equals the energy needed to create and heat the fusion plasma.

deuterium-deuterium fusion (D-D fusion) Fusion of two nuclei of the heavy hydrogen isotope deuterium (^2_1H). The two possible products are tritium (^3_1H) and ordinary hydrogen (^1_1H), or helium-3 (^3_2He) and a neutron.

deuterium-tritium fusion (D-T fusion) Fusion of the heavy hydrogen isotopes deuterium (^2_1H) and tritium (^3_1H) to produce an alpha particle (^4_2He) and a neutron.

ignition The condition when a fusion plasma produces enough energy to keep the plasma at fusion temperatures, thus ensuring sustained fusion.

inertial confinement An approach to fusion that relies on compression of a fuel pellet to fusion conditions, usually with lasers, with those conditions lasting for so short a time that fusing nuclei don't have time to escape the fusion region.

magnetic confinement An approach to fusion that relies on magnetic fields to confine hot plasma.

magnetic field line A line that traces out the direction of a magnetic field. The magnetic field is stronger where field lines are closer together.

muon-catalyzed fusion A scheme that replaces electrons in hydrogen isotopes with heavier muons, making the atoms much smaller and allowing their nuclei to get close enough to fuse. Still under investigation, but a long shot for fusion energy.

plasma A gas consisting of electrically charged particles, characterized by long-range interactions among the particles that result in collective behaviors not seen in ordinary gases.

proton-boron fusion Fusion of a proton (${}^{1}_{1}H$) with boron-11 (${}^{11}_{5}B$), yielding three alpha particles (${}^{4}_{2}He$).

tokamak A toroidal fusion device characterized by electric current flowing within the plasma itself to create a twisted magnetic field that helps reduce instabilities.

toroid A doughnut-shaped structure that is commonly used in magnetic-confinement fusion devices because its magnetic field lines form closed loops that don't begin or end.

Notes

1. That's degrees Celsius (or kelvins; the difference is negligible at these temperatures). If you prefer Fahrenheit, you can roughly double any temperatures given here. But physicists don't use either scale; they give the temperature directly in energy units, since temperature is just a measure of particles' energy.

2. A. Glaser and R. J. Goldston, "Proliferation Risks of Magnetic Fusion Energy: Clandestine Production, Covert Production, and Breakout," *Nuclear Fusion* 52, no. 4, (March 2012).

3. Figure prepared from calculations using parameterizations of reaction rates: p-^{11}B from Equations 13, 14, 15 and Table 2 of W. M. Nevins and R. J. Swain, *Nuclear Fusion* 40, no. 4 (2000): 865–872; others from Equations 12, 13, 14 and Table VII of H.-S. Bosch and G. M. Hale, *Nuclear Fusion* 32, no. 4 (2000): 611–631.

4. H. Hora et al., "Road Map to Clean Energy Using Laser Beam Ignition of Boron-Hydrogen Fusion," *Laser and Particle Beams* 35, no. 4 (December 2017): 730–740.

5. C. P. Burlinguette et al., "Revisiting the Cold Case of Cold Fusion," *Nature* 570 (June 6, 2019): 540.

Further Reading

Chen, Francis F. *An Indispensable Truth: How Fusion Power Can Save the Planet* (New York: Springer, 2011). Despite its effusive title, this is an authoritative account of fusion science and technology by a plasma physicist who spent a career in fusion research and other applications of plasma physics. Especially strong on magnetic confinement.

Chen, Francis F. *Introduction to Plasma Physics and Controlled Fusion*, 3rd ed. (Cham, Switzerland: Springer, 2015). Update of a classic undergraduate textbook. All the details, with no math spared.

Clery, Daniel. *A Piece of the Sun: The Quest for Fusion Energy* (New York: Overlook, 2013). A well-respected science journalist shares his perspectives on the history, physics, economics, and politics of the quest for controlled fusion.

Huizenga, John R. *Cold Fusion: The Scientific Fiasco of the Century* (Oxford: Oxford University Press, 1994). A thorough and authoritative account of cold fusion in the half-decade after the Fleischmann-Pons announcement, by the chair of the U.S. Department of Energy's panel investigating cold fusion.

National Academies of Science, Engineering, and Medicine. *Final Report of the Committee on a Strategic Plan for U.S. Burning Plasma Research* (Washington, DC: National Academies Press, 2018). Available for download at https://www.nap.edu/download/25331. A prestigious group of scientists recommends stepping up U.S. funding for fusion research, including continued involvement with ITER, even as the political leadership threatens withdrawal from the ITER project. Also recommends a U.S.-specific program to produce a pilot fusion power plant.

United States Department of Energy. "Report of the Review of Low Energy Nuclear Reactions," 2004. Available at https://www.lenr-canr.org/acrobat/DOEreportofth.pdf. This brief summary reaffirms the DOE's earlier conclusion that evidence for cold fusion is lacking—although the report doesn't rule out support for careful research by individual scientists.

III

Nuclear Weapons

12

Nuclear Weapons: History and Technology

Our nation can be destroyed in half an hour, and there's nothing we can do to prevent it. That statement appeared in the predecessor to this book nearly three decades ago and, unfortunately, it's still true. The potential for national and even global devastation lies fundamentally in the nuclear difference, that million-fold increase in the energy associated with nuclear reactions. The energy stored in the thousands of nuclear weapons in the world's arsenals exceeds by many orders of magnitude the destructive capability that humankind has invoked in all its wars.

Quantifying Destruction

Nothing in our experience compares even remotely to the destructive effects of nuclear weapons. Relatively few human beings—residents of Hiroshima and Nagasaki—have directly experienced those effects. Most of them are now dead. Even those who survived the effects of the small, crude nuclear weapons used on those two cities cannot imagine warfare with today's multiple-warhead missiles, each of which might carry the destructive power of more than 500 Hiroshima bombs.

How are we to quantify these weapons, whose effects lie so far beyond experience or even imagination? The scientists who worked on the first nuclear bombs were well aware of the million-fold nuclear difference and knew that a few pounds of nuclear explosive could yield the equivalent of thousands of tons of chemical explosive. Thus was born the **kiloton** (kt), a unit of nuclear explosive energy (also called *yield*) equal to that of 1,000 tons of the chemical explosive TNT. By the 1950s, technology had outpaced the kiloton, and the **megaton** (Mt)—1,000 kilotons, or the equivalent of 1 million tons of TNT—became the appropriate unit for the largest weapons.

How can we make these abstract numbers meaningful? Only by comparisons with realized destruction. The medium-sized city of Hiroshima was essentially destroyed by a single 15-kiloton bomb; figure 12.1 thus

Figure 12.1
A 15-kiloton bomb, small by today's standards, damaged or destroyed over 90 percent of Hiroshima's buildings. At least 100,000 people died immediately; by the end of 1945, the toll had reached 140,000, and by 1950, 200,000 had died from the bombing. (U.S. Department of Defense)

provides a yardstick for nuclear destruction. The largest nuclear weapon ever tested yielded nearly 60 megatons or 4,000 Hiroshimas. Today's arsenals range from megaton-range weapons down to battlefield nuclear explosives with sub-Hiroshima yields. The strategic nuclear weapons that the superpowers have aimed at each other's homelands range from a few hundred kilotons to over one megaton. Russia's SS-18 Mod 5 intercontinental ballistic missile, for example, carries 10 warheads with yields in the range from 500 kt to 1 Mt.

How are we to visualize entire arsenals of nuclear weapons? In all of World War II, the explosive energy released in all weapons used by all sides was the equivalent of 6 million tons of TNT.[1] Most of that destructive power was by chemical explosives, from the gunpowder driving individual bullets to bombs weighing thousands of pounds. Those 6 megatons also include the two nuclear bombs dropped on Japan. If we visualize 6 megatons as being "one World War II" of destruction, we have an appropriate unit for describing the destructive potential of today's nuclear arsenals. In those terms, a single United States nuclear missile submarine—armed with 24 Trident missiles manufactured by Lockheed Martin and each carrying up to eight 475-kt warheads—has as much firepower as 15 World War IIs!

Today's Nuclear Arsenals

A year after World War II ended, the world's nuclear arsenal consisted of nine Nagasaki-type bombs possessed by the United States; their combined explosive yield of 200 kt was the equivalent of 1/30 World War II. By 1950, just after the Soviet Union exploded its first nuclear weapon, the global arsenal held the nuclear equivalent of 12 World War IIs. The world's arsenal peaked in 1973 at 25,000 megatons or more than 4,000 World War IIs. Today it's about 7 percent of that value—still almost 300 World War IIs. With some 4,000 nuclear weapons deployed worldwide in 13 countries, and another 6,000 in storage, there's a grave risk of miscalculation during international crises, of accidental detonations from storage sites, or even of deliberate detonations by nonstate actors (more on that possibility in chapter 16). Figure 12.2 shows the evolution of the world's nuclear arsenals since 1945. Our data on weapons numbers are based in part on government reports, including those required under treaty obligations, but often they're best estimates by experts who track nuclear arsenals.

With the stakes so high it behooves us to ask what, exactly, do these weapons do for us and how do they influence our security? What could they do *to* us? How might we intend to use them? How many do we really need? Can we protect ourselves from them? Will nuclear weapons spread to other nations? What's the connection between nuclear power and nuclear weapons? What can be done to control these instruments of unparalleled destruction? We'll explore these nuclear questions in subsequent chapters. First, let's look briefly at the history of nuclear weapons and at the technology that makes them work.

A Brief History

Hiroshima's destruction occurred less than seven years after Lise Meitner and Otto Frisch first recognized the occurrence of nuclear fission in neutron-bombarded uranium. Scientists around the world immediately saw the potential for a weapon of unprecedented destructiveness, although its technical feasibility remained far from certain. That the actual fission experiments had first occurred in Germany alarmed many, including refugee scientists who had fled Hitler for safer havens in England and the United States. Frightened by the prospect of a nuclear-armed Third Reich, scientists lobbied the American and British governments to begin all-out efforts to develop fission weapons. The most famous of these was a 1939

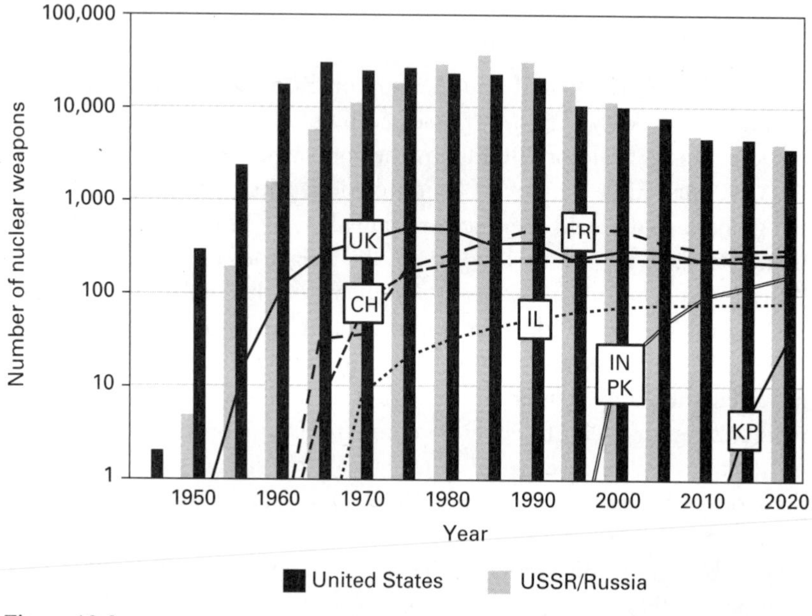

Figure 12.2
Evolution of nuclear arsenals, as measured by the number of nuclear weapons plotted on the vertical axis in decades (that is, logarithmically) and on the horizontal axis in five-year increments starting in 1945. Note the dominance by the United States and Russia (plotted with columns). The substantial decrease in nuclear weapons (by a factor of 10, given the logarithmic scale) since the 1980s is a result of arms control treaties we'll discuss in chapter 18. Arsenals of rivals India and Pakistan track each other, so they're shown as a double line. Also clear are the outlier status of Israel (IL) and North Korea (KP), and the fact that France (FR) and China (CH) have roughly the same number of nuclear weapons while the U.K. has about 30 percent fewer—but still 200 highly lethal weapons. (Data source: SIPRI Yearbook 2018)

letter to President Franklin Roosevelt, carrying Albert Einstein's signature. The letter was actually the idea of three Hungarian refugee physicists—Leo Szilard, Eugene Wigner, and Edward Teller—and the American businessman Alexander Sachs, an associate of Roosevelt's. The Hungarians informed Einstein of the fission work and prepared draft letters; Einstein's signature lent his prestige and probably contributed, along with $E = mc^2$, to the mistaken view of Einstein as a progenitor of the bomb.[2]

Despite the urgings of famous scientists, including also the Danish physicist Niels Bohr, the Allied governments and their military establishments were slow to act. Germany launched a more vigorous nuclear weapons program, and the Japanese military began nuclear explorations. In the Soviet Union, a nuclear weapons program begun in 1939 made only slow progress during the war years. Meanwhile, basic research on nuclear fission continued at laboratories in the United States and England. It soon became apparent that only the rare isotope uranium-235 could sustain a fission chain reaction, but that the greater effectiveness of slow neutrons in causing U-235 fission might nevertheless permit a slow-neutron chain reaction in uranium with U-235 at its low natural abundance. This possibility led to the search for effective moderators that would make a chain reaction possible in natural uranium. The American efforts soon focused primarily on graphite, while, fatefully, the Germans chose heavy water. In December 1942, the Italian-born physicist Enrico Fermi and his team at the University of Chicago achieved the first self-sustaining chain reaction in their uranium-and-graphite "pile" (which we showed in figure 7.6).

A moderated reaction employing slow neutrons could never release its energy fast enough for the catastrophic explosiveness of a weapon. It became obvious that nuclear fission bombs must be fast-neutron devices, highly enriched in U-235 so the less effective fast neutrons could sustain the chain reaction. By mid-1941, a British group had arrived at a rough design for a simple U-235 fission bomb. The proposed weapon was to use 25 pounds of U-235 and would have an explosive yield estimated at 1.8 million pounds (just under 1 kt). Although different in some details, the British design anticipated the larger Hiroshima bomb.

Because a uranium bomb must contain a high proportion of U-235, it seemed that uranium enrichment—separation of the uranium isotopes—was essential to the bomb project, although it might not be necessary for a reactor. Experimental enrichment schemes produced small quantities of enriched uranium in the laboratory, but large-scale production of enough bomb-grade uranium seemed a daunting task.

Another realization dawned on nuclear scientists in the early 1940s: the element with atomic number 94, not yet named and not occurring naturally, might be fissile. In early 1941, University of California chemist Glenn Seaborg first identified element 94 and began producing microscopic samples. Within months, Seaborg and his colleagues had strong evidence that element 94 was fissile. In 1942, Seaborg named the new element *plutonium*, after what was then considered the solar system's outermost planet (since downgraded to a dwarf planet). Now there were two approaches to a nuclear fission weapon: separate enough fissile uranium-235 from natural uranium, or produce fissile plutonium in a moderated reactor powered by natural uranium.

By 1942, the U.S. nuclear weapons program was finally moving, with efforts underway at the University of Chicago, Berkeley, Columbia University, and elsewhere. The following year saw the establishment of major government centers for nuclear weapons work: a huge uranium-enrichment complex at Oak Ridge, Tennessee (recall figure 9.3); plutonium-production reactors at Hanford, Washington; and the famous weapons laboratory at Los Alamos, New Mexico. Army general Leslie Groves oversaw the entire operation, now code-named the Manhattan Engineer District (widely shortened to "Manhattan Project"). Groves chose Berkeley physicist J. Robert Oppenheimer for the project's scientific leader (figure 12.3). Oppenheimer brought many of the world's leading scientists to the hastily assembled, isolated laboratory in the mountains of northern New Mexico.

Figure 12.3
J. Robert Oppenheimer and General Leslie Groves at the Trinity test site. Metal rods are all that remain of the 100-foot tower on which the first nuclear device exploded. (Los Alamos National Laboratory)

The Los Alamos scientists pursued two bombs: a simpler U-235 device and a more sophisticated plutonium weapon. By the summer of 1944, intelligence had determined that a German nuclear bomb was highly unlikely, but the Manhattan Project pushed on. Early in 1945, sizable quantities of bomb-grade uranium from Oak Ridge and plutonium from Hanford began arriving at Los Alamos. On May 8, 1945, Germany's unconditional surrender ended the war in Europe. At Los Alamos, momentum continued toward the first nuclear test, and scientists met with government officials to advise them on the possible use of the bomb against Japan.

By July 1945, the uranium and plutonium bomb designs were complete. The scientists were confident that the uranium device would work if any nuclear weapon would, and in any event they had only enough enriched uranium for one bomb, so they chose to test the plutonium design. In the early morning of July 16, at the Trinity site in southern New Mexico, the first nuclear weapon was detonated, with a yield equivalent to nearly 20,000 tons of TNT (figure 12.4). For many who witnessed it, the Trinity explosion ushered in a new and different world. Oppenheimer is quoted saying: "We waited until the blast had passed, walked out of the shelter and then it was extremely solemn. We knew the world would not be the same. ... I remembered the line from the Hindu scripture, the Bhagavad-Gita: ... 'Now I am become Death, the destroyer of worlds.'"[3]

Figure 12.4
Fireball of the Trinity test, shown just 0.016 second after detonation. Height of the fireball at this point is about 200 meters or 650 feet. (Berlyn Brixner/Los Alamos National Laboratory)

The successful test spurred a vigorous debate among government, military, and scientific leaders. Should a demonstration explosion be staged to persuade the Japanese to surrender? Or should the bomb be used against the enemy? Should they be warned? Should the target be military or civilian? Would dropping the bomb save the lives of American soldiers, poised for a bloody invasion of Japan? And what about the Soviet Union? Might a quick, nuclear end to the war forestall Soviet involvement? President Harry Truman made the nuclear choice, probably motivated to end the war as quickly as possible, and the bomb was dropped—not once, but twice. The uranium bomb Little Boy exploded over Hiroshima on August 6, and the plutonium bomb Fat Man devastated Nagasaki three days later.

Many nuclear scientists shared Oppenheimer's view that a world with nuclear weapons "would not be the same," and they fervently believed their creation would put an end to war and lead to an era of openness and global cooperation. Their view did not prevail. In the climate of worsening relations with the Soviet Union, military and political opinion embraced the notion that nuclear superiority would make the United States invulnerable. The idea of a fusion-based "superbomb," first suggested by Enrico Fermi and Edward Teller early in World War II, gained new support after the Soviet Union's 1949 fission test. By the mid-1950s, both the United States and the Soviet Union had fusion weapons with yields in the 10-megaton range—nearly 1,000 times that of the Hiroshima bomb. The nuclear arms race was underway, and for several decades nuclear weapons proliferated, in number and in variety, within and beyond the American and Soviet arsenals. Since the 1980s, those arsenals have shrunk (recall fig 12.2), but proliferation is ongoing as additional states acquire nuclear weapons and the variety of weapons and delivery systems continues to grow.

How Do Nuclear Weapons Work?

How do nuclear weapons achieve their devastating explosive yields? How easy is it to make a nuclear bomb? How are "atomic" and "hydrogen" bombs different? These nuclear questions are best not left just to scientists, since they have fundamental implications for global security. Your survival hinges on the nuclear weapons policies of your government and of other governments, and to understand and influence those policies it helps to know how nuclear weapons work, how they're made, and what they can do.

Fundamentally, of course, nuclear weapons exploit the nuclear difference—that million-fold energy advantage of nuclear over chemical reactions. A nuclear weapon simply entails a rapid nuclear reaction. The nuclear difference ensures that even a small quantity of nuclear material can yield enormous explosive energy. The fission and fusion reactions we considered in previous chapters in regard to power generation are also suitable for weapons. For the remainder of this chapter, we'll take a closer look at those weapons.

Fission Weapons

In principle, making a fission weapon is simple: Just assemble a supercritical mass of fissile material, and a chain reaction will grow rapidly as each generation of neutrons begets more and more additional fission. You could do that by bringing together two subcritical chunks of fissile material. The trick is to assemble them fast (less than 1 millisecond for highly enriched uranium); otherwise, energy released in the initial fission events will drive the chunks apart and halt the reaction.

An obvious and simple approach is the **gun-type fission weapon**, in which a uranium-235 "bullet" is fired into a U-235 "target" (figure 12.5). Surrounding the fissioning mass with a heavy *tamper* holds the exploding

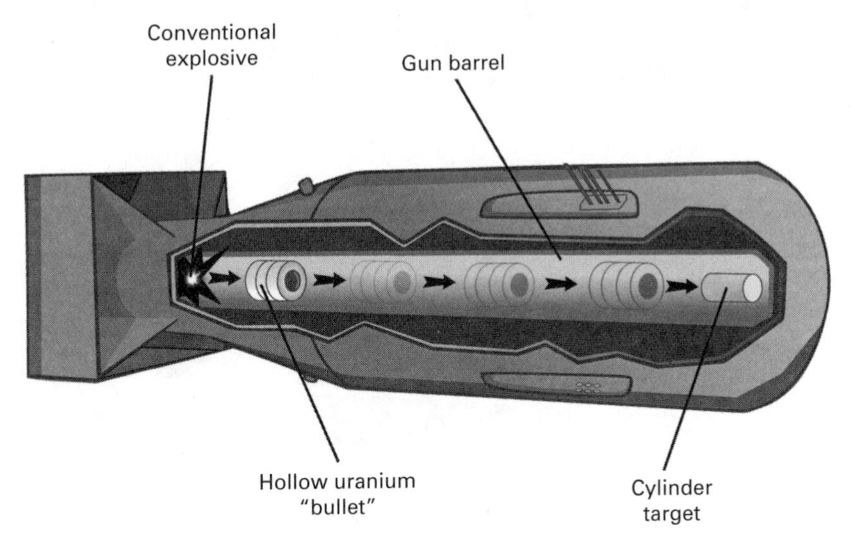

Figure 12.5
Diagram of a gun-type fission weapon. Both the bullet and the target are subcritical masses of uranium-235. Igniting the gunpowder fires the bullet into the target, creating a critical mass that undergoes a fission explosion. A heavy steel tamper holds the exploding mass together briefly, ensuring more complete fission. (Wikimedia Commons, CC BY-SA 3.0)

mass together for a fleeting instant, allowing more complete fission before the weapon blows itself apart. That's basically all there is to it.

Figure 12.6 shows Little Boy, the gun-type fission weapon dropped on Hiroshima. Despite the fact that it destroyed a city and killed 200,000 people, Little Boy was not particularly efficient. The huge, 9,700-pound bomb contained just over 140 pounds (64 kilograms) of uranium, enriched to an average 83 percent U-235. This material was so scarce that varying enrichments were used for different components of the bomb. Of the fissile U-235, less than 2 pounds (1 kilogram) actually underwent fission. The gun design's simplicity made it a good choice for the first combat nuclear weapon, but because of its inefficient use of scarce fissile material it fell out of favor among weapons designers. In addition, gun-type fission bombs tend to be unsafe, in that having so much fissile material makes the weapon vulnerable to stray neutrons that can accidentally trigger the bomb or change conditions to inadvertently make the uranium go critical. Furthermore, the gun-type bomb is triggered by firing its chemical high

Figure 12.6
Little Boy, the gun-type fission bomb that destroyed Hiroshima. The bomb weighed 9,700 pounds. Of that, only about 140 pounds was fissile uranium-235, and of this less than 2 pounds actually underwent fission. (Los Alamos National Laboratory)

explosive at just a single point rather than the more complex multiple-point initiation used with implosion weapons, discussed next.

The world may not have seen its last gun-type fission weapon, however. Its extreme simplicity makes the gun design a logical choice for first-time nuclear nations or terrorist groups. The only significant barrier to the construction of a gun-type fission bomb is the acquisition of highly enriched uranium. As physicist and Nobel laureate Luis Alvarez warned: "With modern weapons-grade uranium, the background neutron rate is so low that terrorists, if they had such material, would have a good chance of setting off a high-yield explosion simply by dropping one half of the material onto the other half. Most people seem unaware that if separate HEU [highly enriched uranium] is at hand it's a trivial job to set off a nuclear explosion ... even a high school kid could make a bomb in short order."[4] The actual amount of HEU can be somewhat less than Little Boy's 140 pounds and in any event is roughly equivalent to a grapefruit-sized uranium sphere. In chapter 16, we'll consider the risk posed by terrorist groups manufacturing nuclear weapons.

What starts the chain reaction in a fission bomb? We addressed this question abstractly in chapter 5, noting that stray neutrons from cosmic-ray interactions or from relatively rare spontaneous fission events guarantee a chain reaction in a supercritical mass. But a bomb requires precise timing, down to a fraction of a millionth of a second. Will there be enough neutrons just as the subcritical pieces come together? Or will there be too many, initiating the chain reaction before the pieces are fully joined? That would *preignite* the bomb, blowing it apart before much fission could occur and greatly reducing its explosive yield. **Preignition** could also be hazardous to a first-time bomb maker experimenting with fissile material, for it blurs the transition to criticality and could result in unanticipated detonation.

For uranium bombs, the available neutrons are sufficient to start the chain reaction in a gun-type design, without being so prolific as to cause preignition. But this isn't true of plutonium, for a subtle reason that we alluded to earlier. In chapter 9, you saw how plutonium bred in fission reactors is a mix of mostly fissile Pu-239 with smaller quantities of Pu-240; the longer nuclear fuel sits in a reactor, the greater the portion of Pu-240. But Pu-240 happens to undergo spontaneous fission at a rate 67,000 times greater than Pu-239, giving rise to neutrons that make preignition a real problem in plutonium fission weapons. This is why light-water reactor fuel, typically three years in a reactor, is a poor source of plutonium for weapons—but, as you saw in chapter 9 and as we'll discuss more in chapter 16, it can still be used as the explosive fuel of a nuclear weapon. This is also why continuously refueled reactors, such as the Canadian CANDU or the Soviet RBMK, have greater potential for making weapons-grade plutonium. The difference, though, has sometimes been exaggerated with the statement that light-water reactors are "safe" because they can't make bomb-grade plutonium. Preignition would reduce the yield of a bomb made from reactor-grade plutonium, but the device would still be a decidedly *nuclear* weapon with a substantial explosive yield. A report from the American Nuclear Society has made this sobering point abundantly clear: "We are aware that a number of well-qualified scientists in countries that have not developed nuclear weapons question the weapons-usability of reactor-grade plutonium. While recognizing that explosives have been produced from this material, many believe that this is a feat that can be accomplished only by an advanced nuclear-weapon state such as the United States. *This is not the case. Any nation or group capable of making a nuclear explosive from weapons-grade plutonium must be considered capable of making one from reactor-grade plutonium.*"[5] (Italics added.)

Even with weapons-grade plutonium, the preignition problem requires more rapid assembly of the critical mass than is possible with the gun-type design. The alternative, originated in 1943 and used in nearly every nuclear weapon today, involves surrounding a subcritical plutonium sphere with conventional (that is, chemical) high explosives. When fired, the explosives compress the plutonium to a higher density, making it supercritical and initiating the nuclear explosion. The inward compression of the plutonium core gives this means of nuclear ignition its name: the **implosion method.**

The challenge with implosion is to compress the plutonium uniformly from all directions. In practice, detonators (32 in the Fat Man design) are placed around the sphere and detonated simultaneously to compress the inner plutonium sphere. Figure 12.7 illustrates the workings of an implosion bomb, and the box describes a major challenge and how it was overcome.[6]

Several improvements greatly increase the yield and the reliability of implosion-type fission weapons. An *initiator* supplies extra neutrons at the precise moment of detonation. Surrounding the plutonium core are materials that reflect neutrons back into the fissioning mass, increasing the efficiency of fission. Surrounding materials may also act as a tamper, holding the plutonium together briefly to ensure more complete fission. If that tamper is depleted uranium, it may also add to the weapon's yield as

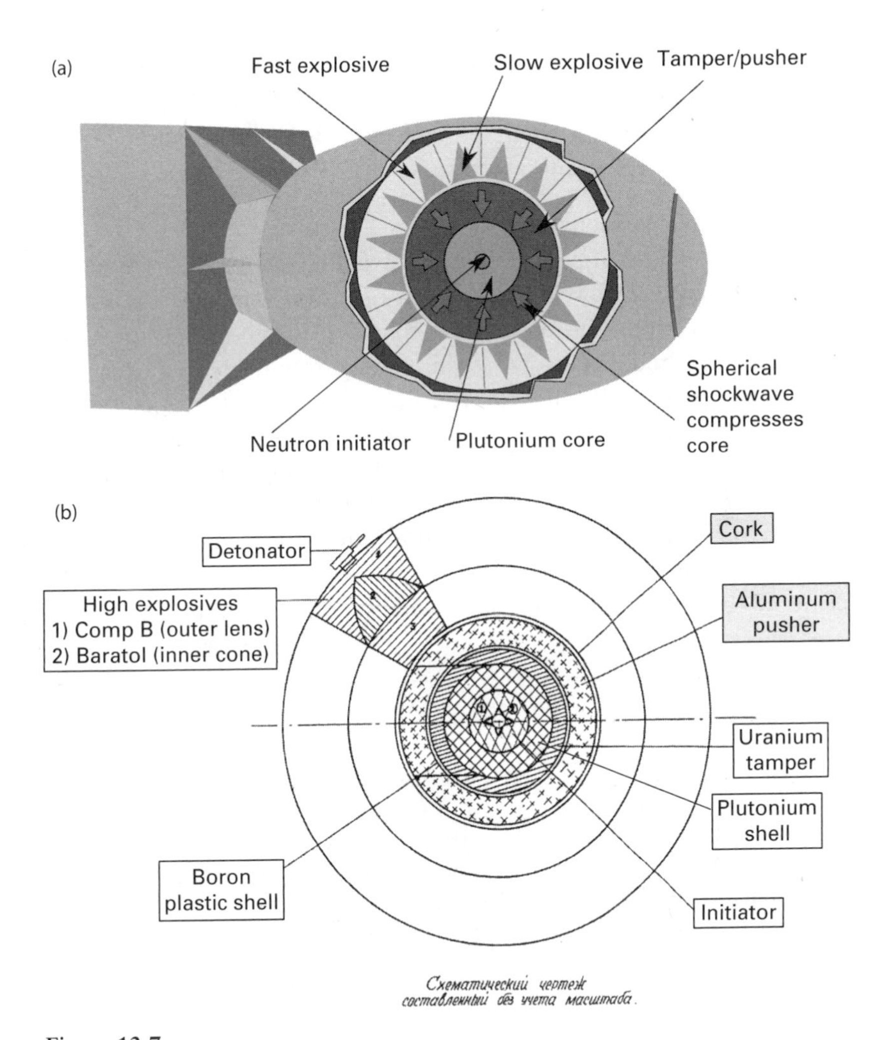

Figure 12.7
In the implosion method, high explosives compress a plutonium sphere to supercritical density. In practice, a tamper would surround the plutonium. (a) Simplified diagram of a complete bomb of the Fat Man type. (Wikimedia Commons, CC BY-SA 2.5.) (b) Drawing of an implosion bomb leaked by Soviet spies embedded in the Manhattan Project and published in a Soviet book. Grayed-out labels designate bomb components whose functions aren't discussed here. Only one sector of high explosive is shown; additional sectors surround the core to provide spherically symmetric compression. (Reproduced by permission of Alex Wellerstein; see note 6 for source details.)

faster neutrons from the fissioning plutonium are able to induce fission in U-238 (that's right: fissionable but not fissile U-238). In fact, the Trinity and Fat Man bombs got about 30 percent of their yield from U-238 fission in their depleted uranium tampers.[7]

Boosted Fission Weapons

Another technique employed in today's fission explosives was developed in the early 1950s in anticipation of fusion weapons. Recall that the deuterium-tritium fusion reaction considered in chapter 11 produces helium, energy, and a neutron. Getting energy from fusion isn't easy; you saw that in chapter 11 when we considered the feasibility of controlled fusion for electric power generation, and in the next section you'll see how the inventors of fusion weapons cleverly surmounted that difficulty. But the neutrons from fusion can be useful even when the energy isn't, as they help sustain *fission* reactions. In a so-called fusion-boosted fission weapon, a small quantity of deuterium-tritium mixture at the center of the imploding plutonium reaches such a high temperature that it

Explosive Lenses

Picture a spherical shell of plutonium surrounded by chemical high explosives with detonators on the outside. The problem is that shock waves from detonation of the high explosive travel spherically from the detonators, like ripples from a rock thrown into a pond. When the shock waves reach the plutonium, they'll strike at the same number of points as the number of detonators—which won't compress the plutonium symmetrically. Manhattan Project scientists were perplexed by this fundamental problem. In the official history of the project they're cited as worrying that they might never get a working plutonium design. The implosion group's team leader, Ukrainian-American scientist George Kistiakowsky, famously predicted that if the test of the implosion bomb were to fail, he would "go nuts" and "be locked up."

Then James Tuck, a talented scientist from the British mission who was working with Hungarian mathematician John von Neumann, hit on a solution. Tuck's concept was based on the fact that shock waves from explosions travel at different speeds in different chemical explosives, making it possible to steer the waves so that, instead of hitting the fissile plutonium at only specific points, they "hug" the plutonium to compress it. The scientists knew that just as light changes direction when it enters another medium (that's how the lenses in your glasses and eyes work, and why a straw looks bent where it enters water), so will the shock waves change direction when they enter different explosive materials. Using two types of explosive material to compress the plutonium made the explosives act as lenses, shaping

Explosive Lenses (*continued*)

the shock waves (see figure 12.8) so the compression is even across the surface of the plutonium. The two explosives with different shock-wave speeds were known as Comp B (speed 7.8 km/second) and Baratol (speed 5 km/second). This explosive-lens innovation was a game changer, allowing plutonium to be used in nuclear weapons.

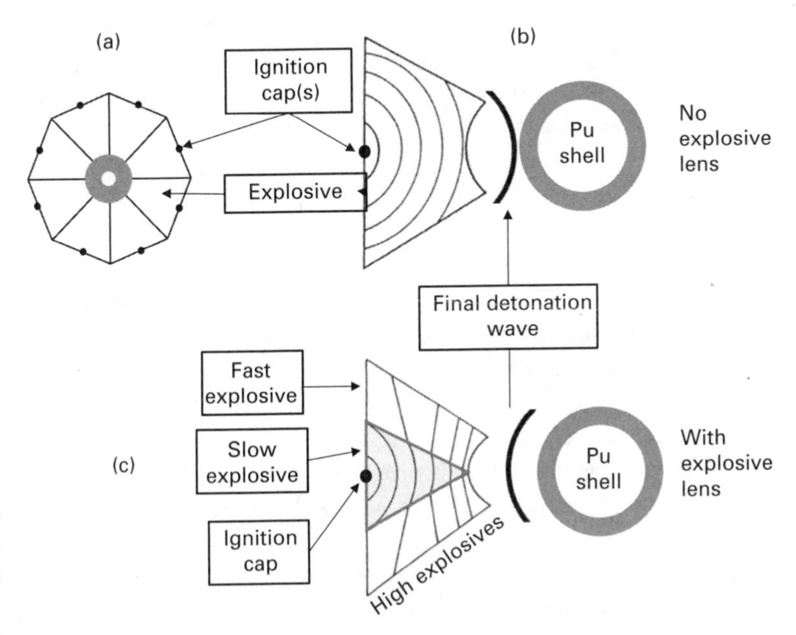

Figure 12.8
Simplified depiction of the explosive lens "trick" to symmetrically compress a spherical shell of plutonium. In the top panel (a and b) the case without explosive lenses is shown. Here a shock wave travels spherically from the detonation point (the ignition cap) and hits the plutonium shell at one point. However, in the bottom panel (c), two explosives are used, one that transmits a shock wave very rapidly and one more slowly. In the end, the original spherical wave is shaped so it compresses the plutonium uniformly.

undergoes fusion. The resulting burst of neutrons causes more complete fission of the surrounding plutonium, greatly increasing the weapon's explosive yield. Together with the other enhancements discussed above, fusion boosting helps modern fission explosives utilize far more of their fissile material than the Hiroshima bomb's roughly 1 percent. Fusion-boosted weapons, however, require a source of tritium, which is usually produced in specialized reactors. In the United States, tritium for weapons was recently produced in civilian power reactors owned by the Tennessee Valley Authority—a clear compromise of the U.S. goal of keeping civilian and military nuclear activity separate.

Why all the fuss about fission weapons, tritium-boosted or otherwise? Aren't today's nuclear weapons "hydrogen bombs," and doesn't that make them fusion weapons? Generally, yes. The larger weapons in today's nuclear arsenals do employ fusion, but all fusion weapons require fission triggers to generate the enormous temperature necessary to ignite thermonuclear fusion. You can't have a fusion weapon without fission. And, as we'll see in the next section, most of today's nuclear weapons are actually complex hybrids whose explosive yield generally involves both fission and fusion.

Fusion Weapons

In September 1941, Enrico Fermi and Edward Teller were strolling in New York when Fermi suddenly asked if the fission bomb they were beginning to conceive might be used to ignite nuclear fusion. Teller became obsessed with the notion of a fusion bomb, an obsession that decades later held him among the United States' foremost advocates of continuing nuclear weapons development. A modest program of theoretical fusion research continued at Los Alamos through the war years and advanced slowly after the war. The Soviet Union's first fission explosion, in late 1949, intensified debate on the moral and practical issues surrounding the development of fusion superbombs. Early in 1950, the U.S. government authorized a full-scale fusion effort.

Igniting nuclear fusion proved complicated. Blowing up a fission bomb amidst fusion fuel would just disperse the fusion material. The problem was to channel the immense energy of the fission bomb to ignite fusion without the fission explosion's simultaneously destroying the fusion fuel. The solution came in a famous meeting in 1951 between Edward Teller and talented Polish mathematician Stanislaw Ulam. Ulam realized, as a recent and revealing book has quoted, that "radiation at sufficiently high temperature is a 'substance' with remarkable properties. It can flow like a liquid and then push like a giant steel piston ... radiation that can be channeled until it surrounds a container of thermonuclear fuel and implodes it."[8] Teller realized that it would be possible to compress fusion fuel using X-rays emitted from the fission bomb, heating it sufficiently to ignite nuclear fusion. This Teller-Ulam invention is at the heart of today's **thermonuclear weapons**. Although the invention remains classified, journalist Howard Morland pieced together most of the details from unclassified sources in the late 1970s. After a celebrated legal case in which the government sought to halt publication of Morland's findings, the "H-bomb secret" was published in *The Progressive* in 1979.[9]

The Teller-Ulam weapon—the "hydrogen bomb"—involves a carefully orchestrated sequence of events, including both fission and fusion reactions. Independently, the prominent Soviet physicist and dissident Andrei Sakharov came up with the same concept, known in Russia as Sakharov's Third Idea. Figure 12.9 is a schematic diagram of the entire device. At one end is the fission trigger known as the *primary*—a fusion-boosted fission bomb as described in the preceding section. The entire trigger is about the size of a soccer ball, and it alone probably has an explosive yield comparable to the Hiroshima bomb. When it explodes, the fission trigger produces intense X-rays that travel ahead of its physical blast, vaporizing the plastic foam that occupies most of the inside of the weapon. Since the X-rays and gamma rays spread out in all directions

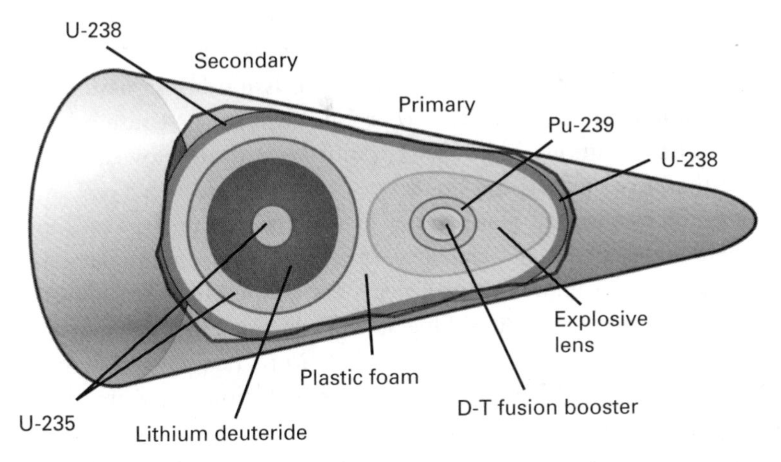

Figure 12.9
Diagram of a W-88 thermonuclear warhead, which uses the Teller-Ulam configuration. X-rays from the fusion-boosted fission primary are absorbed and reemitted by the plastic foam, compressing the lithium deuteride and uranium-235, causing the latter to fission. Compression heats the fusion fuel, while fission neutrons convert lithium to tritium. Deuterium-tritium fusion ensues, giving the weapon roughly half its explosive yield. Neutrons from fusion reactions cause fission in the uranium-238 that surrounds the entire package, producing most of the remaining explosive energy and nearly all the radioactive fallout.

from the fission trigger, the nearer side of the fusion secondary would normally be more compressed than the far side. But the secondary needs to be compressed uniformly to reach sufficient temperature for fusion. Modern weapons likely use a type of plastic foam that absorbs X-rays and reemits them in all directions to achieve uniform compression.

The complex secondary structure probably consists of a uranium-235 shell surrounding solid lithium deuteride (a blend of lithium and deuterium) that in turn surrounds a U-235 sphere. X-rays reemitted by the vaporized foam compress this entire structure, initiating fission in the U-235. Neutrons from fission convert lithium to tritium in the lithium deuteride, and the high temperature from compression and fission ignites deuterium-tritium fusion. The neutrons produced in this reaction convert more lithium to tritium, and D-T fusion quickly engulfs the mass of fusion fuel, giving the weapon about half its explosive yield. But the bang isn't over yet. High-energy neutrons from D-T fusion now induce fission in the uranium-238 jacket, producing the remaining explosive energy and nearly all the weapon's radioactive fallout. Fission in U-238? Shouldn't that read U-235? No. U-238 will undergo fission when struck by neutrons of high enough energy. It won't sustain a chain reaction, but that isn't necessary when an external neutron source is available. In a thermonuclear weapon, that source is the D-T fusion reaction. So a "hydrogen bomb" is really a **fission-fusion-fission** bomb: fission initiates fusion, and fusion-produced neutrons cause additional fission. The invention of the Teller-Ulam mechanism allowed yields far higher than from any fission bomb.

In 1952, the United States tested its first thermonuclear weapon—a cumbersome device that weighed 62 tons and occupied an entire building. It yielded an impressive 10 megatons (some 600 Hiroshimas) and vaporized the Pacific island of Elugelab, leaving an underwater crater two miles wide and half a mile deep. By 1954, a practical design, deliverable by aircraft, yielded 15 megatons. The Soviet Union tested its first thermonuclear weapon in 1955, and within a few years both nations were deploying megaton-range thermonuclear weapons, first on bombers and later on missiles. Figure 12.10 shows a thermonuclear explosion.

In the race to develop new nuclear weapons, the Soviets were never far behind. That's to be expected given brilliant Soviet physicists such as Andrei Sakharov, Yakov Zeldovich, Vitaly Ginzburg, and Igor Kurchatov, who were on par with the best Western scientists.

Figure 12.10
A French thermonuclear test explosion in 1970 brightens the night sky over an uninhabited Pacific island. The weapon's explosive yield was just under 1 megaton. (CTBTO/Science Source)

Advances in Nuclear Weaponry

The development of thermonuclear weapons brought a thousand-fold increase in explosive yield over simpler fission weapons. The requirements for a critical chain reaction put a limit—probably several hundred kilotons—on the yield of fission weapons. In fact, the largest pure fission bomb ever detonated was the United States' Ivy King test in 1952, which yielded 550 kilotons or 0.55 megatons. But there's no such thing as a critical mass for fusion, so fusion-based weapons can be made arbitrarily large. In the early years of thermonuclear weaponry, both the United States and the Soviet Union experimented with very large weapons; in 1961, the Soviets exploded the largest ever, a 58-megaton device.[10] That bomb, known as the Tsar Bomba (king of bombs), would have been even larger, but for fear of devastating fallout the Soviets replaced some uranium with lead. Even so, the bomb's explosion broke windows more than 500 miles distant. The bomb's global fallout potential led Andrei Sakharov to become an advocate for banning atmospheric nuclear testing. (Sakharov also became a critic of the human rights situation in the Soviet Union, for which he was ultimately awarded the Nobel Peace Prize in 1975.)

In the 1950s, interest in so-called **tactical nuclear weapons** had led to small, relatively low-yield fission weapons that could be placed in artillery shells for battlefield use. By 1955, a 12-kiloton device was available that could fit into an 8-inch-diameter cannon. Later innovations included a hand-carried "special atomic demolition munition" with a subkiloton yield. A host of small-scale fission weapons proliferated, many of them deployed on the front lines of a divided Europe during the 45 years of the Cold War.

The development of intercontinental missiles in the late 1950s brought pressure to decrease the size and weight of thermonuclear weapons as well. This led to modern weapons whose ratio of explosive yield to weight is up to 1,000 times greater than that of the nuclear bombs used in World War II (see figure 12.11). The development of multiple-warhead missiles in the 1970s continued the emphasis on smaller warheads, causing an actual drop in the total explosive yield of the superpowers' nuclear

Figure 12.11
A model of the Nagasaki Fat Man bomb alongside the W-78 warhead used in Minuteman III missiles. Fat Man weighed nearly 11,000 pounds and yielded the equivalent of 22 kilotons of TNT. The W-78 weighs less than 800 pounds and yields 335 kilotons. Until 2014, some Minuteman III missiles carried three W-78 warheads; today they're single-warhead missiles. (Los Alamos National Laboratory)

arsenals. However, increased accuracy and the ability to target individual warheads meant there was no corresponding reduction in destructive capability.

Nuclear weapons can be tailored for special purposes. Incorporating materials that become highly radioactive when they absorb neutrons produces an abundance of lethal fallout. Replacing uranium-238 in a thermonuclear weapon with a nonfissionable material leads to a relatively "clean" bomb whose fallout is limited to fission products from the small fission trigger. The **neutron bomb**, or enhanced-radiation weapon (ERW), is a thermonuclear device designed to minimize blast and maximize lethal effects of high-energy neutrons produced in fusion. ERWs would be used against tanks, killing or disabling their occupants in a burst of neutron radiation without producing much physical damage. After a protracted moral debate, the United States developed enhanced-radiation weapons in the 1980s, but they were never deployed. They were retired in 1992, at the end of the Cold War, and subsequently dismantled. Another nuclear innovation is the variable-yield weapon (known as **dial-a-yield**), whose explosive violence is selected with the turn of a dial.

Trending: Low-Yield Nuclear Weapons

Nuclear weapons nowadays have a much lower yield than in the early decades of the nuclear era. Very few megaton-range weapons remain in the U.S. and Russian arsenals. There are several reasons why large-yield weapons are no longer considered necessary. Primary is the increased accuracy of weapons delivery. Quantitatively, accuracy is expressed as the **circular error probable** (CEP), which describes the minimum distance by which a warhead would miss its target half the time. If the CEP is 100 meters, then half the time a warhead would miss its target by more than 100 meters.

As accuracy improves, warheads with less yield are sufficient to destroy the target. Improving the CEP by a factor of 3 reduces the required yield by a factor of 25. For example, a new version of the B-61 gravity bomb, the B-61–12, will be have a CEP of 30 meters, compared with its predecessor's 100 meters. That allows a decrease in maximum yield to 50 kilotons—although the B-61–12 is customizable to much lower yields, increasing its strategic versatility.

In addition, a group of smaller weapons can destroy a larger area than the equivalent yield in one large weapon. But smaller yields don't necessarily make us safer. Some analysts argue that if yields get too small, then weapons will be perceived as more "usable," leading to greater chance of a nuclear exchange.

Hard Targets and Earth Penetrators

Nations often site high-value military facilities deep underground to protect from attack and to hide from satellites. The capability to destroy these facilities with conventional or nuclear weapons is a priority for adversaries.

Earth-penetrating warheads known as *bunker busters* operate essentially as powerful needles that penetrate the ground and then detonate at the desired depth. Higher impact speed results in deeper penetration. These warheads, even those with conventional explosives, may be encased in depleted uranium (nearly all U-238, left over from enrichment), which is heavy and stronger than steel. This construction ensures that, even under the immense pressure of impact, the warhead will penetrate, intact, deep into the ground.

A nuclear version of the bunker buster, the B-61–11, can penetrate 5 or 6 meters (15 to 20 feet) and subsequently detonate to destroy underground targets. These bombs don't have to tunnel as deep as their targets, because energy from the explosion produces a seismic wave that greatly magnifies the explosive effect—making a 100-kiloton explosion feel like 2 megatons underground. Penetrating deeper than 1 or 2 meters doesn't greatly amplify the effect, so the warhead doesn't have to drill deep to have a devastating effect.

Because bunker busters don't go deep underground, they can cause massive casualties on the surface, especially if weapons detonate near urban areas. Nuclear weapons tests, in contrast, are conducted at depths of hundreds of meters, not the several meters associated with bunker busters (more on nuclear testing in chapter 19).

Where Are We Now?

The good news is that, as of 2020, there's a marked decrease in the sheer number of nuclear weapons worldwide even compared to a few years ago. In just a year, the world's arsenals dropped by 600 nuclear weapons, about 4 percent and the equivalent of many World War IIs. But don't celebrate just yet: 2,000 of the remaining weapons are on high alert, ready for launch on a moment's notice. More reductions in the U.S. and Russian arsenals are possible as long as the 2011 Treaty on Measures for the Further Reduction and Limitation of Strategic Offensive Arms (New START) is followed, but the treaty is up for renewal and no serious effort is currently underway for a follow-on treaty. We'll discuss arms control treaties further in chapter 18.

Worrying Trends

The Federation of American Scientists (FAS) notes that from 2010 to 2018 the United States declared publicly the size of its nuclear stockpile, but as of 2019 the Trump administration has decided to not disclose this information. As Hans Kristensen of the FAS has said: "This is curious since the Trump administration had repeatedly complained about secrecy in the Russian and Chinese arsenals. Instead, it now appears to endorse their secrecy." Russia has also not publicly declared its nuclear weapons although it does reveal their numbers to the U.S. government as obligated under the New START treaty. A further concern is that while the total number of nuclear weapons is decreasing in the United States and Russia, both countries' weapons and delivery systems are being upgraded (more in chapter 14). Rivals India and Pakistan continue to produce more fissile material and to increase their nuclear weapons inventory. China has announced new nuclear weapons delivery systems and is increasing the size of its nuclear arsenal. Finally, North Korea is likely increasing its nuclear weapons inventory. As long as nuclear weapons aren't abolished outright, the history of these weapons is unfortunately still being written.

What about nuclear testing? Since the 1990s, no one has tested nuclear weapons except for North Korea. Most countries have observed a testing moratorium since 1996, except for North Korea and, in 1998, India and Pakistan—although in 2020 the Trump administration reportedly discussed a possible resumption of U.S. nuclear testing. The testing moratorium is a challenge because countries like the United States need to verify that their nuclear weapons are safe, secure, and effective. A dedicated program known as the *Science Based Stockpile Stewardship* has redirected the priorities of the U.S. nuclear weapons labs to simulate nuclear weapons with computers and physics-based experiments but without actual nuclear explosions. We'll discuss nuclear testing and stewardship further in chapter 19.

Summary

You've now had a quick survey of the history of nuclear weapons, and you understand the scientific basis of their operation. That history, beginning in the 1940s, saw growth in the world's nuclear arsenal to the equivalent of 3,000 World War IIs at the height of the Cold War, subsequently dropping to about 300 WWIIs. The offspring of the simple fission weapons that ended World War II have multiplied in number, variety, explosive yield, and countries of ownership. Today's arsenals contain not only kiloton-range fission devices but also thermonuclear weapons employing a complex sequence of fission and fusion to deliver yields into the megaton range. Smaller, more versatile, and more accurate nuclear weapons add to uncertainty about the likelihood of their use. The existence and continuing production of nuclear weapons pose grave moral and political questions.

Glossary

circular error probable (CEP) The minimum distance from a weapon's target beyond which the weapon will miss the target half the time.

dial-a-yield Option available on some nuclear weapons allowing the explosive yield of a nuclear weapon to be varied.

fission-fusion-fission weapon A high-yield nuclear weapon in which fission initiates fusion, whence fusion-generated neutrons cause additional fission.

gun-type fission weapon A nuclear fission weapon in which a subcritical U-235 "bullet" is fired into a U-235 target to form a critical mass, which then undergoes an explosive chain reaction.

implosion method A technique for detonating a nuclear weapon that uses conventional explosives to compress a subcritical sphere of fissile material to critical density.

kiloton (kt) A unit of explosive yield equivalent to that of 1,000 tons of the chemical explosive TNT.

megaton (Mt) A unit of explosive yield, equivalent to that of 1 million tons of the chemical explosive TNT, or 1,000 kilotons.

neutron bomb A nuclear weapon designed to produce relatively little blast damage but to kill people with intense neutron radiation.

preignition Premature detonation of a nuclear weapon due to excess neutrons, usually from spontaneous fission. Preignition blows apart the fissile material before it undergoes complete fission, greatly reducing explosive yield.

tactical nuclear weapon A nuclear weapon designed for use on the battlefield, typically characterized by small size and modest explosive yield.

thermonuclear weapon A nuclear weapon involving thermonuclear fusion—the fusion of light nuclei at high temperature to produce heavier nuclei and energy.

Notes

1. See Arthur H. Westing, "Misspent Energy: Munition Expenditures Past and Future: The World Arsenal of Nuclear Weapons," *Bulletin of Peace Proposals* 16 (1985): 9.

2. These and other details of nuclear history are chronicled in Richard Rhodes's book *The Making of the Atomic Bomb* (New York: Simon & Schuster, 1986) and its 2011 reissue.

3. J. Robert Oppenheimer, quoted in Len Giovannitti and Fred Freed, *The Decision to Drop the Bomb* (New York: Coward-McCann, 1965), 197.

4. Luis Alvarez, *Adventures of a Physicist* (New York: Basic Books, 1987), 125.

5. American Nuclear Society, *Special Report on the Protection and Management of Plutonium*, 1995.

6. Material No. 464 (January 28, 1946), Document 338 in L. D. Riabev, ed., *Atomnyi proekt SSSR: Documenti i materiali*, vol. 2, book 6 (Sarov, Russia: VNIIEF, 2006), 823–829, diagram on 829. As sourced on Alex Wellerstein, "Soviet Drawings of an American Bomb," *Restricted Data: The Nuclear Secrecy Blog*, November 30, 2012, http://blog.nuclearsecrecy.com/2012/11/30/soviet-drawings-of-an-american-bomb/. Labels suggested by Alex Wellerstein; reproduced with permission of Alex Wellerstein.

7. See B. Cameron Reed, "Composite Cores and Tamper Yield: Lesser-Known Aspects of Manhattan Project Fission Bombs," *American Journal of Physics* 88, no. 2 (February 2020). Recall that U-238, while not fissile, can fission when struck by neutrons of sufficiently high energy.

8. Kenneth Ford, *Building the H Bomb: A Personal History* (Singapore: World Scientific Publishing, 2015), 8.

9. Howard Morland, "The H-Bomb Secret," *The Progressive* 43 (November 1979): 14.

10. In 2020, Russia released previously classified video showing preparations for the Tsar Bomba test and the subsequent explosion. It's available at https://www.youtube.com/watch?v=XJhZ3i-HXS0.

Further Reading

Federation of American Scientists Nuclear Notebook. https://fas.org/issues/nuclear-weapons/nuclear-notebook/. This website reports the inventories of nuclear weapons worldwide. The team at FAS annually assesses the numbers of nuclear weapons through open source intelligence. The FAS website is the main clearinghouse for information on nuclear weapons and their delivery systems. Follow FAS on Twitter at @FAScientists.

Ford, Kenneth W. *Building the H Bomb: A Personal History*. Singapore: World Scientific Publishing, 2015. This is a personal account of the development of the H-bomb, including technical issues on the bomb's construction. The U.S. Department of Energy asked Ford to redact portions of the book, claiming it contained government secrets, but Ford refused and it was published as he wrote it.

Gerber, Michele. *On the Home Front: The Cold War Legacy of the Hanford Nuclear Site*, 3rd ed. Lincoln, NE: Bison, 2007. The history of nuclear weapons includes the distressing environmental legacy of early weapons production. In *On the Home Front*, Gerber leverages her work on the National Academy of Sciences Committee on Declassification to provide an authoritative account plutonium production and its aftermath at Hanford, Washington.

Kelly, Cynthia, ed., and Richard Rhodes. *The Manhattan Project: The Birth of the Atomic Bomb in the Words of Its Creators, Eyewitnesses, and Historians*, rev. ed. New York: Black Dog and Leventhal, 2020. A compendium of works on the Manhattan Project by writers from all walks of life.

Montillo, Roseanne. *Atomic Women: The Untold Stories of the Scientists Who Helped Create the Nuclear Bomb*. New York: Little, Brown, 2019. You've seen the crucial role that Lise Meitner played in the discovery of fission, but here, in Montillo's book for teenage readers, are other women who contributed to the development of nuclear weapons.

Reed, B. Cameron. *The Physics of the Manhattan Project*, 3rd ed. Heidelberg: Springer, 2014. This book is part nuclear physics textbook and part history. It describes in a great detail the Little Boy and Fat Man bombs, and also features useful technical appendices.

Rhodes, Richard. *Making of the Atomic Bomb*. New York: Simon & Schuster, 1986; reissued 2011. A Pulitzer Prize–winning history of nuclear weapons through the early 1950s. A gripping blend of history and technology. Although a classic, Rhode's book missed out on the extensive declassification of initially secret nuclear history that occurred in the 1990s.

Serber, Robert. *The Los Alamos Primer: The First Lectures on How to Build an Atomic Bomb, Updated with a New Introduction by Richard Rhodes*. Berkeley: University of California Press, 2020. Robert Serber was Oppenheimer's protégé on the Manhattan Project. In his *Primer*, which was declassified in 1965 and published as a book in 1992, Serber gives a first-person account of the physics behind the development of the fission bomb. The *Primer* derived from lectures on nuclear weapons given to new personnel when they arrived at Los Alamos.

Walker, Samuel J. *Prompt and Utter Destruction: Truman and the Use of Atomic Bombs Against Japan*, 3rd ed. Chapel Hill: University of North Carolina Press, 2016. The former historian of the Nuclear Regulatory Commission gives an authoritatively researched account of the decision to drop nuclear bombs on Japan. Walker draws on recently released materials from both Japan and the U.S. to provide a thorough and nuanced view of Truman's consequential nuclear choice.

U.S. Department of Defense. *The Nuclear Matters Handbook*. https://www.acq.osd.mil/ncbdp/nm//NMHB/index.htm (2016 version). This is an introduction to the concepts and terms of the U.S. nuclear weapons program.

U.S. Department of Energy, Office of History and Heritage Resources. *The Manhattan Project: An Interactive History*. This website lets users explore eras and events associated with the project. Interactive is a relative term! Those used to flashy websites with true interactivity may not be impressed with the presentation, but the history is solid. https://www.osti.gov/opennet/manhattan-project-history/Events/1945/retrospect.htm.

Wellerstein, Alex. *Restricted Data: The Nuclear Secrecy Blog*. http://blog.nuclearsecrecy.com/. Alex Wellerstein is a well-known historian of science and nuclear weapons and a professor at the Stevens Institute of Technology. He is also the creator of NUKEMAP. This blog began in 2011. For more, follow on Twitter @wellerstein.

13

Effects of Nuclear Weapons

What can nuclear weapons do? How do they achieve their destructive purpose? What would a nuclear war be like? In this chapter, we will explore these and related questions that reveal the most horrifying manifestations of the nuclear difference.

A Bomb Explodes: Short-Term Effects

The preceding chapter outlined the carefully orchestrated sequence of events involved in the detonation of a modern nuclear weapon. Those events are over in a millionth of a second, but their effects continue for seconds, minutes, hours, days, and even weeks and more. The most immediate effect of a nuclear explosion is an intense burst of nuclear radiation, primarily gamma rays and neutrons. This **direct radiation** is produced in the weapon's nuclear reactions themselves, and lasts well under a second. Lethal direct radiation extends nearly a mile from a 10-kiloton explosion. With most weapons, though, direct radiation is of little significance because other lethal effects generally encompass greater distances. An important exception is the enhanced-radiation weapon, or neutron bomb, which maximizes direct radiation and minimizes other destructive effects.

An exploding nuclear weapon instantly vaporizes itself. What was cold, solid material microseconds earlier becomes a gas hotter than the Sun's 15-million-degree core. This hot gas radiates its energy in the form of X-rays, which heat the surrounding air. A **fireball** of superheated air forms and grows rapidly; 10 seconds after a 1-megaton explosion, the fireball is a mile in diameter. The fireball glows visibly from its own heat—so visibly that the early stages of a 1-megaton fireball are many times brighter than the Sun even at a distance of 50 miles (figure 13.1). Besides light, the glowing fireball radiates heat.

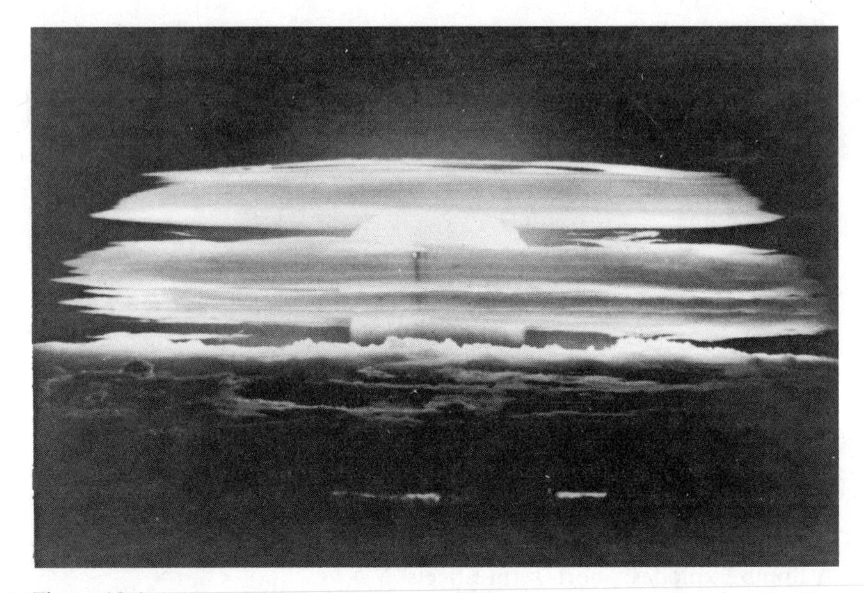

Figure 13.1
The fireball of a megaton-range nuclear explosion, photographed from 50 miles away. (U.S. Department of Energy/National Atomic Museum)

This **thermal flash** lasts many seconds and accounts for more than one-third of the weapon's explosive energy. The intense heat can ignite fires and cause severe burns on exposed flesh as far as 20 miles from a large thermonuclear explosion. Two-thirds of injured Hiroshima survivors showed evidence of such flash burns (figure 13.2). You can think of the incendiary effect of thermal flash as analogous to starting a fire using a magnifying glass to concentrate the Sun's rays. The difference is that rays from a nuclear explosion are so intense that they don't need concentration to ignite flammable materials.

As the rapidly expanding fireball pushes into the surrounding air, it creates a **blast wave** consisting of an abrupt jump in air pressure. The blast wave moves outward initially at thousands of miles per hour but slows as it spreads. It carries about half the bomb's explosive energy and is responsible for most of the physical destruction. Normal air pressure is about 15 pounds per square inch (psi). That means every square inch of your body or your house experiences a force of 15 pounds. You don't usually feel that force, because air pressure is normally exerted equally in all directions, so the 15 pounds pushing a square inch of your body one way is counterbalanced by 15 pounds pushing the other way. What you do feel is **overpressure**, caused by a greater air pressure on one side of an

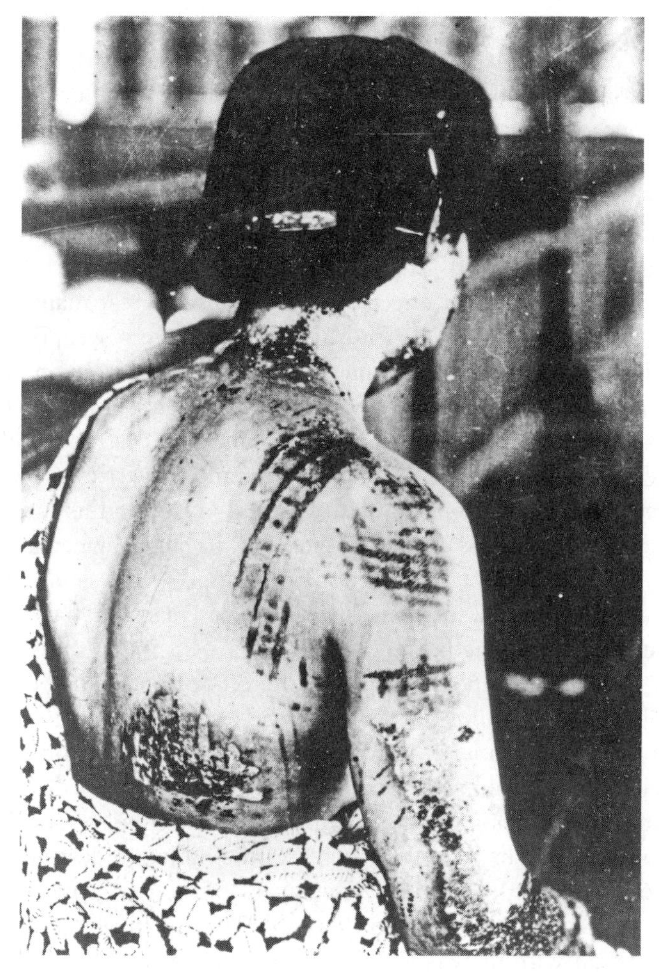

Figure 13.2
Flash burns on a Hiroshima victim's skin. The skin wasn't burned where dark clothing blocked the thermal flash. (Defense Nuclear Agency)

object. If you've ever tried to open a door against a strong wind, you've experienced overpressure. An overpressure of even 1/100 psi could make a door almost impossible to open. That's because a door has lots of square inches—about 3,000 or more. So 1/100 psi adds up to a lot of pounds. The blast wave of a nuclear explosion may create overpressures of several psi many miles from the explosion site. Think about that! There are about 50,000 square inches in the front wall of a modest house—and that means 50,000 pounds or 25 tons of force even at 1 psi overpressure.

Overpressures of 5 psi are enough to destroy most residential buildings, as shown dramatically in figure 13.3. An overpressure of 10 psi collapses most factories and commercial buildings, and 20 psi will level even reinforced concrete structures.

People, remarkably, are relatively immune to overpressure itself. But they aren't immune to collapsing buildings or to pieces of glass hurtling through the air at hundreds of miles per hour or to having themselves hurled into concrete walls—all of which are direct consequences of a blast wave's overpressure. Blast effects therefore cause a great many fatalities (figure 13.4). Blast effects depend in part on where a weapon is detonated. The most widespread damage to buildings occurs in an **air burst**, a detonation thousands of feet above the target. The blast wave from an air burst reflects off the ground, which enhances its destructive power. A **ground burst**, in contrast, digs a huge crater and pulverizes everything in the immediate vicinity, but its blast effects don't extend as far. Nuclear attacks on cities would probably employ air bursts, whereas ground bursts would be used on hardened military targets such as underground missile silos. As you'll soon see, the two types of blasts have different implications for radioactive fallout.

How far do a weapon's destructive effects extend? That distance—the **radius of destruction**—depends on the explosive yield. The volume encompassing a given level of destruction depends directly on the weapon's yield. Because volume is proportional to the radius cubed, that means the destructive radius grows approximately as the cube root of the yield. A 10-fold increase in yield then increases the radius of destruction by a factor of only a little over 2. The area of destruction grows faster but still not in direct proportion to the yield. That relatively slow increase in destruction with increasing yield is one reason why multiple smaller weapons are more effective than a single larger one. Twenty 50-kiloton warheads, for example, destroy nearly three times the area leveled by a numerically equivalent 1-megaton weapon.

What constitutes the radius of destruction also depends on the level of destruction you want to achieve. Roughly speaking, though, the distance at which overpressure has fallen to about 5 psi is a good definition of destructive radius. Many of the people within this distance would be killed, although some wouldn't. But some would be killed beyond the 5-psi distance, making the situation roughly equivalent to having everyone within the 5-psi circle killed and everyone outside surviving. Figure 13.5 shows how the destructive zone varies with explosive yield for a hypothetical explosion. This is a simplified picture; a more careful calculation

Figure 13.3
Destruction of a wood-frame house during a nuclear test in the 1950s. First the house is illuminated by visible radiation from the fireball; note the shadow cast by the house as it blocks the bomb's light. Almost immediately the house begins to smolder from the intense heat of the thermal flash. Two seconds later the blast wave arrives, bringing total destruction. (U.S. Department of Energy)

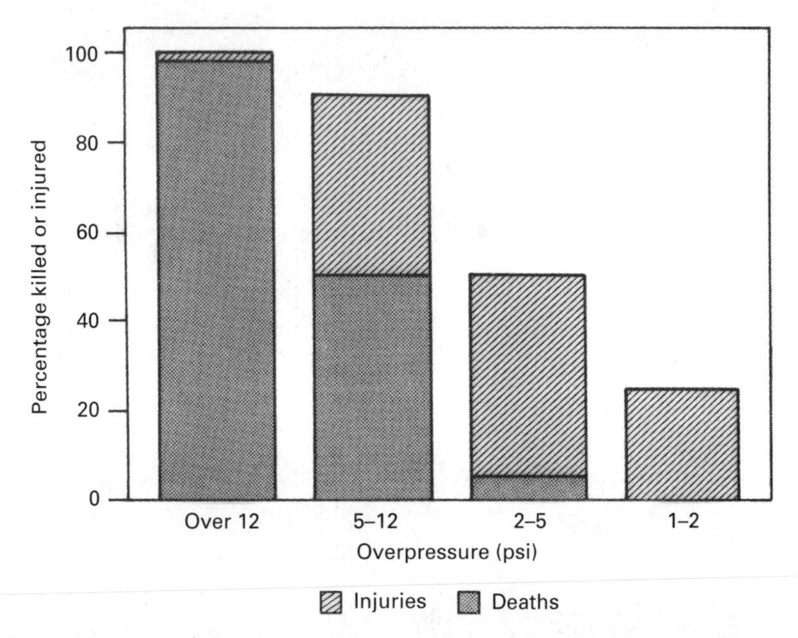

Figure 13.4
Injuries and fatalities caused by blast from a nuclear explosion. The four categories represent different ranges of blast-wave overpressure; for a 1-megaton weapon, the 5–12 psi overpressure zone would extend over four miles from ground zero. (U.S. Congress Office of Technology Assessment, *The Effects of Nuclear War*)

of the effects of nuclear weapons on entire populations requires detailed simulations that include many environmental and geographic variables.

The blast wave is over in a minute or so, but the immediate destruction may not be. Fires started by the thermal flash or by blast effects still rage, and under some circumstances they may coalesce into a single gigantic blaze called a **firestorm** that can develop its own winds and thus cause the fire to spread. Hot gases rise from the firestorm, replaced by air rushing inward along the surface at hundreds of miles per hour. Winds and fire compound the blast damage, and the fire consumes enough oxygen to suffocate any remaining survivors.

During World War II, bombing of Hamburg with incendiary chemicals resulted in a firestorm that claimed 45,000 lives. The nuclear bombing of Hiroshima resulted in a firestorm; that of Nagasaki did not, likely because of Nagasaki's rougher terrain. The question of firestorms is important not only to the residents of a target area: firestorms might also have significant long-term effects on the global climate, as we'll discuss later in this chapter.

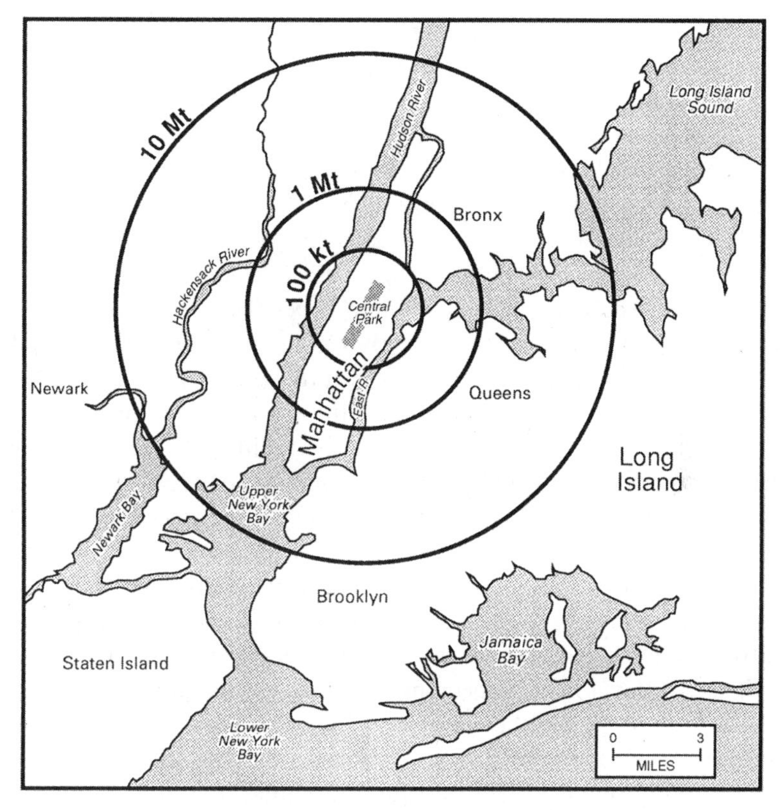

Figure 13.5
Destructive radii of 100-kiloton, 1-megaton, and 10-megaton weapons superimposed on a map of the New York City area. The destructive radius is defined as the distance within which blast overpressure exceeds 5 pounds per square inch, and it measures 2 miles, 4.4 miles, and 9.4 miles for the weapon yields shown. These values assume air-burst explosions at optimum altitudes over Central Park. (U.S. Office of Technology Assessment, *The Effects of Nuclear War*)

Fallout

Both nuclear and conventional weapons produce destructive blast effects, although of vastly different magnitudes. But radioactive **fallout** is unique to nuclear weapons. Fallout consists primarily of fission products, although neutron capture and other nuclear reactions contribute additional radioactive material. The term *fallout* generally applies to those isotopes whose half-lives exceed the time scale of the blast and other short-term effects. Although fallout contamination may linger for years and even decades, the dominant lethal effects last from days to weeks, and contemporary civil defense recommendations are for survivors to stay inside for at least 48 hours while the radiation decreases.

The fallout produced in a nuclear explosion depends greatly on the type of weapon, its explosive yield, and where it's exploded. The neutron bomb, although it produces intense *direct* radiation, is primarily a fusion device and generates only slight fallout from its fission trigger. Small fission weapons like those used at Hiroshima and Nagasaki produce locally significant fallout. But the fission-fusion-fission design used in today's thermonuclear weapons introduces the new phenomenon of *global fallout*. Most of this fallout comes from fission of the U-238 jacket that surrounds the fusion fuel (recall figure 12.9). The global effect of these huge weapons comes partly from the sheer quantity of radioactive material and partly from the fact that the radioactive cloud rises well into the stratosphere, where it may take months or even years to reach the ground. Even though we've had no nuclear war since the bombings of Hiroshima and Nagasaki, fallout is one weapons effect with which we have experience. Atmospheric nuclear testing before the 1963 Partial Test Ban Treaty resulted in detectable levels of radioactive fission products across the globe, and some of that radiation is still with us.

Fallout differs greatly depending on whether a weapon is exploded at ground level or high in the atmosphere. In an air burst, the fireball never touches the ground, and radioactivity rises into the stratosphere. This reduces local fallout but enhances global fallout. In a ground burst, the explosion digs a huge crater and entrains tons of soil, rock, and other pulverized material into its rising cloud. Radioactive materials cling to these heavier particles, which drop back the ground in a relatively short time. Rain may wash down particularly large amounts of radioactive material, producing local hot spots of especially intense radioactivity. A hot spot in Albany, New York, thousands of miles from the 1953 Nevada test that produced it, exposed area residents to some 10 times their annual background radiation dose. The exact distribution of fallout depends crucially on wind speed and direction; under some conditions, lethal fallout may extend several hundred miles downwind of an explosion. However, it's important to recognize that the lethality of fallout quickly decreases as short-lived isotopes decay.

Recommended Response to a Nuclear Explosion

The United States government has recently provided guidance on how to respond to a nuclear detonation.[1] One recommendation is to divide the region of destruction due to blast effects into three separate damage zones. This division provides guidance for first responders in assessing the

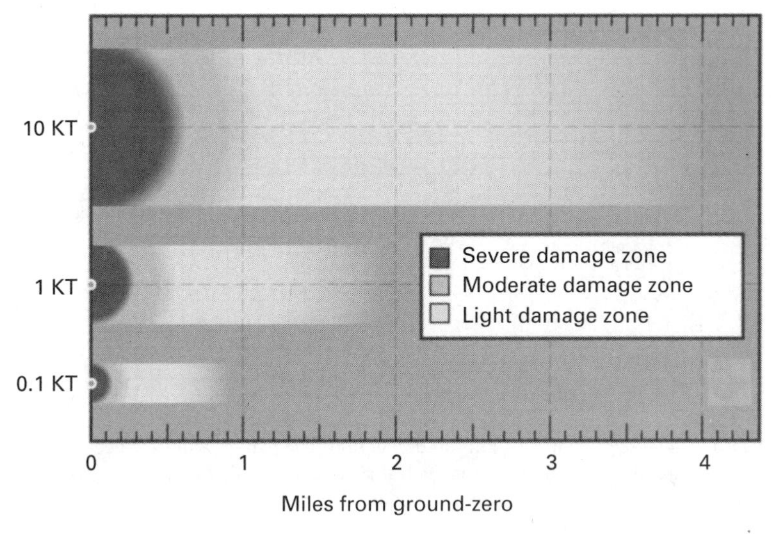

Figure 13.6
Radii of damage zones for 0.1, 1, and 10 kiloton explosions.

situation. Outermost is the *light damage zone*, characterized by "broken windows and easily managed injuries." Next is the *moderate damage zone* with "significant building damage, rubble, downed utility lines and some downed poles, overturned automobiles, fires, and serious injuries." Finally, there's the *severe damage zone*, where buildings will be completely collapsed, radiation levels high, and survivors unlikely (see figure 13.6).[2]

The recommendations also define a *dangerous fallout zone* spanning different structural damage zones (figure 13.7).[3] This is the region where dose rates exceed a whole-body external dose of about 0.1 Sv/hour. First responders must exercise special precautions as they approach the fallout zone in order to limit their own radiation exposure. The dangerous fallout zone can easily stretch 10 to 20 miles (15 to 30 kilometers) from the detonation depending on explosive yield and weather conditions.

Electromagnetic Pulse

A nuclear weapon exploded at very high altitude produces none of the blast or local fallout effects we've just described. But intense gamma rays knock electrons out of atoms in the surrounding air, and when the explosion takes place in the rarefied air at high altitude this effect may extend hundreds of miles. As they gyrate in Earth's magnetic field, the electrons generate an intense pulse of radio waves known as an **electromagnetic pulse** (EMP).

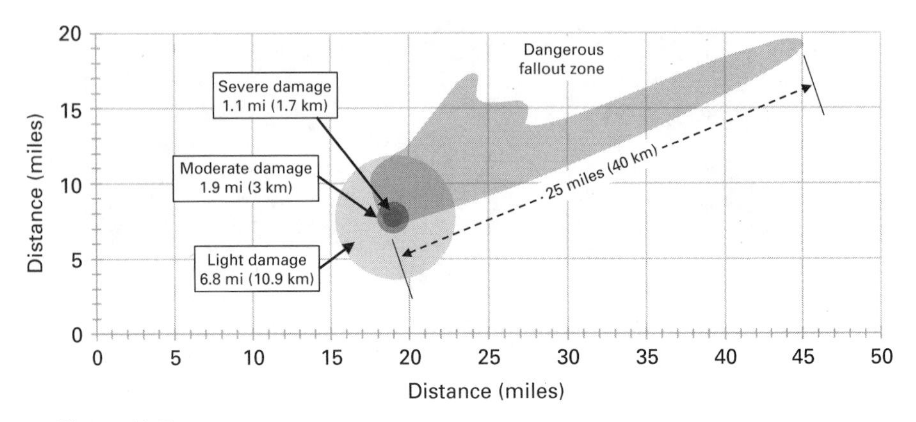

Figure 13.7
Dangerous fallout zone for a 10-kiloton detonation. The pattern will vary depending on wind speed and direction, and the zone will shrink substantially over a day or so. Also shown are the structural damage zones of figure 13.6.

A single large weapon exploded some 200 miles over the central United States could blanket the entire country with an electromagnetic pulse intense enough to damage computers, communication systems, and other electronic devices. It could also affect satellites used for military communications, reconnaissance, and attack warning. The EMP phenomenon thus has profound implications for a military that depends on sophisticated electronics. In 1962, the United States detonated a 1.4-megaton warhead 250 miles above Johnston Island in the Pacific Ocean. People as far as Australia and New Zealand witnessed the explosion as a red aurora appearing in the night sky. Hawaiians, only 800 miles from the island, experienced a bright flash followed by a green sky and the failure of hundreds of street lights. In total, the Soviet Union and the United States conducted 20 tests of EMP from nuclear detonations. However, it's unclear how to extrapolate the results to today's more sensitive and more pervasive electronic equipment.

Since the Partial Test Ban Treaty of 1963 it has been virtually impossible to study EMP effects directly, although elaborate devices have been developed to mimic the electronic impact of nuclear weapons. Increasingly, crucial electronic systems are "hardened" to minimize the impact of EMP. Nevertheless, the use of EMP in a war could wreak havoc with systems for communication and control of military forces.

Many countries are around the world are developing high-powered microwave weapons which, although not nuclear devices, are designed to produce EMPs. These *directed-energy weapons*, also called *e-bombs*, emit large pulses of microwaves to destroy electronics on missiles, to stop cars, to detonate explosives remotely, and to down swarms of drones. Despite these EMP weapons being nonlethal in the sense that there's no bang or blast wave, an enemy may be unable to distinguish their effects from those of nuclear weapons.

Would the high-altitude detonation of a nuclear weapon to produce EMP or the use of a directed-beam EMP weapon be an act of war warranting nuclear retaliation? With its electronic warning systems in disarray, should the EMPed nation launch a nuclear strike on the chance that it was about to be attacked? How are nuclear decisions to be made in a climate of EMP-crippled communications? These are difficult questions, but military strategists need to have answers.

Nuclear War

So far we've examined the effects of single nuclear explosions. But a nuclear war would involve hundreds to thousands of explosions, creating a situation for which we simply have no relevant experience. Despite decades of arms reduction treaties, there are still thousands of nuclear weapons in the world's arsenals. Detonating only a tiny fraction of these would cause mass casualties.

What would a nuclear war be like? When you think of nuclear war, you probably envision an all-out holocaust in which adversaries unleash their arsenals in an attempt to inflict the most damage. Many people—including your authors—believe that misfortune to be the likely outcome of almost any use of nuclear weapons among the superpowers. But nuclear strategists have explored many scenarios that fall short of the all-out nuclear exchange. What might these limited nuclear wars be like? Could they really remain *limited*?

Limited Nuclear War

One form of limited nuclear war would be like a conventional battlefield conflict but using low-yield tactical nuclear weapons. Here's a hypothetical scenario: After its 2014 annexation of Crimea, Russia attacks a Baltic country with tanks and ground forces while the United States is distracted by a domestic crisis. NATO responds with decisive counterforce, destroying Russian tanks with fighter jets, but this doesn't quell Russian resolve. Russia responds with even more tanks and by bombing NATO installations, killing several hundred troops. NATO cannot tolerate such aggression and to prevent further Russian advance launches

Hawaii in the Crosshairs

On January 13, 2018, during the height of a missile crisis when President Trump tweeted that hostile action by North Korea will be "met with fire and fury like the world has never seen," Hawaiians awoke to a terrifying alert: "Ballistic missile threat inbound to Hawaii. Seek immediate shelter. This is not a drill." (Figure 13.8.)

It's not unrealistic to expect that Hawaii might be in the crosshairs of North Korean missiles. After all, the island of Kauai hosts the Pacific Missile Range—the very site that tested antimissile missiles that would be used to shoot down North Korean missiles. Well-informed residents knew that it takes only 20 to 25 minutes from a North Korean launch to impact on a Hawaiian target. For 38 long minutes, the missile alert terrified the residents of Hawaii. People didn't know whether to seek shelter or to just remain with their loved ones to wait for the inevitable.

Figure 13.8
Screenshot of the missile alert sent to cell phones across Hawaii by the Hawaii Emergency Management Agency. (Ryan Ozawa)

Hawaii in the Crosshairs (*continued*)

In the end, the alert was a false alarm. The reason for the error is still uncertain, but the Hawaii Emergency Management Agency claims that an unstable operator at their emergency department inadvertently pressed a button. However, the employee has claimed that he was told the scenario was real and that he responded accordingly.

We may never know the true chain of events, but there's still a lot we can learn from this event. First, there should never have been just one person authorized to release the statewide alert. Second, it was far too easy for the operator to issue the alert using software with no easy way to opt out. Finally, the fact that it took more than 38 minutes to follow up with a false alarm response shows a serious problem. While it was easy for the Hawaii Emergency Management Agency to blame the employee who pushed the button, the blame also lies with the agency for a system riddled with flaws.

The Hawaii incident is just one of many examples of false alarms in nuclear emergency alert and warning systems. Most of these never reached the public, so we aren't as aware of them. Daniel Ellsberg, of Pentagon Papers fame and himself a former military analyst, has documented many such incidents,[4] as has the U.K.'s Royal Institute of International Affairs.[5]

low-yield tactical nuclear weapons with their *dial-a-yield* positions set to the lowest settings of only 300 tons TNT equivalent. The goal is to signal Russia that it has crossed a line and to deescalate the situation. NATO's actions are based on fear that if the Russian aggression weren't stopped the result would be all-out war in northern Europe.

This strategy is actually being discussed in the higher echelons of the Pentagon. The catchy concept is that use of a few low-yield nuclear weapons could show resolve, with the hoped-for outcome that the other party will back down from its aggressive behavior (this concept is known as *escalate to deescalate*). The assumption is that the nuclear attack would remain limited, that parties would go back to the negotiating table, and that saner voices would prevail. However, this assumes a chain of events where everything unfolds as expected. It neglects the incontrovertible fact that, as the Prussian general Carl von Clausewitz observed in the nineteenth century, "Three quarters of the factors on which action in war is based are wrapped in a fog of greater or lesser uncertainty."[6] Often coined *fog of war*, this describes the lack of clarity in wartime situations on which decisions must nevertheless be based. In the scenario described, sensors could have been damaged or lines of communication severed that would have reported the low-yield nature of the nuclear weapons. As a result, Russia might feel its homeland threatened and respond with an all-out attack using strategic nuclear weapons, resulting in millions of deaths.

There is every reason to believe that a limited nuclear war wouldn't remain limited. A 1983 war game known as Proud Prophet involved top-secret nuclear war plans and had as participants high-level decision makers including President Reagan's Secretary of Defense Caspar Weinberger. The war game followed actual plans but unexpectedly ended in total nuclear annihilation with more than half a billion fatalities in the initial onslaught—not including subsequent deaths from starvation. The exercise revealed that a limited nuclear strike may not achieve the desired results! In this case, that was because the team playing the Soviet Union responded to a limited U.S. nuclear strike with a massive all-out nuclear attack.

What about an attack on North Korea? In 2017, some in the U.S. cabinet advocated for a "bloody nose" strategy in dealing with North Korea's flagrant violations of international law. This is the notion that in response to a threatening action by North Korea, the U.S. would destroy a significant site to "bloody Pyongyang's nose." This might employ a low-yield nuclear attack or a conventional attack. The "bloody nose" strategy relies on the expectation that Pyongyang would be so overwhelmed by U.S. might that they would immediately back down and not retaliate. However, North Korea might see any type of aggression as an attack aimed at overthrowing their regime, and could retaliate with an all-or-nothing response using weapons of mass destruction (including but not necessarily limited to nuclear weapons) as well as their vast conventional force.

In September 2017, during the height of verbal exchanges between President Trump and the North Korean dictator Kim Jong-un, the U.S. flew B-1B Lancer bombers along the North Korean coast, further north of the demilitarized zone than the U.S. had ever done, while still staying over international waters. However, North Korea didn't respond at all, making analysts wonder whether the bombers were even detected. Uncertainty in North Korea's ability to discriminate different weapon systems might exacerbate a situation like this one and could lead the North Koreans viewing any intrusion as an "attack on their nation, their way of life and their honor." This is exactly how the Soviet team in the Proud Prophet war game interpreted it.

What about a limited attack *on* the United States? Suppose a nuclear adversary decided to cripple the U.S. nuclear retaliatory forces (a virtual impossibility, given nuclear missile submarines, but a scenario considered with deadly seriousness by nuclear planners). Many of the 48 contiguous states have at least one target—a nuclear bomber base, a submarine support base, or intercontinental missile silos—that would warrant destruction in such an attack. The attack, which would require only a tiny fraction of the strategic nuclear weapons in the Russian arsenal, could kill millions of civilians. Those living near targeted bomber and submarine bases would suffer blast and local radiation effects. Intense fallout from groundburst explosions on missile silos in the Midwest would extend all the way to the Atlantic coast. Fallout would also contaminate a significant fraction of U.S. cropland for up to year and would kill livestock. On the other hand, the U.S. industrial base would remain relatively unscathed, *if* no further hostilities occurred.

In contrast to attacking military targets, an adversary might seek to cripple the U.S. economy by destroying a vital industry. In one hypothetical attack considered by the congressional Office of Technology Assessment, ten Soviet SS-18 missiles, each with eight 1-megaton warheads, attack United States' oil refineries. The result is destruction of two-thirds of the U.S. oil-refining capability. And even with some evacuation of major cities in the hypothetical crisis leading to the attack, 5 million Americans are killed.

Each of these "limited" nuclear attack scenarios kills millions of Americans—many, many times the 1.2 million killed in all the wars in our nation's history. Do we want to entertain limited nuclear war as a realistic possibility? Do we believe nuclear war could be limited to "only" a few million casualties? Do we trust the professional strategic planners who prepare our possible nuclear responses to an adversary's threats? What level of nuclear preparedness do we need to deter attack? We'll examine these difficult nuclear questions further when we explore nuclear strategies in chapter 15.

All-Out Nuclear War

Whether from escalation of a limited nuclear conflict or as an outright full-scale attack, an all-out nuclear war remains possible as long as nuclear nations have hundreds to thousands of weapons aimed at one another. What would be the consequences of all-out nuclear war?

Within individual target cities, conditions described earlier for single explosions would prevail. (Most cities, though, would likely be targeted with multiple weapons.) Government estimates suggest that over half of the United States' population could be killed by the prompt effects of an all-out nuclear war. For those within the appropriate radii of destruction, it would make little difference whether theirs was an isolated explosion or part of a war. But for the survivors in the less damaged areas, the difference could be dramatic.

Consider the injured. Thermal flash burns extend well beyond the 5-psi radius of destruction. A single nuclear explosion might produce 10,000 cases of severe burns requiring specialized medical treatment; in an all-out war there could be several million such cases. Yet the United States has facilities to treat fewer than 2,000 burn cases—virtually all of them in urban areas that would be leveled by nuclear blasts. Burn victims who might be saved, had their injuries resulted from some isolated cause, would succumb in the aftermath of nuclear war. The same goes for fractures, lacerations, missing limbs, crushed skulls, punctured lungs, and myriad other injuries suffered as a result of nuclear blast. Where would be the doctors, the hospitals, the medicines, the equipment needed for their treatment? Most would lie in ruin, and those that remained would be inadequate to the overwhelming numbers of injured. Again, many would die whom modern medicine could normally save.

In an all-out war, lethal fallout would cover much of the United States (see figure 13.9). Survivors could avoid fatal radiation exposure only

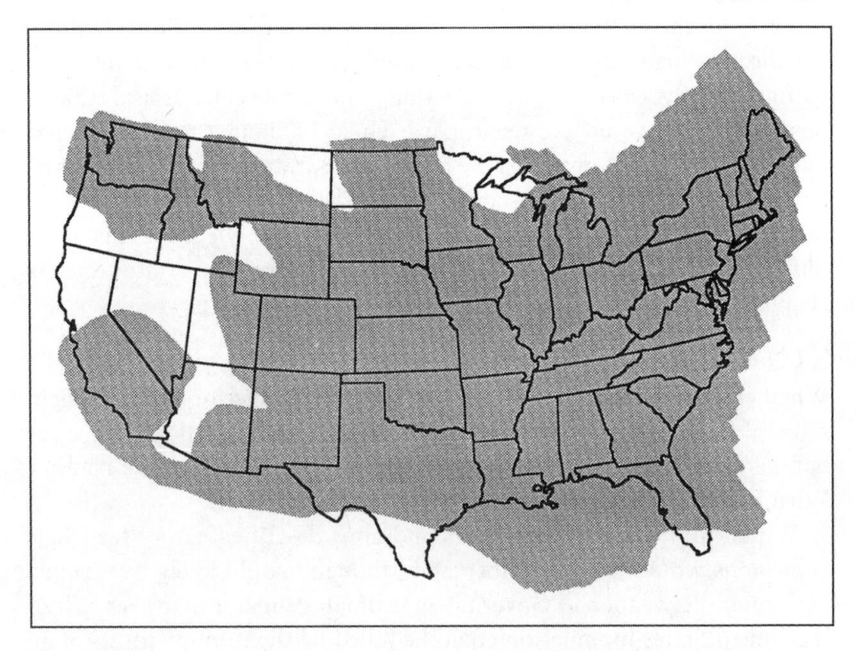

Figure 13.9
After all-out nuclear war, unsheltered persons in the shaded areas would receive weekly radiation doses in excess of 1 Sv (1,000 mSv or 100 rem), some 200 times annual background; 60 percent would receive fatal doses in excess of 10 Sv.

when sheltered with adequate food, water, and medical supplies. Even then, millions would be exposed to radiation high enough to cause lowered disease resistance and greater incidence of subsequent fatal cancer. Lowered disease resistance could lead to death from everyday infections in a population deprived of adequate medical facilities. And the spread of diseases from contaminated water supplies, nonexistent sanitary facilities, lack of medicines, and the millions of dead could reach epidemic proportions. Small wonder that the international group Physicians for Social Responsibility has called nuclear war "the last epidemic."

Attempts to contain damage to cities, suburbs, and industries would suffer analogously to the treatment of injured people. Firefighting equipment, water supplies, electric power, heavy equipment, fuel supplies, and emergency communications would be gone. Transportation into and out of stricken cities would be blocked by debris. The scarcity of radiation-monitoring equipment and of personnel trained to operate it would make it difficult to know where emergency crews could safely work. Most of all, there would be no healthy neighboring cities to call on for help; all would be crippled in an all-out war.

Is Nuclear War Survivable?

We've noted that more than half the United States' population might be killed outright in an all-out nuclear war. What about the survivors?

Recent studies have used detailed three-dimensional, block-by-block urban terrain models to study the effects of 10-kiloton detonations on Washington, D.C. and Los Angeles. The results settle an earlier controversy about whether survivors should evacuate or shelter in place: Staying indoors for 48 hours after a nuclear blast is now recommended. That time allows fallout levels to decay by a factor of 100. Furthermore, buildings between a survivor and the blast can block the worst of the fallout, and going deep inside an urban building can lower fallout levels still further (figure 13.10).[7] The same shelter-in-place arguments apply to survivors in the nonurban areas blanketed by fallout as shown in figure 13.9.

These new studies, however, consider only single detonations as might occur in a terrorist or rogue attack. We'll discuss those situations further in chapter 16. Here, in considering all-out nuclear war, we have to ask a further question: Then what?

Individuals might survive for a while, but what about longer term, and what about society as a whole? Extreme and cooperative efforts would be needed for long-term survival, but would the shocked and weakened survivors be up to those efforts? How would individuals react to watching their loved ones die of radiation sickness or untreated injuries? Would

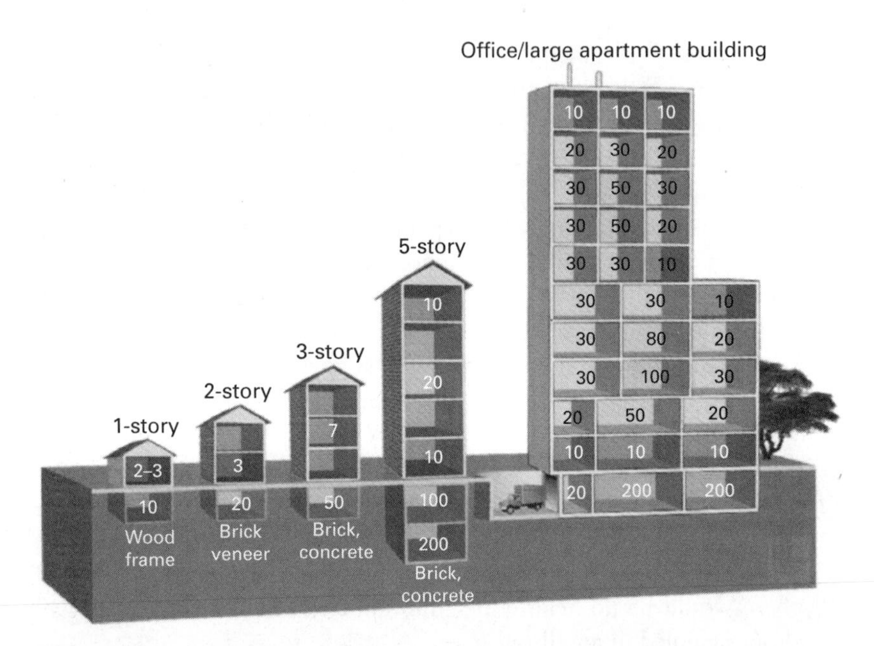

Figure 13.10
Fallout shielding by buildings. Numbers represent factors by which the radiation dose is reduced relative to outdoor exposure. A factor of 20, for example, means that a person in that area would receive 1/20th the dose of someone in the open.

an "everyone for themselves" attitude prevail, preventing the cooperation necessary to rebuild society? How would residents of undamaged rural areas react to the streams of urban refugees flooding their communities? What governmental structures could function in the postwar climate? How could people know what was happening throughout the country? Would international organizations be able to cope?

Some students of nuclear war see postwar society in a race against time. An all-out war would have destroyed much of the nation's productive capacity and would have killed many of the experts who could help guide social and physical reconstruction. The war also would have destroyed stocks of food and other materials needed for survival.

On the other hand, the remaining supplies would have to support only the much smaller postwar population. The challenge to the survivors would be to establish production of food and other necessities before the supplies left from before the war were exhausted. Could the war-shocked survivors, their social and governmental structure shattered, meet that challenge? That is a very big nuclear question—so big that it's best left unanswered, since only an all-out nuclear war could decide it definitively.

Climatic Effects

A large-scale nuclear war would pump huge quantities of chemicals and dust into the upper atmosphere. Humanity was well into the nuclear age before scientists took a good look at the possible consequences of this. What they found was not reassuring.

The upper atmosphere includes a layer enhanced in ozone gas, an unusual form of oxygen that vigorously absorbs the Sun's ultraviolet radiation. In the absence of this *ozone layer*, more ultraviolet radiation would reach Earth's surface, with a variety of harmful effects. A nuclear war would produce huge quantities of ozone-consuming chemicals, and studies suggest that even a modest nuclear exchange would result in unprecedented increases in ultraviolet exposure. Marine life might be damaged by the increased ultraviolet radiation, and humans could receive blistering sunburns. More UV radiation would also lead to a greater incidence of fatal skin cancers and to general weakening of the human immune system.

Even more alarming is the fact that soot from the fires of burning cities after a nuclear exchange would be injected high into the atmosphere. A 1983 study by Richard Turco, Carl Sagan, and others (the so-called TTAPS paper[8]) shocked the world with the suggestion that even a modest nuclear exchange—as few as 100 warheads—could trigger drastic global cooling as airborne soot blocked incoming sunlight. In its most extreme form, this **nuclear winter** hypothesis raised the possibility of extinction of the human species. (This is not the first dust-induced extinction pondered by science. Current thinking holds that the dinosaurs went extinct as a result of climate change brought about by atmospheric dust from an asteroid impact; indeed, that hypothesis helped prompt the nuclear winter research.)

The original nuclear winter study used a computer model that was unsophisticated compared to present-day climate models, and it spurred vigorous controversy among atmospheric scientists. Although not the primary researcher on the publication, Sagan lent his name in order to publicize the work. Two months before *Science* would publish the paper, he decided to introduce the results in the popular press. This backfired, as Sagan was derided by hawkish physicists like Edward Teller who had a stake in perpetuating the myth that nuclear war could be won and the belief that a missile defense system could protect the United States from nuclear attack. Teller called Sagan an "excellent propagandist" and suggested that the concept of nuclear winter was "highly speculative." The damage was done, and many considered the nuclear winter phenomenon discredited.

But research on nuclear winter continued. Recent studies with modern climate models show that an all-out nuclear war between the United States and Russia, even with today's reduced arsenals, could put over 150 million tons of smoke and soot into the upper atmosphere. That's roughly the equivalent of all the garbage the U.S. produces in a year! The result would be a drop in global temperature of some 8°C (more than the difference between today's temperature and the depths of the last ice age), and even after a decade the temperature would have recovered only 4°C. In the world's "breadbasket" agricultural regions, the temperature could remain below freezing for a year or more, and precipitation would drop by 90 percent. The effect on the world's food supply would be devastating.[9]

Even a much smaller nuclear exchange could have catastrophic climate consequences. The research cited above also suggests that a nuclear exchange between India and Pakistan, involving 100 Hiroshima-sized weapons, would shorten growing seasons and threaten annual monsoon rains, jeopardizing the food supply of a billion people. Figure 13.11 shows the global picture one month after this hypothetical 100-warhead nuclear exchange.[10]

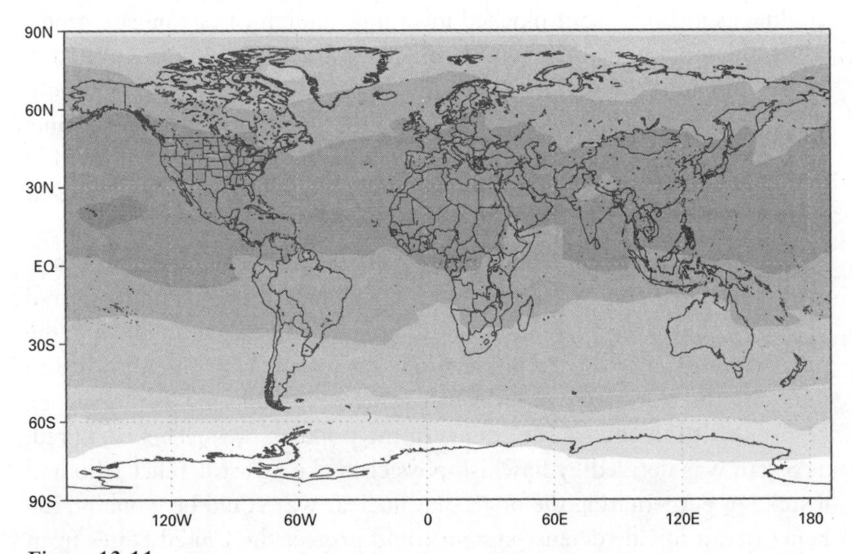

Figure 13.11
Global smoke distribution from a 100-warhead nuclear exchange between India and Pakistan one month after the event. Darker shading indicates greater sunlight absorption. This map assumes that 5 million tons of soot was injected into the air, almost the same mass as the Great Pyramid of Giza. (See note 10.)

Summary

Nuclear weapons have devastating effects. Destructive blast effects extend miles from the detonation point of a typical nuclear weapon, and lethal fallout may blanket communities hundreds of miles downwind of a single nuclear explosion. An all-out nuclear war would leave survivors with few means of recovery, and could lead to a total breakdown of society. Fallout from an all-out war would expose most of the belligerent nations' surviving populations to radiation levels ranging from harmful to fatal. And the effects of nuclear war would extend well beyond the warring nations, possibly including climate change severe enough to threaten much of the planet's human population.

Debate about national and global effects of nuclear war continues, and the issues are unlikely to be decided conclusively without the unfortunate experiment of an actual nuclear war. But enough is known about nuclear war's possible effects that there is near universal agreement on the need to avoid them. As the great science communicator and astronomer Carl Sagan once said, "It's elementary planetary hygiene to clean the world of these nuclear weapons." But can we eliminate nuclear weapons? Should we? What risks might such elimination entail? Those are the real issues in the ongoing debates about the future of nuclear weaponry.

Glossary

air burst A nuclear explosion detonated at an altitude—typically, thousands of feet—that maximizes blast damage. Because its fireball never touches the ground, an air burst produces less radioactive fallout than a ground burst.

blast wave An abrupt jump in air pressure that propagates outward from a nuclear explosion, damaging or destroying whatever it encounters.

direct radiation Nuclear radiation produced in the actual detonation of a nuclear weapon and constituting the most immediate effect on the surrounding environment.

electromagnetic pulse (EMP) An intense burst of radio waves produced by a high-altitude nuclear explosion, capable of damaging electronic equipment over thousands of miles.

fallout Radioactive material, mostly fission products, released into the environment by nuclear explosions.

fireball A mass of air surrounding a nuclear explosion and heated to luminous temperatures.

firestorm A massive fire formed by coalescence of numerous smaller fires.

ground burst A nuclear explosion detonated at ground level, producing a crater and significant fallout but less widespread damage than an air burst.

nuclear winter A substantial reduction in global temperature that might result from soot injected into the atmosphere during a nuclear war.

overpressure Excess air pressure encountered in the blast wave of a nuclear explosion. Overpressure of a few pounds per square inch is sufficient to destroy typical wooden houses.

radius of destruction The distance from a nuclear blast within which destruction is near total, often taken as the zone of 5-pound-per-square-inch overpressure.

thermal flash An intense burst of heat radiation in the seconds following a nuclear explosion. The thermal flash of a large weapon can ignite fires and cause third-degree burns tens of miles from the explosion.

Notes

1. National Security Staff Interagency Policy Coordination Subcommittee for Preparedness and Response to Radiological and Nuclear Threats, *Planning Guidance for Response to a Nuclear Detonation*, 2nd ed. (2010). Available at https://www.fema.gov/media-library-data/20130726-1821-25045-3023/planning_guidance_for_response_to_a_nuclear_detonation___2nd_edition_final.pdf.

2. Ibid.

3. Ibid., figure 1.6.

4. Daniel Ellsberg, *The Doomsday Machine: Confessions of a Nuclear War Planner* (New York: Bloomsbury, 2017).

5. Patricia Lewis, Heather Williams, Benoit Pelopidas, and Sasan Aghlani, *Too Close for Comfort: Cases of Near Nuclear Use and Options for Policy* (London: Chatham House, 2014).

6. Carl von Clausewitz, *On War*, reprint edition translated and edited by M. E. Howard and P. Paret (Princeton, NJ: Princeton University Press, 1989). (Original published in 1873.)

7. National Security Staff Interagency Policy Coordination Subcommittee for Preparedness and Response to Radiological and Nuclear Threats, *Planning Guidance for Response to a Nuclear Detonation*, 2nd ed. (2010).

8. R. P. Turco, O. B. Toon, T. P. Ackerman, J. B. Pollack, and Carl Sagan, "Nuclear Winter: Global Consequences of Multiple Nuclear Explosions," *Science* 222, no. 4630 (December 23, 1983): 1283–1292.

9. Alan Robock and Owen Toon, "Self-Assured Destruction: The Climate Impacts of Nuclear War," *Bulletin of the Atomic Scientists* 68, no. 5 (September 1, 2012): 66–74.

10. Adapted from A. Robock, L. Oman, G. L. Stenchikov, O. B. Toon, C. Bardeen, and R. P. Turco, "Climatic Consequences of Regional Nuclear Conflicts," *Atmospheric Chemistry and Physics* 7, no. 8 (2007): 2003–2012, https://doi.org/10.5194/acp-7-2003-2007. We are grateful to Luke Oman for supplying this image.

Further Reading

Buddemeier, Brooke. "Nuclear Detonation in a Major City." YouTube, June 21, 2011. https://www.youtube.com/watch?v=ttv1NLf6Cs4&t=209s. This presentation details the findings of a research study on the effects of the detonation of a nuclear weapon on a major U.S. city.

Craig, Paul P., and John A. Jungerman. *The Nuclear Arms Race: Technology and Society*, 2nd edition. New York: McGraw-Hill, 1990. Chapters 13–18 detail the physical effects of nuclear weapons; chapters 19 and 22 describe the physical and psychological effects of nuclear war.

Glasstone, Samuel, and Philip J. Dolan. *The Effects of Nuclear Weapons*. Washington, DC: U.S. Department of Defense and Energy Research and Development Administration, 1977. Old, but still the authoritative reference on nuclear weapons effects. The entire book is freely available at https://www.dtra.mil/Portals/61/Documents/NTPR/4-Rad_Exp_Rpts/36_The_Effects_of_Nuclear_Weapons.pdf.

Helfand, Ira. *Nuclear Famine: Two Billion People at Risk*, 2nd edition. International Physicians for the Prevention of Nuclear War and Physicians for Social Responsibility, 2013. Available at https://www.ippnw.org/pdf/nuclear-famine-two-billion-at-risk-2013.pdf. This article describes the effects of food shortage, lack of resources, and shortened growing season for crops following to a small nuclear exchange.

Lewis, Jeffrey. *The 2020 Commission Report on the North Korean Nuclear Attacks Against the United States: A Speculative Novel*. Boston: Mariner Books, 2018. A fictional account of a hypothetical 48-hour nuclear war initiated by North Korea, by an expert on nuclear nonproliferation and colleague of your author FDV at the James Martin Center for Nonproliferation Studies.

National Security Staff Interagency Policy Coordination Subcommittee for Preparedness and Response to Radiological and Nuclear Threats. Planning guidance for response to a nuclear detonation. 2010. Available at https://www.fema.gov/media-library/assets/documents/24879. This is a highly recommended publication describing guidance for first responders to nuclear incidents.

Robock, Alan. Robock's website at Rutgers University is an excellent resource for information on climatic effects of nuclear war. Many recent and older papers are archived here. http://climate.envsci.rutgers.edu/nuclear/.

Robock, Alan, and Owen Brian Toon. "Local Nuclear War, Global Suffering." *Scientific American* 302, no. 1 (January 2010): 74–81. This highly influential article describes recent research on the climatic effects of a relatively small nuclear exchange.

Solomon, Fred, Robert Q. Marston, and Lewis Thomas. *The Medical Implications of Nuclear War*. Washington, DC: National Academies Press, 1986. A compendium of authoritative papers covering not only medical but also physical, climatic, and psychological effects of nuclear warfare.

Southard, Susan. *Nagasaki: Life After Nuclear War*. New York: Penguin Books, 2016. The author, a creative fiction writer, recounts conversations with a Nagasaki survivor as well as her own visits to Japan.

Wellerstein, Alex. Nukemap. This is an excellent tool developed by historian Wellerstein, where the effects of a nuclear detonation of a chosen yield, including casualty estimates, can be calculated for any place on Earth. https://nuclearsecrecy.com/nukemap/.

U.S. Congress, Office of Technology Assessment. *The Effects of Nuclear War*. Washington, DC: U.S. Government Printing Office, 1979. Clear, straightforward discussions of the effects of nuclear weapons and nuclear warfare. Includes scenarios of attacks on Detroit and Leningrad (modern-day Saint Petersburg), limited and all-out nuclear war, and long-term effects. A fictional account envisions life after a nuclear war. Available for download at https://ota.fas.org/reports/7906.pdf.

14

Delivering Nuclear Weapons

Nuclear weapons aren't of much military use without some means of delivering them to their targets. **Nuclear delivery systems** also serve as key elements in the international balance of political and military power. Delivery systems range from "suitcase bombs" that might be carried by nuclear terrorists to land- and submarine-based intercontinental ballistic missiles. They also include aircraft to drop bombs or launch nuclear-tipped cruise missiles. Here we'll focus primarily on **strategic** delivery systems—those designed to carry nuclear weapons to an adversary's homeland.

Features of Nuclear Delivery Systems

Features of delivery systems depend on what a nation wants to do with its weapons; in turn, delivery systems may give clues to a nation's nuclear intentions. Important aspects of delivery systems include range, time to target, and accuracy.

Range describes the distance a delivery vehicle can travel from base to target. Today's longest-range delivery vehicles are truly intercontinental and can reach virtually anywhere on the globe. **Flight time** is the time it takes from base to target. Shorter flight times would be especially useful in a surprise attack—a **first strike**, in nuclear parlance. Flight times of today's strategic weapons range from about 10 minutes for submarine-launched missiles to hours for bombers. **Payload** is the maximum weight a vehicle can deliver. There's a tradeoff between range and payload, with payload increasing at some cost in range. **Accuracy** describes how close to its target a weapon will get. Accuracy matters less for **soft targets** such as cities but becomes increasingly important to ensure destruction of **hard targets** such as underground missile silos and command centers. If a 1-megaton warhead lands a mile from its center-city target, the city will still be destroyed, but only a direct hit will destroy a hardened missile silo. Accuracy is quantitatively expressed in terms of the *circular error probable* (CEP), introduced in chapter 12. Today's intercontinental missiles have CEPs of about 100 meters, meaning that a warhead has a 50–50 chance of landing within this distance of its target after a flight of some 10,000 miles. There's a tradeoff between accuracy and warhead yield: with its huge destructive radius, a 9-megaton weapon needs less accurate targeting than one with a 300-kiloton yield. Extreme accuracy is a first-strike requirement, since destruction of an adversary's missiles in hardened silos would be important in a first strike. A delivery system is no good if it's destroyed before reaching its target. **Penetration** characterizes the ability to penetrate enemy territory in the face of defensive systems. Slow-moving aircraft have lower penetration than high-speed missiles, although even aircraft penetration is improving with technological advances.

A delivery system might also be destroyed before being launched. That possibility, a nightmare for nuclear strategists, demands that nuclear delivery systems have a measure of **invulnerability**. If the primary purpose of nuclear weapons is to deter an adversary's attack by threatening nuclear retaliation, then that threat is hollow if your adversary can destroy your retaliatory forces in a first strike. Invulnerable weapons keep the retaliatory threat credible, helping preserve the nuclear peace. The quest for invulnerability has been a significant driver of the nuclear arms race.

What about accidental nuclear war or war provoked inadvertently in time of crisis? Can the designs of nuclear weapons and their delivery systems affect these frightening possibilities? Yes. High-speed systems with short flight times are more dangerous. They can't be recalled if an accidental or mutinous launch occurs, and their short flight time gives the nation under attack little opportunity to make rational decisions about how to respond. Other features, including vulnerability of communications systems and deployment of multiple-warhead missiles, also have significant bearing on the likelihood of nuclear conflict.

Today's nuclear balance hinges on the delivery systems the nuclear powers have deployed. This chapter examines technical details of those delivery systems and their implications for international security. We'll focus primarily on the United States' nuclear forces, which, along with Russia's, still dominate the global nuclear balance. However, we face a multipolar world with nuclear weapons also in China, the United Kingdom, France, Pakistan, India, North Korea, and Israel. Despite a near-global nuclear testing moratorium, the nuclear capabilities of these nations aren't static, and most continue to modernize their warheads and delivery

Figure 14.1
The B-52 Stratofortress bomber with some of the munitions it can carry. B-52s have been flying since the early 1950s, but ongoing modernization keeps them a vital part of the U.S. nuclear delivery arsenal. (U.S. Air Force)

systems. Something as mundane as the U.S. Air Force's planned installation of new engines in the B-52 bomber (figure 14.1) to increase its range by 40 percent may upset an already uneasy strategic balance.

The Strategic Triad

Since about 1960, each superpower's nuclear forces have comprised a **triad** of delivery systems that differ in **basing schemes**—where weapons are deployed—and in actual delivery vehicles. The triad includes aircraft, land-based **intercontinental ballistic missiles** (ICBMs), and submarine-launched ballistic missiles (SLBMs). One rationale for the triad is redundancy: if a surprise attack destroys one leg, the others remain capable of retaliation. The triad's legs differ in the delivery-system qualities discussed above; since those qualities involve tradeoffs, a diversified triad may give more security than would "putting all one's eggs in one basket." Another reason for the triad is political: each major branch of the military would like its own share of nuclear forces, and the triad affords just such a sharing. Here we'll focus on the United States and Russian nuclear triads but will look briefly at other nations' nuclear forces and delivery systems. We'll address broader questions of nuclear strategy in chapter 15.

The American and Russian nuclear triads are equivalent but not identical. Accidents of geography and technology have given them different emphases. The Soviet Union's vast land area and poor ocean access favored land-based ICBMs. Strength in high-payload rocketry and lagging guidance technology further tipped the Soviets (and now the Russians) toward large, land-based missiles with high-yield warheads. Figure 14.2 shows the results: the U.S. and Soviet triads, while comparable in total yield, show distinctly different balances among the legs of the triad.[1]

The triad doesn't include nonstrategic nuclear weapons—short-range weapons intended for battlefield use. Russia vastly outnumbers the United States in nonstrategic warheads, with some 1,800 versus 200 for the U.S.—with all those U.S. weapons deployed in Europe.

Who's ahead? The United States has more strategic warheads deployed, but Russia's land-based ICBMs carry more than twice the total megatonnage while the reverse is true for the other two legs of the triad—giving the

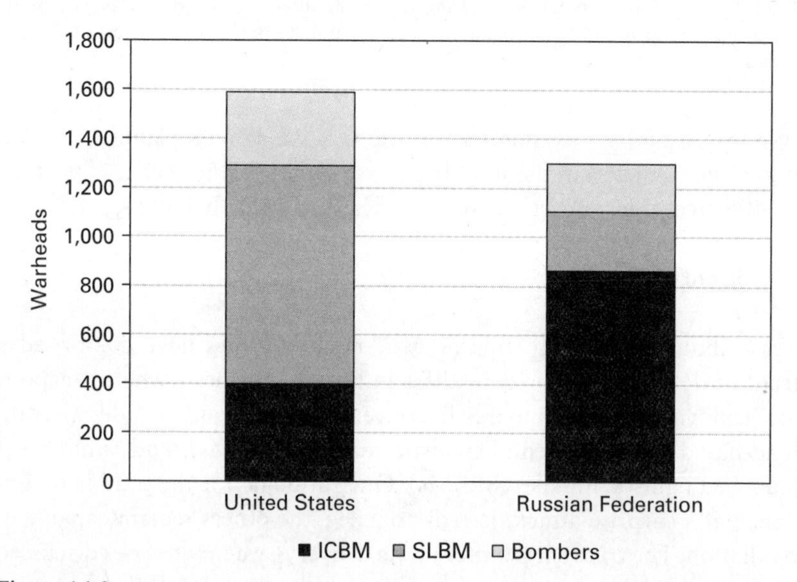

Figure 14.2
Compositions of the U.S. and Russian strategic nuclear triads in 2018, as measured by warheads actually deployed on the three legs of the triad. Not included are additional warheads available but in storage and tactical nuclear weapons. As discussed in chapter 13, only 1 to 2 percent of each arsenal can cause catastrophic damage as well as lasting climate change that could put billions of people at risk. (Data source: *Bulletin of the Atomic Scientists*; see note 1 for details.)

two countries about the same total yield. Considering the huge destructive power of both arsenals, the question of who's ahead is essentially meaningless. It's more important to ask whether strategic weapons systems enhance or diminish international stability, especially in time of crisis. Sheer destructive power does not buy security!

Ballistic Missiles

If you've thrown a baseball, you intuitively know how a **ballistic missile** works. First your arm moves to set the ball on the desired trajectory. In missile parlance this is the **boost phase**—the time when the rocket engines burn until they've exhausted their fuel, also called **propellant**. In fact, most of the missile's initial weight is propellant. The payload is only a few percent of the total weight (that's why it's expensive to send objects into space).

During the boost phase, the exhaust plume is hot, and the missile is easily tracked with plume-searching satellites. Back to your baseball: once it leaves your hand, it has its maximum speed and moves (neglecting air resistance) under the influence of gravity alone. Its path—called a **ballistic trajectory**—is fully determined by the ball's speed, direction, and position when it leaves your hand. The faster it's going, the further it will go. A ballistic missile works much the same way.

The rocket motors on today's intercontinental ballistic missiles (ICBMs) burn for only about three minutes, leaving their payloads at just over 100 miles altitude and traveling at 15,000 miles per hour (over 4 miles per second). Missiles are first launched vertically upward, then steered onto the appropriate trajectory for their targets. An **inertial guidance system** senses every change in the missile's motion, and at the end of the boost phase small rockets fire to send one or more warheads on precise trajectories toward their targets. The period from the end of the boost phase to when the warheads reenter Earth's atmosphere marks the **midcourse phase**. In midcourse the warheads travel thousands of miles under the influence of gravity alone, with no further propulsion or guidance. This is the longest phase of flight and in the baseball analogy it lasts from when the baseball leaves the hand until it's caught. Then comes the **terminal phase**, when warheads plunge through the atmosphere toward their targets. The total intercontinental flight time is about 20 to 30 minutes. That relatively long flight without guidance makes the extreme accuracy of today's ICBMs all the more impressive. Critical to that accuracy are those first few minutes of achieving the correct speed and orientation: too fast or too slow and the warhead will miss its target.

Missiles use two principal types of fuel, liquid or solid. Either way, the fuel needs oxygen to burn. Missiles can't get oxygen fast enough from the air; hence they need an **oxidizer** to provide oxygen. A liquid-fueled rocket has two tanks, one for fuel and one for oxidizer. A pump-and-valve system combines the two liquids and introduces them into a combustion chamber. In solid-fueled rockets, fuel and oxidizer are combined and cast into a solid compound that's ignited like a giant firecracker. Liquid-fueled rockets are more maneuverable, but it takes time to load the fuel prior to launch. Solid-fueled rockets can be launched instantly and are therefore preferred for military applications.

Missiles are classified according to their range (see table 14.1). However, there's an important caveat. Just as you can throw a baseball farther than a bowling bowl, so a missile's range depends on its payload. Thus range isn't an inherent characteristic of a missile but depends on the weight of the warhead or warheads it's delivering.

Did that last paragraph imply more than one warhead on a missile? Yes. In the late 1960s, American nuclear strategists worried that Soviet advances in missile-defense technology might prevent U.S. missiles from reaching their targets. The United States responded by developing multiple-warhead missiles, with the goal of overwhelming Soviet missile defenses. Multiple-warhead missiles were a way to maximize damage at minimum cost—literally getting more bang for the buck! The multiple-warhead concept is called **MIRV,** for **multiple independently targetable reentry vehicles** (figure 14.3). The uppermost stage of a MIRVed missile is the **bus.** It carries the individual warheads and maneuvers to release each one (hence the term *independently*) on a separate trajectory to its designated target. The warheads from a single missile can be targeted to land in a so-called *footprint* measuring roughly 100 by 300 miles.[2]

Are MIRVs a good thing? In retrospect, many nuclear strategists argue that they aren't, and that MIRV technology should have been banned by

Table 14.1
Classification of missiles by range

Designation	Range	Example
Short	< 1,000 km (~650 miles)	Scud (Russian origin)
Medium	1,000–3,000 km	Sejil (Iranian)
Intermediate	3,000–5,500 km	Musudan (North Korea)
ICBM	> 5,500 km (~3,400 miles)	Minuteman III (U.S.)

Figure 14.3
Technicians working on a Minuteman III's MIRV system. Since 2014, Minuteman missiles have carried only a single warhead. (U.S. Air Force)

treaty. Its multiple warheads give one MIRVed missile a very good chance of destroying one or more of an adversary's missiles before launch. And destruction of a single MIRVed missile means destruction of multiple warheads. For both these reasons, a nation armed with MIRVs might be tempted to strike first in an attempt to destroy its adversary's MIRVs. Combined with increased accuracy, MIRVed missiles have made today's U.S. silo-based missiles much more vulnerable, although the air force believes that currently only Russia would be capable of destroying the U.S. ICBM force.

While the United States no longer deploys land-based MIRVs, Russia still MIRVs both land- and submarine-based ballistic missiles. For example, Russia's new Sarmat SS-28 missile can carry as many as 16 warheads. A single SS-28's total firepower is more than all the explosives and gunpowder used in World War II, while its range is believed to be as much as 11,000 miles (18,000 km) with an accuracy on the order of 10 meters (30 feet). Those facts should really terrify you!

In addition to multiple warheads, missiles may carry **penetration aids**, including **decoys** that, in the vacuum of space, are difficult to distinguish from real warheads. Decoys can be as simple as balloons that inflate to produce a cloud of hundreds or thousands of objects that can all look, to the target country, like incoming nuclear warheads. Another penetration aid is *chaff*, small strips of aluminum foil deployed to create a reflective cloud that confuses defensive radars. We'll discuss penetration aids further in chapter 17.

Our discussion so far has emphasized intercontinental missiles. Until 2019, intermediate-range ballistic missiles were banned by the 1987 Intermediate-Range Nuclear Forces Treaty (INF), which prohibited the United States and Russia from testing or fielding land-based ballistic missiles with ranges from 500 to 5,500 kilometers (300–3,400 miles). In 2019, the Trump administration pulled the U.S. out of the treaty and Russia's Putin administration followed suit. Each country had accused the other of violating of the treaty. Withdrawal from the INF is an ominous sign for future arms control (more in chapter 18).

Cruise Missiles

Ballistic missiles follow a ballistic trajectory—that is, a trajectory due to gravity alone. Such a trajectory is entirely predictable, giving the nation under attack a fighting chance of either destroying the missile (with interceptor missiles, to be discussed in chapter 17) or at least a few minutes warning.

This isn't the case with cruise missiles or with hypersonic glide vehicles (discussed later in this chapter). In contrast to a rocket-boosted ballistic missile, the **cruise missile** is a small, pilotless airplane that usually flies subsonically at low altitude with conventional wings and a jet engine. Unlike ballistic missiles, cruise missiles can be retargeted during flight. They hug the ground to avoid radar, which can't see well over the horizon. This is especially true if cruise missiles include **stealth technology**, which makes them hard to detect. Cruise missiles have been used countless times in conventional conflicts, but they can also carry nuclear weapons or other unconventional armaments. They're classified according to the launch

Figure 14.4
A U.S. AGM-86 air-launched cruise missile in flight. Different versions carry conventional or nuclear warheads. The missile is launched from B-52 bombers and flies at 550 miles per hour. (U.S. Air Force)

platform: cruise missiles launched from aircraft are **air-launched cruise missiles** (ALCMs) and essentially extend the aircraft's attack range while decreasing risk to the aircraft itself (figure 14.4).

Two important developments led to the modern cruise missile: small, inexpensive jet engines and sophisticated computerized guidance systems. Cruise-missile guidance involves sensors such as terrain-contour matching (TERCOM) radar, in which an onboard computer compares radar images of the terrain over which it's flying against internal maps. Another technology, digitized scene-mapping area correlator, compares real-time camera images with stored images from prior aerial surveillance. Both guidance systems require a point of reference. During the 1991 Gulf War, Baghdad freeway intersections were used as reference points—which the Iraqis knew and used to shoot down U.S. Tomahawk cruise missiles.

An upgrade to both of these technologies is to use the Global Positioning System (GPS) to determine the missile's location. But GPS is subject to jamming using inexpensive devices that, although illegal in the United States, are readily available. GPS antennas like the ones in your smartphone comb the sky for satellites, and if a jammer happens to broadcast in the line of sight of the antenna, it will pick up this signal and confuse the location. Sophisticated new GPS receivers used in cruise missiles have narrow directional antennas that know where the satellites are and thus help deter jamming. In addition to these automated systems, some cruise missiles use real-time human direction to guide the missile to its target (known as *man-in-the-loop*). Soon, quantum-sensor-based inertial guidance may be used both in missiles and submarines when GPS is inoperative.

Cruise missiles' extreme accuracy makes them valuable delivery systems for conventional as well as nuclear explosives, and it's not obvious to the targeted country whether an incoming cruise missile carries a nuclear or a conventional warhead. This is a fundamental problem with any delivery system that can carry both conventional and nuclear weapons.

Drones

You've surely seen small drones operated by private individuals or read of military drones used in precision attacks on suspected terrorists. Could these craft also deliver nuclear weapons?

We probably don't need to worry about the small drones anyone can purchase these days. But those are at one end of a continuum of **unmanned aerial vehicles** (UAVs) that range through craft used commercially for remote sensing, aerial photography, small package delivery, law enforcement, and a host of other applications, to "aerial trucks" capable of lifting several tons, and on to military drones that are essentially pilotless fighter aircraft. Even cruise missiles fall into the category of UAVs. As you've seen, some cruise missiles are designed to carry nuclear weapons. But there's no reason a beefy enough drone couldn't be adapted to do the same.

A 2019 drone and cruise missile attack on Saudi Arabian oil facilities served as a wake-up call to the dangers of burgeoning drone technology. It's unclear whether the attack came from a nonstate actor or from Iran, but either way it showcased drones' offensive military capabilities and raised fears of nuclear-armed drones—something the U.S. military is actively exploring. And the attack further blurred the distinction between drones and cruise missiles, which is sometimes taken as the difference between reusable vehicles (drones) and craft that fly a single attack mission (missiles). The Saudi attack used "suicide drones" whose explosive payloads destroyed the drones along with their targets—just as happens with cruise missiles.

Drones might play other roles in the nuclear equation. Using artificial intelligence and "swarming technology," hordes of drones could target an adversary's nuclear delivery systems or, alternatively, be used to protect those systems. In fact, thousands of drones and cruise missiles might work in unison and communicate with each other. Camera-equipped sensor drones would identify potential targets, environmental hazards, or defenses, and transmit this information to the rest of the swarm. Swarms of drones with decoy warheads could overwhelm a missile defense system. In 1982, Israel used drones as decoys to manipulate Syrian air defenses into believing the drones were Israeli planes. As Syrian radar locked on to those drones and attempted to fire interceptors, the Israelis detected this and destroyed the radar installations.[3]

Drones and their uses continue to proliferate, and some of those uses will not be benign. In our nuclear age, that's a worry.

Submarine-Launched Ballistic Missiles

Submarine-launched ballistic missiles (SLBMs) constitute another leg of the nuclear triad. SLBMs are launched from submerged submarines off an adversary's coast, and their guidance systems allow considerable accuracy despite submarines' changeable locations. SLBMs are similar to their land-launched ballistic missile counterparts except that they're generally given a cold launch where they're ejected from the submarine before the rocket engines start so as not to damage the submarine. Typical flight times of 10 to 15 minutes make defense or even warning of incoming missiles impractical.

If any nuclear weapons system is invulnerable, it's the SLBM—or, rather, the submarines that carry SLBMs. Despite continued research in undersea detection and tracking, ballistic-missile submarines remain virtually impossible to find once they're at sea, where they can stay hidden for months. For some nuclear strategists, SLBMs' penetrability makes them alone an entirely adequate retaliatory force.

The invulnerability of SLBMs is made possible by the nuclear reactor. (As noted in chapter 9, today's pressurized-water reactors for electric power generation are scaled-up versions of PWRs first developed in the 1950s for submarine propulsion.) The nuclear difference means that a submarine can travel a long way on a small quantity of nuclear fuel. A modern nuclear-powered ballistic missile submarine (**SSBN**, for ship, submersible, ballistic missile, nuclear-powered) is refueled at 15-year intervals and may travel half a million miles on less than 100 pounds of highly enriched uranium. Equally important is the fact that fission doesn't require oxygen as does chemical fuel; consequently, nuclear submarines can remain submerged almost indefinitely. In fact, nuclear-generated electricity is used to dissociate seawater into hydrogen and oxygen, providing oxygen for the crew to breathe. In practice, a nuclear missile submarine carries a crew of about 150 on patrol missions lasting three to six months.

Strategic Bombers

B-29 bombers dropped nuclear bombs on Hiroshima and Nagasaki. Today, bombers constitute the third leg of the nuclear triad. They've been vastly improved over their World War II counterparts, thanks to lighter materials and fuel-efficient jet engines. They can carry bombs and cruise missiles (recall figure 14.1). The bombers themselves maximize lift, minimize drag, and have high engine thrust to achieve great range. Bombers must be able to penetrate enemy territory and get away with minimal risk of being shot down. To maximize range, they fly at very high altitudes where air resistance is lowest.

However, high-altitude aircraft are at greater risk of detection because they're in the line of sight of ground-based radars. Therefore, the general approach of bombers is to fly a high-low-high profile where the bomber first flies as high as possible and then, close to enemy territory, drops low enough to avoid detection. It then drops its bombs and rises to high altitude for its return. Another quandary is that bombers must fly fast to increase survivability, but that increases fuel consumption and thus decreases range. Also, at high speeds it becomes more challenging to hug the terrain to remain below ground-based radar, especially for supersonic aircraft. Therefore, bombers are usually subsonic or only slightly supersonic.

An improvement akin to MIRVed missiles was the deployment of air-launched cruise missiles on bombers and strike aircraft. This meant the aircraft themselves didn't need to penetrate enemy airspace. This extended airplanes' strike range and allowed a single aircraft with several ALCMs to strike multiple targets.

Delivery Systems of the United States and Russia

The United States and Russia both maintain full, robust nuclear triads. As we discussed previously, they're militarily equivalent but not identical. Here we describe both countries' triads in detail.

Intercontinental Ballistic Missiles

Since the late 1960s, the mainstay of the United States' land-based ICBM force has been the **Minuteman III**, a solid-fuel rocket with an 8,000-mile range (13,000 km). Some 400 Minutemen are deployed in silos in Wyoming, Montana, and North Dakota. They're armed with single warheads in the

Figure 14.5
A Minuteman missile in its silo. (U.S. National Park Service)

300-kiloton range (see figure 14.5). Minuteman missiles can be MIRVed with three warheads each. Since 2014, however, all have been de-MIRVed to meet treaty obligations.

The Minuteman III's accuracy (its CEP) is approximately 120 meters. That's equivalent to throwing a dart from seven feet to hit a point the size of a human hair! Together, the deployed Minuteman III missiles total a yield of some 130 megatons—more than 20 World War IIs on this single leg of the U.S. strategic triad. Despite their lethality, Minuteman III missiles are decades old. In 2015, a modernization program extended the Minuteman's life to 2030, when a new ICBM, the *Ground-Based Strategic Deterrent*, is scheduled to replace them.

As you've seen, the introduction of accurate, MIRVed missiles made fixed, land-based ICBMs increasingly vulnerable to first-strike attack. Although a treaty banning MIRVs might have been an effective response to that threat, nuclear strategists have instead sought technological advances to restore the invulnerability of ICBMs. One solution is to deploy mobile missiles whose location changes frequently. Russia uses this strategy, with missiles not only in fixed silos but also on massive trucks called *transporter erector launchers* (figure 14.6). The transporters move the missiles, raise them to vertical, and then launch. Another mobile-missile tactic is to use railroads. For a while the Soviet Union had a dozen *doomsday trains* that roamed around loaded with three SS-23 missiles carrying 10 warheads of 550 kilotons each, but these were decommissioned under the 1993 START II treaty. In 2013, Russia announced plans to resurrect the rail-mobile idea, but the project, dubbed Barguzin, appears to be on hold.

Russia does not publish unclassified descriptions of its strategic nuclear forces, but analysts at the Federation of American Scientists (FAS) have tracked the world's nuclear armaments for decades. Russia has a formidable ICBM program with as many as 10 different missile types and over 1,138 warheads. More than 30 percent of all active Russian ICBMs

Figure 14.6
Russian mobile ICBM on display during a military parade. The missile itself is inside the large tube atop the transporter erector launcher truck. The tube rises to a vertical position for launch. The missile itself is probably a 10-warhead Yars. (Vitaly Kuzmin/CC BY-SA 4.0)

have been deployed since 2010, indicating a marked increase in ICBM deployments.

The Russian ICBM program is undergoing extensive modernization, replacing old Soviet-era missiles. The new SS-27 mod 2, also called Yars, is MIRVed with up to 10 independently targetable 300-kiloton warheads. The Yars reentry vehicle can change direction to evade missile defense. This type of missile is known as *quasi-ballistic* because its trajectory isn't fixed by gravity but can change in flight like a cruise missile. This makes Yars a formidable part of Russia's nuclear triad.

The Russian SS-18 (Satan) missile, unlike the U.S. Minuteman III, can be cold launched, meaning that the silo can be quickly reloaded, which some U.S. strategists fear would give Russia an advantage in a prolonged nuclear exchange. SS-18s are being replaced by the RS-28 (Sarmat or Satan 2). After multiple delays in testing, this missile is expected to go into production in 2021. It can reportedly carry up to 16 warheads or 24 Yu-71 Avanguard hypersonic glide vehicles, which, with their unpredictable trajectory and global reach, will challenge ballistic missile defense systems.

Submarine-Launched Ballistic Missiles (SLBMs) and Submarines

The current U.S. SLBM fleet comprises 14 *Ohio*-class ballistic missile submarines (SSBNs) and 4 *Ohio*-class cruise missile submarines (see figure 14.7). Four of originally 24 missile tubes were removed from each SSBN to comply with the New START treaty's ceiling of 280 missiles in the submarine leg of the triad. All *Ohio*-class SSBNs carry 20 Trident D5 ICBMs. Each missile is MIRVed with the capability to carry up to 14 W76 (100 kt) or W88 (455 kt) warheads, but they average typically 3 to 6 warheads per missile. To put that in a perspective, that's a maximum total firepower ranging from 6 to over 100 megatons—the equivalent of at least 1 and at most 20 WWII's on each submarine! The other four *Ohio*-class submarines were converted to carry 154 Tomahawk cruise missiles armed with conventional weapons. Roughly two-thirds of U.S. submarines are at sea at a given time.

Effective missiles alone aren't enough for an invulnerable submarine-based deterrent. Stealth is also required. During the Cold War, both the Soviet Union and the United States deployed sophisticated ocean-based acoustic sensors to detect submarines. Expendable sensors known as *sonobuoys* were deployed from aircraft; their ultrasensitive microphones would begin transmitting 90 seconds after they dropped into the ocean. In response to the resulting increased vulnerability, submarines had to become even more stealthy. The *Ohio*-class submarines were modified to

Figure 14.7
The *Ohio*-class ballistic missile submarine USS *Nebraska* (SSBN-739) returning to its home base in Washington State. The sub routinely patrols the Pacific Ocean. (U.S. Navy photo/Specialist 1st Class Amanda Gray)

circulate the reactor coolant using natural convection rather than noisy pumps. This and other advances have diminished submarine noise and have therefore lowered detectability.

The oceans' impenetrability to common signaling mechanisms is what gives submarines their invulnerability, but it also rules out easy communication with submerged submarines. The difficulty of communication puts the submarine leg of the triad in a unique position: submarine-launched ballistic missiles are the only nuclear weapons that can be launched without first receiving special codes from high government authorities. That fact enhances SLBM invulnerability since no amount of disruption back home can disable them. But it also raises the specter that an accidental launch, a mutiny, or an act of insanity confined to a single submarine could initiate a nuclear war. During the Cuban Missile Crisis the difficulty in communication by a submerged Soviet submarine almost led to disaster (see this chapter's box "You're Alive Because ...").

The U.S. Navy plans to replace the *Ohio*-class submarines with the next generation *Columbia*-class submarines. The new submarine will have a lifetime reactor core that won't need refueling for its full 42 years. One major advance will be electric drive, quieter than the mechanical drive of

Ohio-class submarines. *Columbia* submarines will carry 16 Trident D5 missiles and will regain surface maneuverability that was lost in earlier generations when propellers were modified to make them quieter. (Propeller designs are big secrets and photos or diagrams of propellers are rare. That's why you'll usually see photos featuring of the front of submarines.[4])

The Russian nuclear missile fleet consists of 11 submarines in three classes designated Delta III, Delta IV, and Borei. Each submarine can launch as many as 16 MIRVed SLBMs. As with the U.S. fleet, not all subs are at sea concurrently. The total number of warheads, according to FAS open-source estimates, is around 600. There are three principal SLBMs deployed on Russian submarines. The two newer models have a range of 5,200 miles (8,300 km) and can carry multiple warheads. Russia claims that these missiles can carry decoys and that their reentry vehicles are capable of maneuvering in flight to escape missile defenses.

Strategic Bombers

In 1952, the United States began deploying the B-52 Stratofortress long-range bomber, with eight jet engines, intercontinental range, and in-flight refueling capability. In 2020, this workhorse of the bomber fleet still made up about half of the bomber leg of the U.S. strategic triad. B-52s are no longer armed with nuclear gravity bombs but with air-launched cruise missiles. Use of ACLMs allows the slow, not-very-stealthy aircraft to stand off from enemy territory when launching their missiles. The current version, the B-52H, is based at Barksdale Air Force Base in Louisiana and Minot Air Force Base in North Dakota. The United States retains some 76 B-52 aircraft in its strategic forces, but only 44 are focused on nuclear missions. You may have seen B-52s pass overhead, and they may even have been carrying nuclear weapons. In fact, in August 2007, six W-80 nuclear warheads were inadvertently loaded onto a B-52 and flown, unknown to the air force, from one base to another—including bases that aren't supposed to host nuclear weapons. The warheads weren't reported missing for one-and-a-half days nor were security precautions for nuclear weapons instituted. Investigation of this incident resulted in steps to improve morale and quality of life for air force personnel in the nuclear sector.

Bombers' penetrability can be improved by decreasing radar visibility and increasing speed. The B-2, also known as the stealth bomber, has been in service since 1997 and is the United States' only other long-range strategic bomber (figure 14.8). Like the B-52, it's subsonic, with maximum speed 630 miles per hour. It incorporates carbon composite materials that absorb rather than reflect radar signals. This and other

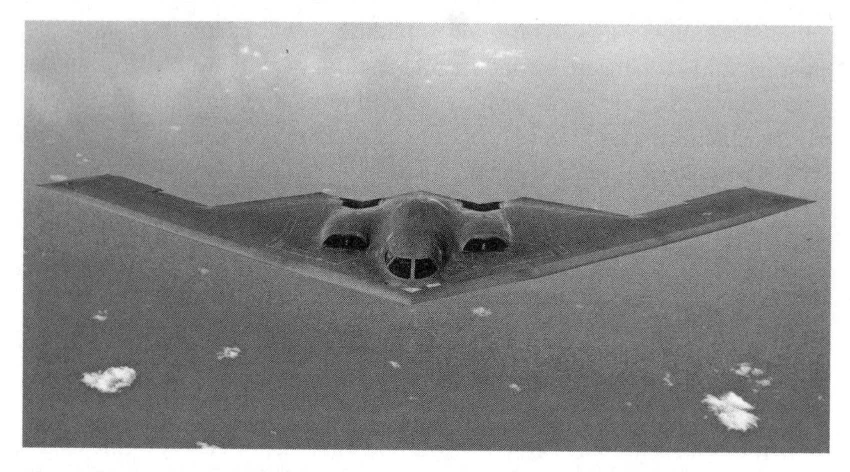

Figure 14.8
A B-2 stealth bomber over the Pacific Ocean. (U.S. Air Force photo/Staff Sgt. Bennie J. Davis III)

features—including "flying wing" design and angular contours—give the 172-foot-wingspan plane the same radar profile as a pigeon! With in-air refueling, the B-2 can attack targets anywhere in the world. It flies as high as 50,000 feet and has sensors that alert the pilot to change altitude in order to eliminate the contrails that normally form behind jet airplanes. The B-2 is painted black to blend in with the dark sky at high altitudes. It's been reported that this sophisticated aircraft even has a sensor that alerts the pilots to change their altitude in order to match the sky illuminance. The B-2 can also fly at near-treetop level, using sophisticated terrain-following systems.

The B-2 bomber can carry 16 B61–7 gravity bombs, each with a variable yield up to 340 kilotons, or 16 B-83 gravity bombs with a maximum yield of 1.2 megatons each. Or it can carry 8 B61–11 earth-penetrating bunker-buster bombs with a single yield of 400 kilotons, or a mixture of these several bombs.

The B-52 and B-2 will be in service for a long time. However, they're scheduled to be joined and eventually replaced by the B-21, a stealth design similar to the B-2. Although details are sparse, the B-21 will probably be a subsonic aircraft with even less radar detectability than the B-2. It could go into service in the mid- to late 2020s.

How does the U.S. bomber fleet stack up against our criteria for nuclear delivery systems? The B-52, B-2, and future B-21 all have intercontinental range with in-flight refueling. Each carries a substantial payload—about one-half World War II's worth of destruction per airplane—and can strike with good to excellent accuracy. Flight times of these subsonic bombers are measured in hours, a disadvantage in a first strike but a definite safety feature that mitigates against accidental or inadvertent nuclear attack.

What about vulnerability? It takes two to four minutes for a bomber crew on alert to board its aircraft, and another minute or so to start the engines and take off. During the Cold War, nuclear-armed bombers were on 24-hour "strip alert," with bombers sitting on the runway ready for takeoff. This is no longer the case, but that could change if the security situation worsens and the United States feels threatened.

Their long flight times make bombers primarily second-strike weapons, intended for retaliatory nuclear attack. Whether bombers could survive a first strike depends on their ability to get off the ground quickly and on their being hardened enough to withstand blast and EMP effects from nuclear weapons exploding on the bases that they had just left.

Although enemy missile submarines probably couldn't approach the United States coast without detection, they can operate undetected 500 or more miles out. From that distance, an SLBM would take 8 to 15 minutes to reach its target. Even though radars might spot the missile, it's doubtful that bombers could become airborne before they're destroyed on the ground.

These days the primary role of U.S. bombers is to signal to adversaries resolve in defending itself and allies. For example, in 2017 the U.S. flew B1-B bombers off the coast of North Korea as a response to North Korean missile tests. These aircraft don't carry nuclear weapons but pack massive conventional firepower.[5]

On the other hand, the virtual invisibility of the B-2 or B-21 might make it useful in a first strike. How is a potential adversary to perceive the evolution of the U.S. bomber fleet? Is the introduction of stealth technology an indication of potential U.S. belligerence, or is even the B-2 deployed solely to deter others' nuclear aggression? An adversary's answers to those questions can be as important as the actual technical capabilities of a nuclear weapons system.

Russia has two types of strategic bombers in service, the Tupolev Tu-160 Blackjack and the Tu-95MS Bear. Unlike U.S. stealth bombers, the Russian bombers are very much visible and therefore they would launch cruise missiles from a large standoff distance—meaning that the bombers' presence might indicate to the enemy that an attack is imminent.

Cruise Missiles

Today's cruise missiles are slow, so they're not an obvious first-strike choice. But as we discussed, they fly at nearly treetop level, below the line of sight of ground-based radars, making them practically undetectable. And they can deliver high-yield nuclear warheads with silo-killing accuracy. Currently, the United States deploys air-launched cruise missiles (the AGM-86B) on B-52 bombers at Minot, Whitman, and Barksdale Air Force Bases. These ALCMs can travel more than 1,500 miles and carry the W80–1 warhead, with variable yield from 5 to 150 kilotons. Nuclear-tipped ALCMs have been operational for decades, but many have been converted to carry conventional warheads—and they've seen plenty of use in conventional conflicts.

The air force plans to replace nuclear-armed ALCMs with the LRSO (long-range standoff weapon) cruise missile set to deploy in 2025. Few details have been released about this missile except that it will use state-of-the-art stealth techniques to penetrate air defenses. The missile will be retargetable in that it could receive remotely generated instructions (man-in-the-loop, as discussed earlier) to alter its target while in flight. The LRSO will be deployable on multiple platforms, including the B-52, B-2, and B-21 aircraft. It will carry the W80–4 nuclear warhead, a life-extended but functionally identical version of the W80–1.

Both Russian strategic bombers can launch AS-15 cruise missiles with warheads of 200 to 250 kilotons from a standoff distance of up to 1,500 miles (2,500 km). The Blackjack and Bear bombers can carry 12 and 6 Kh-55 missiles, respectively.

Russia has developed a new cruise missile known as the Kalibr-NK that can carry both conventional and nuclear armaments and can be deployed on newer Russian missile submarines and other naval platforms. The Kalibr travels close to the sea or ground to prevent detection. It has already been used in combat when, in 2015, Russia launched a salvo of conventionally armed missiles against Islamic State targets in Syria.

Since 2012, Russia has also deployed the KH-102 (nuclear) and KH-101 (conventional) air-launched cruise missiles for which Russia claims a maximum range of 4,500 kilometers (2,800 miles) with a 250-kiloton warhead. Like the sea-launched Kalibr, the KH-101 was used in 2015–2017 against multiple Islamic State targets in Syria.

The Developing Chinese Triad

You've seen that the Russian and U.S. strategic nuclear triads are robust and that substantial national resources are devoted to maintain and modernize them. China has also decided on a nuclear triad to ensure second-strike

You're Alive Because of This Man You've Never Met

You probably haven't heard of Soviet naval officer Vasily Alexandrovich Arkhipov, but you probably exist because of his quick thinking, resolve, and steady hand (figure 14.9). He's one of two people that we know have prevented nuclear war (the other was Soviet Air Force officer Stanislav Petrov, who, in 1983, disobeyed orders to prevent a possible retaliatory strike in response to a false alarm about incoming U.S. missiles). In 1962, President John F. Kennedy blockaded Cuba while a Soviet submarine was hiding on the ocean floor. This sub was meant to operate in the cool North Atlantic and didn't have air conditioning. Conditions inside were like a sauna! The coolest interior temperature was 113°F, and there was no fresh drinking water. The sub's crew hadn't been able to communicate with Moscow for a long time and weren't up to date on the political and military situation. Furthermore, they were overwhelmed by the cacophony of explosions from U.S. practice depth charges intended to force them to surface.

The submarine's commander tried to contact Moscow, but "Moscow was totally jammed. There was nothing ... emptiness. It was like Moscow doesn't exist."[6] Not having situational awareness and presuming that a nuclear war might already be under way, the commander ordered the launch of a 15-kiloton nuclear torpedo at the U.S. ships nearby. He reportedly screamed, "We're gonna blast them now! We will die, but we will sink them all—we will not become the shame of the fleet." The officer in charge of the torpedo agreed to launch. However, Arkhipov, being commodore of the entire Soviet submarine flotilla, outranked the commander and refused.

There are different accounts of what transpired, but all suggest that if Arkhipov had not intervened the nuclear launch would have occurred and the U.S. would likely have responded with nuclear weapons. Many issues contributed to this incident, but foremost was the sub's inability to communicate with Moscow or with the Americans. So despite the leadership of both governments working to diffuse the Cuban Missile Crisis, world catastrophe was actually avoided by one level-headed naval officer.

Figure 14.9
Vasily Arkhipov and wife Olga Arkhipova. (M. Yarovskaya, A. Labunskaya, from the personal archive of Olga Arkhipova)

capability. China has a sophisticated program of short- and intermediate-range ballistic missiles that could deliver nuclear weapons. The well-known DF-31 missile is solid fueled and has a range exceeding 6,800 miles (11,000 km)—enough to reach most of the United States. It may carry multiple warheads and penetration aids to increase the probability of a warhead reaching its target. The DF-31 is deployed on large mobile transporter erector launchers. A newer missile, the DF-41, has a range up to 9,300 miles (15,000 km), which puts the entire U.S. in its crosshairs. China also has intermediate-range missiles that can hit U.S. bases in the Pacific, as well as a mature program of cruise missiles.

To further develop its nuclear triad, China has fielded four JIN-class ballistic missile nuclear submarines, all of which can field a submarine-launched version of the DF-31 with a range of more than 4,400 miles (7,000 km) but only one warhead. However, these subs are rather noisy,[7] so it's not clear whether they're sufficiently invulnerable to provide a viable second-strike threat. Their vulnerability calls into question whether the motive for these subs was a second-strike capability or some other purpose.

China also has bombers, like the H-6K, an upgraded copy of an older Soviet aircraft, with a range of 4,500 miles (7,200 km). It's not clear what role these bombers play or whether they carry nuclear weapons. Yet another variant, the H-6N, carries an air-launched version of the formidable DF-21D medium-range ballistic missile, which can deliver both conventional and nuclear warheads. The DF-21D is probably the fastest intermediate-range missile currently deployed. It's capable of speeds faster than Mach 10 and can evade all current missile-defense systems. In addition, China has unveiled, in a flashy video, hints of a new stealth bomber. Not much is known about this aircraft except that it will have a range greater than 6,000 miles and a payload exceeding 10 tons.

India and Pakistan

India's Growing Triad

India operates land-, sea- and air-based nuclear delivery systems. The mainstay of the Indian nuclear forces are French Mirage 2000H and Franco-British Jaguar fighter-bombers, which together number 48 aircraft. Both were used in their countries of origin to serve a nuclear role, so could do so for India. India has a variety of land-based missiles (Agni-1, Agni-2, and Agni-3) that carry single nuclear warheads and have both short and intermediate ranges. India is now testing the road-mobile Agni-4 and three-stage Agni-5 missiles; the latter's range exceeds 3,100 miles (5,000 km).

India has also developed the 700-kilometer-range Nirbhay nuclear-capable cruise missile. At sea, India has the Arihant nuclear missile submarine, based on the Akula-class Russian subs, known to be among the quietest submarines. After lengthy trials and an accident that flooded the submarine, Arihant finally conducted its first patrol in 2018. However, Arihant's current missiles aren't capable of reaching Islamabad or Beijing. But successors in the Arihant class will be armed with 6,000-kilometer-range missiles that will allow India to deter a Chinese attack while staying far from Chinese territory.

Pakistan's Future Triad

Pakistan does not have a well-developed nuclear triad but has multiple platforms from which to enforce its deterrence relationship with India. Pakistani missiles include the short-range Hatf-3 and Hatf-4, the longer range Hatf-5 (1,250 km), and the newer Shaheen-3 with 2,750-kilometer range. With China's help Pakistan has also developed the Ababeel, a 2,200-kilometer solid-fueled missile whose extra-large nosecone can accommodate multiple warheads—probably the first MIRVed missile in Southeast Asia. It's expected that India will respond with their own MIRV. Nuclear policy expert Michael Krepon warns that "modest perturbations in the triangular competition among China, India, and Pakistan can have undesirable and unintended consequences" and should be avoided.[8]

Pakistani cruise missiles include the Babur, a land- or sea-launched model designed to carry a 10-kiloton or 35-kiloton warhead, and the Ra'ad, an air-launched cruise missile with stealth technology, maneuverability, and terrain-hugging capability that can deliver a 5–12 kiloton warhead. Pakistan's motivation, unsurprisingly, is to have a credible second-strike capability to respond to India's indigenous nuclear submarines. Pakistan's submarines are diesel-powered and therefore easily tracked, so they're not good platforms for second-strike weapons. Pakistan can also deliver nuclear weapons using aircraft including modified U.S.-built F-16 fighters and French Mirage fighter-bombers.

What about North Korea?

North Korea has had long-standing philosophical principles of self-reliance (*juche*) and of prioritizing the military (*songun*) over all else. These principles have been strengthened since Kim Jong Un took over in 2011. North Korea believes that maintaining a strong nuclear deterrent is the only way to ensure its survival. The country has an advanced nuclear weapons program, including ballistic missiles that can target not only neighbors South Korea and Japan but also Hawaii and the U.S. mainland. Since Kim Jong Un's tenure, production and development of ballistic missiles have greatly accelerated, with well over 100 missile tests (figure 14.10).

North Korea has produced ballistic missiles of all ranges, including short- and medium-range missiles like Soviet-era SCUD variants. But it also has ICBMs such as the Hwasong-14 and Hwasong-15. The former is a two-stage liquid-fueled missile whose range may exceed 6,200 miles (10,000 km) depending on payload. These ranges are estimated from the maximum height reached during tests in which the missile is fired nearly vertically, rather than at the 45-degree angle that would maximize range. We don't know whether the test payload was equivalent in weight to a nuclear weapon, so range estimates are approximate. North Korea probably chose the vertical test trajectory so as to avoid overflying Japanese or Russian territory.

North Korea's Hwasong-15 is a true ICBM whose estimated range of 8,800 miles (13,000 km) means it can hit any location in the United States. North Korea tested this missile in 2017 in a near-vertical launch reaching an altitude of 2,780 miles (4,475 km). Again, the payload was

Figure 14.10
August 2019 launch of a North Korean KN-23 quasi-ballistic missile capable of flying at low altitudes to a range of about 250 miles (400 km). (Korean Central News Agency/ Korea News Service via AP)

unknown so we don't know if the range estimate applies when carrying a nuclear weapon. It's also unclear whether the technology is adequate for a nuclear weapon to survive atmospheric reentry. Nor do we know the missile's accuracy. We do know that the Hwasong-15 has what appears to be a spacious nosecone that might admit multiple warheads.

Does North Korea have a triad? Not yet, but there's every indication that they seek such a force configuration. North Korea has a huge submarine fleet with as many as 86 submarines of various classes. However, they're all rather small. The exception is the GORAE-class diesel-electric submarine, which is capable of firing ballistic missiles. This submarine lacks an air-independent propulsion (AIP) system, which would allow it to remain submerged longer—although not indefinitely like a nuclear-powered submarine. In 2018, it was revealed that North Korea is building a new submarine known as the Sinpo-C class. Little is known about this vessel, although it evidently has tubes for missile launching and, while not nuclear powered, does have an air-independent propulsion system. In 2019, North Korea released images of Kim Jung Un inspecting a new submarine, clearly projecting North Korea's interest in a nuclear triad.

North Korea has its land-based missiles on mobile platforms so they can be fired from different locations—essentially taking a similar role to submarines. North Korea has also demonstrated that they can launch at any time of day or night—thus signaling to adversaries that an attack or response could come at any time.

France, United Kingdom, and Israel

Three remaining known nuclear-armed states are France, the United Kingdom, and Israel. Although these are currently stable democracies, a changing political situation could lead to instability in which nuclear weapons use becomes more likely. Furthermore, all three have sizable nuclear arsenals that are far more than enough to do serious harm to other countries and perhaps to the global climate. However, none has a full nuclear triad.

France

As we discussed in chapter 12, France has about 300 deployed nuclear weapons. The majority of French warheads are on the M51 series of submarine-launched ballistic missiles carried on three Triomphant-class nuclear submarines. The M51 SLBMs have ranges exceeding 5,600 miles (9,000 km) and can carry up to six warheads each, although the actual number of warheads is assumed to be less than that. At least one French SSBN is continuously at sea enforcing France's deterrent posture. With 16 missiles per submarine, this implies that about 64 warheads are deployed at once. The remaining French warheads, of approximately 300-kiloton yield each, are on ASMPA 500-kilometer range air-launched cruise missiles that launch from Rafale nuclear strike aircraft.

United Kingdom

British nuclear weapons are deployed solely on four Vanguard-class Trident nuclear missile submarines. As with the French nuclear posture, one SSBN is always on patrol and carries no more than eight D5 missiles with no more than three warheads each. Currently these missiles are on reduced alert, requiring several days' notice to launch. But they're solid-fueled rockets, so they could be put on higher alert with a much shorter launch time—as they were during the Cold War. Unlike the French, and rather oddly, the U.K. does not actually own the missiles but rather leases them from the United States, although the warheads are British made. This means the U.K.'s nuclear deterrent isn't entirely independent because the U.S. could deny access to GPS, gravitational, or weather data, "rendering that form of navigation and targeting data useless if the U.K. were to launch without U.S. approval."[9]

Israel

Israel is known to have a nuclear arsenal of about 80 weapons but officially neither denies nor admits that it possesses nuclear weapons. Israel has not signed the Non-Proliferation Treaty, so it's not bound by the rules ascribed to non-nuclear-weapon states. It's believed that some 30 Israeli weapons are gravity bombs deployed in bunkers near Tel Aviv, with the remainder on Jericho series ballistic missiles. The 1,500-kilometer-range Jericho II can be launched from mobile platforms and is sometimes hidden in caves, making it harder for adversaries to locate. The Jericho III missile has a range that's possibly as high as 4,000 miles (6,500 km).

Modernization Programs and Future Delivery Systems

There's a common perception that the nuclear arms race ended with the dissolution of the Soviet Union. But nothing could be further from the truth. In fact, the United States, Russia, and China are focusing on improving nuclear weapons and their delivery systems, and they do so without violating current arms control treaties. Each of these countries feels the need to upgrade delivery systems and warheads, or to build weapons that are more accurate, less easily defended against, and perhaps safer—the latter in the sense of avoiding inadvertent launch or detonation.

Hypersonic Glide Vehicles

Hypersonic glide vehicles (HGVs, also known as boost glide vehicles) are being developed by the United States, Russia, and China (figure 14.11). These devices share characteristics of both ballistic and cruise missiles. Like ballistic missiles, they achieve hypersonic speeds (greater than five times the speed of sound), far faster than cruise missiles but not generally as fast as ICBMs. HGVs can be launched from multiple platforms, including other missiles, thus acting as maneuverable reentry vehicles. Like cruise missiles, HGVs can fly long distances and maneuver to evade missile defenses.

While Russia, China, and the United States are all developing hypersonic missiles, the objectives, speeds, and ranges appear to be different. The United States has experimental vehicles designated for its *Prompt Global Strike* mission to hit anywhere on the planet in an hour or less.

Figure 14.11
Artist's conception of a U.S. hypersonic glide vehicle. (U.S. Air Force Scientific Advisory Board)

These vehicles have a top speed of Mach 20, or 20 times the speed of sound, and they're intended to carry conventional warheads. In contrast, Russia and China plan for hypersonic technology to deliver nuclear weapons. China's hypersonic missile, the DF-ZF, has undergone multiple tests and became operational in 2020; it travels at about half the speed of the U.S. hypersonics.

Russia's hypersonic vehicle, Avanguard, claims a top speed of Mach 27. In his 2018 State of the Nation address, President Putin announced with great fanfare that the testing phase was complete and the Avanguard was in production. A second Russian weapon is the hypersonic cruise missile known as the Khinzal, which travels at Mach 10 and can cover 1,200 miles. It's released from a modified MiG-31 fighter jet and can maneuver around missile defense systems. The Khinzal, which entered service in 2017, can carry either conventional or nuclear warheads.

Super-Fuze and Second-Strike Capability

As described in chapter 13, a nuclear warhead is usually detonated at a preset altitude in order to maximize its destructive effect. This is accomplished with barometric fuzes that use air pressure to sense altitude. However, if a trajectory overshoots its target, then the greatest damage doesn't necessarily occur at the specific preset altitude. In a glancing trajectory, for example, the trajectory may pass very close to the target, but detonation at the preset altitude occurs far away—too far to destroy hardened targets. New fuzes, dubbed *super-fuzes*, detonate the warhead at a specific altitude customized to the actual trajectory so as to wreak the greatest possible damage. The super-fuze works by calculating the altitude (60–80 km) as the warhead reenters the atmosphere. Since the trajectory is known to very high accuracy, the optimal detonation point relative to the target can then be determined.[10]

Although the super-fuze may seem just a simple modification of the fuze that every warhead carries, it transforms missiles designed for soft targets into weapons that can destroy hardened targets like missile silos. When installed in submarine-based warheads, the super-fuze turns the submarine fleet from a second-strike force into a formidable first-strike arsenal. In addition, it frees up warheads for other targets, thus expanding the destructiveness of a country's nuclear forces.

Russian president Vladimir Putin expressly complained about the United States' new nuclear developments, including the super-fuze: "Your people," referring to the United States and its allies, "do not feel a sense of the impending danger—this is what worries me."[11] This concern may be partly why Russia recently unveiled the new weapon systems to be discussed next. These may well be Potemkin constructs but nonetheless must be taken seriously.

Nuclear-Powered Cruise Missiles

In 2018, Russia unveiled a nuclear-powered cruise missile dubbed Burevestnik (Petrel). The idea is that this missile can fly around for days without refueling. The concept originated in the 1960s with the Soviet Union's prominent physicist and human rights activist Andrei Sakharov but was rejected at the time. The United States had a similar program in the 1960s. Known as Project Pluto, it was never developed because the idea of a nuclear-powered missile seemed too risky and because ICBMs were deemed more practical. The same arguments could certainly apply against the Russian missile. However, it appears that Russia is actively working on its nuclear-powered cruise missile: A mysterious explosion and radiation release in 2019 was, according to U.S. intelligence, the result of an attempt to recover such a missile that had been lost at sea.

Nuclear Torpedo

Russia has also developed a nuclear-powered torpedo known as Poseidon. It's the largest torpedo ever built and can reportedly carry a 2-megaton nuclear warhead. The torpedo is autonomous and can travel long distances using a tiny nuclear reactor. It can also go far deeper than submarines and travel much faster.

There's been a great deal of hype surrounding this weapon, making it sound like a true doomsday device. Some claim that it could be deployed with a 200-megaton warhead, which, when detonated, would produce a 500-meter-high tsunami that could destroy all life more than 600 miles (1,000 km) inland. The Russian navy has plans to acquire at least 30 Poseidons. There's a nuclear choice that surely isn't making the world a safer place.

Summary

Here we've taken a detailed look at the nuclear delivery systems of the United States, Russia, and China. We've also covered India, Pakistan, North Korea, and Israel as well as the United Kingdom and France. Many of these countries have developed or are developing a second-strike capability, often by utilizing a triad that includes nuclear weapons delivered by land, sea, and air. You saw that Russia has a fully developed nuclear triad comparable to that of the United States. China and India have similar triads, while North Korea and Pakistan likely plan to develop them (see figure 14.12 and recall figure 14.2).

You've also had a deeper look at the diverse components comprising the weapons delivery systems of a modern nuclear arsenal. These include bomber aircraft, land-based missiles, and submarine-based missiles. In addition, cruise missiles may appear in each leg of the triad. As triad components, cruise missiles carry nuclear weapons, but they're also available for conventional combat and have been widely used as such.

Important features characterizing nuclear delivery systems include range, payload, accuracy, and penetrability; all these relate to the ability to deliver weapons to their assigned targets. Flight times vary from minutes to hours and set the fundamental timescale for nuclear conflict. Invulnerability of delivery systems is strategically crucial, ensuring the retaliatory force needed to deter an adversary's attack. The ideal nuclear weapons system affords stability during crisis and safety against accidental or mutinous launch.

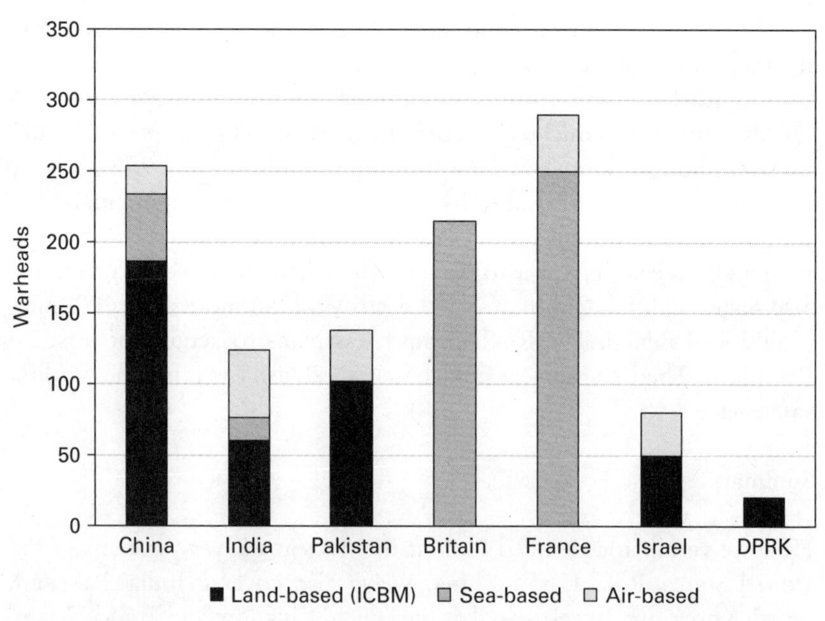

Figure 14.12
Distribution of nuclear warheads other than U.S. and Russia (which were shown in figure 14.2), categorized as land, sea, or air based. The warhead numbers for North Korea (DPRK) are highly uncertain and probably low. (Data source: SIPRI Yearbook 2018, ch. 6)

Those features of weapons that seem important to military planners—destructive force, multiple warheads, warhead accuracy—may actually diminish national security by inviting preemptive attack or encouraging dangerous arms competition. Weighing perceived needs for military strength against the overwhelming necessity of avoiding nuclear war poses some of the most difficult and contested questions.

Glossary

accuracy A measure of how close a warhead gets in relation to its target. Quantified by the circular error probable (CEP), introduced in chapter 12.

air-launched cruise missile (ALCM) A cruise missile launched from an airplane.

ballistic missile A missile that, after being boosted above the atmosphere, travels only under the influence of gravity.

ballistic trajectory A missile trajectory that's determined by gravity alone and doesn't involve rocket power.

basing scheme The way nuclear weapons are deployed in readiness for wartime use. Examples include fixed, land-based deployment; mobile basing; and submarine basing.

boost phase The first part of a ballistic missile's flight, when its rocket engines are firing.

bus The uppermost stage of a multiple-warhead missile, responsible for carrying and aiming the individual warheads.

cruise missile A subsonic, jet-powered missile that flies at treetop level to avoid radar detection.

decoy A penetration aid, often a simple balloon, deployed by a ballistic missile in the midcourse phase to confuse defensive systems.

first strike A nuclear surprise attack aimed at crippling an adversary's retaliatory capability.

flight time The time it takes a weapon to go from its base to its target.

hard target An underground missile silo, military command center, or other structure that has been fortified to resist nuclear blasts.

hypersonic glide vehicle (HGV) A delivery system that combines the speed of a ballistic missile with the maneuvering capabilities of a cruise missile.

inertial guidance system A mechanism that senses changes in motion and from those changes determines accurately its position and motion. Used in high-accuracy ballistic missiles and for submarine navigation.

intercontinental ballistic missile (ICBM) A missile with range exceeding 5,500 kilometers (3,400 miles).

invulnerability A delivery system's ability to withstand an enemy's first strike or surprise attack.

midcourse phase The portion of a ballistic missile's flight when it's traveling unpowered above Earth's atmosphere.

Minuteman III A U.S. intercontinental ballistic missile, currently the only land-based ICBM in the U.S. arsenal.

multiple independently targetable reentry vehicle (MIRV) Individual warhead-bearing reentry vehicles that can be independently targeted. Three to 10 or more MIRVs may be carried on a single missile.

nuclear delivery system Any system designed to deliver nuclear weapons to their targets. Missiles and aircraft are common examples.

oxidizer A substance that supplies oxygen to combust a missile's propellant.

payload The maximum weight a vehicle can deliver to its target.

penetration A delivery system's ability to penetrate enemy territory in order to reach its target.

penetration aid Any object(s), such as a decoy or chaff, or a strategy, that increases the probability of a missile's warhead reaching its target.

propellant A missile's fuel, which also requires an oxidizer to combust.

range The distance a nuclear delivery vehicle can travel from its base to a target.

soft target A target, such as a city or a factory, that isn't fortified against nuclear blast.

SSBN A ballistic missile submarine. The acronym stands for ship, submersible, ballistic missile, nuclear powered.

stealth technology Technology that makes a weapon hard to detect. With aircraft and missiles, stealth includes the use of nonmetallic construction and special designs to minimize radar reflection.

strategic In the context of nuclear weapons, strategic refers to weapons to be used against an adversary's home territory (in contrast to tactical weapons, designed for battlefield use).

submarine-launched ballistic missile (SLBM) A ballistic missile capable of being launched from a submerged submarine.

terminal phase The portion of a ballistic missile's flight between atmospheric reentry and reaching the target.

triad The three-part structure of the superpowers' and other countries' nuclear forces, consisting of bomber aircraft, land-based intercontinental missiles, and submarine-launched missiles.

unmanned aerial vehicle (UAV) A pilotless aircraft that might be used to deliver weapons. Examples include drones and cruise missiles.

Notes

1. U.S. data from Kristensen and Korda, "United States Nuclear Forces, 2020," *Bulletin of the Atomic Scientists* 76, no. 1 (January 13, 2020), Table 1; Russia data from Kristensen and Korda, "Russian Nuclear Forces, 2019," *Bulletin of the Atomic Scientists* 75, no. 2 (March 4, 2020), Table 1. In both cases, footnotes were consulted to get only *deployed strategic* warheads.

2. See rare footage of a bus releasing dummy warheads over Los Angeles in 2015 at https://www.youtube.com/watch?v=cc0_wZat4nI.

3. David Rodman, "Unmanned Aerial Vehicles in the Service of the Israel Air Force," *Meria Journal* 14, no. 3 (2010).

4. A rare exception is this photo (URL below) showing the propeller of the French submarine *Le Redoutable*. Notice that the central axis of symmetry goes into the page and the blades are along a plane perpendicular to this axis. This picture reveals many details that normally wouldn't be disclosed. See https://www.reddit.com/r/drydockporn/comments/38npcn/sevenbladed_propeller_of_french_submarine_snle_le.

5. This was a fact that the South Korean media completely missed and continues to miss. The B1-B bomber has been denuclearized since 1994 and is not part of the nuclear mission. Our concern is that the North Koreans may not know that and may overreact, believing that a B1-B bomber carries nuclear weapons. See https://www.af.mil/About-Us/Fact-Sheets/Display/Article/104500/b-1b-lancer/.

6. Martin J. Sherwin, "The Cuban Missile Crisis Revisited: Nuclear Deterrence? Good Luck!," *Cornerstone 2018*, George Mason University College of Humanities and Social Sciences, https://cornerstone.gmu.edu/articles/4198.

7. R. Medcalf and B. Thoman-Noone, "Nuclear-Armed Submarines in Indo-Pacific Asia: Stabiliser or Menace?," www.lowyinstitute.org/publications/nuclear-armed-submarines-indo-pacific-asia-stabiliser-or-menace.

8. Michael Krepon, Travis Wheeler, and Shane Mason, eds., *The Lure and Pitfalls of MIRVs: From the First to the Second Nuclear Age* (Washington, DC: Stimson Center, 2016).

9. J. Wallis Simons, "How Washington Owns the UK's nukes," *Politico*, April 30, 2015, https://www.politico.eu/article/uk-trident-nuclear-program/.

10. See David Majumdar, "The 'Super-Fuze': The Big Upgrade to America's Nuclear Arsenal That Russia Fears," *National Interest* (Blog), October 17, 2017, https://nationalinterest.org/blog/the-buzz/the-super-fuze-the-big-upgrade-americas-nuclear-arsenal-22765, and Hans M. Kristensen, Matthew McKinzie, and Theodore A. Postol, "How U.S. Nuclear Force Modernization is Undermining Strategic Stability: The Burst-Height Compensating Super-fuze," *Bulletin of the Atomic Scientists* (March 1, 2017), https://thebulletin.org/2017/03/how-us-nuclear-force-modernization-is-undermining-strategic-stability-the-burst-height-compensating-super-fuze/.

11. Tyler Durden, "Hillary's 'War Drums' Confirm Putin's Fears of a World 'Rushing Irreversibly' Towards Nuclear Showdown," *ZeroHedge* (October 17, 2016), https://www.zerohedge.com/news/2016-10-17/hillarys-war-drums-confirm-putins-fears-world-rushing-irreversibly-towards-nuclear-s.

Further Reading

Cochran, Thomas, et al. *Nuclear Weapons Databook* (Cambridge, MA: Ballinger, 1984–1989). An older four-volume compendium of data on nuclear weapons and delivery systems, and on the nuclear weapons industry's research, production, and testing facilities, prepared by the Natural Resources Defense Council. Volume I includes photographs and design specifications of airplanes, missiles, ships, reentry vehicles, and nuclear warheads. Volumes II and III detail U.S. nuclear warhead facilities and production, and volume IV covers Soviet nuclear weapons. Despite its age, the information in these volumes is still very relevant.

Kristensen, Hans, and Matt Korda. "United States Nuclear Forces, 2020." *Bulletin of the Atomic Scientists* 76, no. 1 (January 2020): 46–60. A thorough, up-to-date (through New START) discussion and tabulation of U.S. nuclear forces, including where they're deployed. Published annually.

National Air and Space Intelligence Center. "Ballistic and Cruise Missile Threat." Wright-Patterson Air Force Base, OH, 2017. https://fas.org/irp/threat/missile/bm-2017.pdf. An authoritative look at evolving missile technologies around the world.

Nuclear Notebook. Since 1987, the *Bulletin of the Atomic Scientists* has published the *Nuclear Notebook*, an authoritative accounting of world nuclear arsenals compiled by top experts from the Federation of American Scientists. Today it's prepared by Hans M. Kristensen and Matt Korda of FAS. Because of its importance to researchers, governments, and citizens around the world, the *Nuclear Notebook* is always free to access: https://thebulletin.org/nuclear-risk/nuclear-weapons/nuclear-notebook/. See the second item in this list of readings for a specific article by Kristensen and Korda.

NTI (Nuclear Threat Initiative). *Tutorial on Delivery Systems*. An accessible tutorial on nuclear weapon delivery systems developed by the James Martin Center for Nonproliferation Studies for the Nuclear Threat Initiative website. An excellent overview of nuclear weapon delivery systems. https://tutorials.nti.org/delivery-system/introduction/.

Nuclear Threat Initiative 3D missile models. On these web pages you can virtually inspect particular missiles in three dimensions. https://www.nti.org/analysis/articles/north-korean-ballistic-missile-models/ for North Korean missiles. See also the 3D virtual museum where using VR goggles or Google Cardboard you can walk through a virtual museum containing all North Korean missiles. https://www.nti.org/newsroom/news/evolution-and-scope-north-koreas-nuclear-program-illustrated-new-3d-virtual-museum/.

Stockholm International Peace Research Institute. *SIPRI Yearbook: World Armaments and Disarmament*. Oxford: Oxford University Press, published annually. Authoritative tabulations of weapons systems, military expenditures, weapons proliferation, and related topics, accompanied by summary essays. The nuclear weapons section is authored by experts from the Federation of Atomic Scientists.

15

Nuclear Strategy

How should nuclear weapons be used? In view of the devastation that nuclear war would cause, an obvious answer is that they shouldn't be used. So let's broaden the concept of "use" to include the role of nuclear weapons as instruments of international politics. How, then, should these agents of mass destruction be "used"? How many should a country have, and how should they be deployed in order to enhance national and global security? What moral strictures attach to the possession of nuclear weapons and to the threat of their use? These deep and difficult questions were thrust on humanity with the dawn of the nuclear age, and they remain under vigorous debate. Here we explore briefly some basic issues of nuclear strategy. Since this is a book emphasizing technology, the discussion will necessarily be sketchy; the reader is referred to the many published analyses of nuclear policy issues, some of which are listed at the end of this chapter.

Vision of the Nuclear Scientists

On September 12, 1933, the Hungarian-born physicist Leo Szilard had a sudden, unprecedented vision of the possibility of a nuclear chain reaction. Szilard, a political idealist strongly influenced by the visionary novels of H. G. Wells, saw more than physics. He saw also nuclear bombs and a world order fundamentally altered by their presence.

Over the next decade, Szilard pushed hard to see his chain reaction brought to fruition. He believed the existence of nuclear bombs would mean an end to war and the advent of a new era of enlightened world community. During the war years Szilard's vision was shared by the Danish physicist Niels Bohr, who himself directly influenced J. Robert Oppenheimer. After the war, Oppenheimer became an outspoken advocate for international control of nuclear weapons, and his urging helped the United States pursue that goal in the newly formed United Nations.

But the scientists' vision was not to be. Growing mistrust between the United States and the Soviet Union derailed U.N. negotiations, and led instead to an accelerating nuclear arms race. Disappointment with the course of events made many former bomb makers outspoken and enduring critics of nuclear policy; Szilard himself founded the Council for a Livable World, a group that to this day supports disarmament-minded candidates for the U.S. Congress.

Deterrence

Scientists weren't alone in recognizing that their nuclear offspring had changed the world. In 1946, the strategic analyst Bernard Brodie stated: "Thus far the chief purpose of our military establishment has been to win wars. From now on its chief purpose must be to avert them."[1] In Brodie's statement lies the essence of nuclear **deterrence**, the idea that we possess nuclear weapons in order to prevent their use. In 1983, former U.S. Secretary of Defense Robert McNamara made it even clearer: "Nuclear weapons serve no military purpose whatsoever. They are totally useless—except only to deter one's opponent from using them."[2] A 2013 report by the Obama administration reemphasized deterrence as the only function of nuclear weapons: henceforth U.S. nuclear strategy would "focus on only those objectives and missions that are necessary for deterrence in the 21st century."[3]

Deterrence requires the threat of nuclear retaliation: If you dare to attack me, I will use my nuclear weapons against you. For deterrence to work, that threat must be *credible*; a potential attacker has to believe its opponent *can* and *will* use its nuclear retaliatory force. What makes a nuclear deterrent credible?

First and foremost, a deterrent force must be capable of inflicting "unacceptable" damage on an opponent. Unacceptability depends on both technical and political factors. Technical because retaliatory weapons must be physically capable of sufficient destruction; political because potential adversaries must believe that the level of damage suffered in a retaliatory strike would indeed be unacceptable to their country.

A retaliatory force is no good if it can be destroyed in a first strike; that's why there was so much emphasis on weapon systems' invulnerability in the preceding chapter. Invulnerability is primarily a technical issue, but questions of political will are also involved. For example, land-based

ICBMs may be vulnerable to a first strike—*if* they're in their silos when the attacking warheads hit. But a **launch-on-warning** strategy would have the missiles underway once warning of an impending attack was received, again giving them a measure of invulnerability. Launch-on-warning is risky: it hangs the fateful decision to launch nuclear missiles on electronic warning systems and reduces to almost zero the time for rational thinking. But the mere possibility of launch-on-warning should give a potential aggressor pause, so the possibility alone may contribute to deterrence.

It's not enough that a deterrent force be invulnerable and capable of ample destruction; it's also essential that its possessor show the political will to use its deterrent force if necessary. Therein lies a nuclear dilemma. Deterrent forces exist to deter war, but deterrence works only if a potential attacker believes they would actually be used. Paradoxically, a nation bent on deterring nuclear war must show willingness to engage in such war.

What constitutes a credible deterrent? That's been a central question of nuclear strategy since the 1950s. The debate isn't between those who favor nuclear war and those who oppose it; rather, it's a debate about the best way to prevent nuclear war. That "the best way" isn't obvious is again related to a central dilemma of the nuclear age—that the principal use of nuclear weapons is to prevent their use. In that context, nuclear hawks would argue that we, and indeed the world, are safer with more and better nuclear weapons; nuclear doves counter that huge nuclear arsenals only increase the chance of inadvertent, accidental, or crisis-driven nuclear war, and ensure global catastrophe should deterrence fail.

As long as deterrence holds—and with it the necessity that nations be ready and willing to use their nuclear weapons—nuclear war remains a possibility. How should we plan for that possibility? Should we plan at all, or does planning only make nuclear war more likely? Should we, as some would argue, be prepared to prevail in a nuclear conflict, by maintaining nuclear superiority over potential adversaries? Should we engage in damage limitation measures, such as civil defense and hardening of key industries, to lower nuclear war's impact on our nation? Or would damage limitation only make us more likely to accept nuclear conflict? Another nuclear dilemma: preparing for nuclear war might reduce our losses in the event of war but might at the same time make war more likely.

Is nuclear war winnable? That question is important if we believe deterrence might fail, for an affirmative answer implies nuclear arsenals far in excess of what's needed for deterrence. Today the world's leaders are in consensus that nuclear war is not winnable, but that opinion hasn't always prevailed, even among those in power. And many nuclear doves would argue that continuing technological advances and modernization of nuclear forces reveal a quest for nuclear superiority among military planners, despite what political leaders may say.

The possibility that deterrence could fail leads some hawks to advocate that we prepare to wage, and ultimately to prevail in, a protracted nuclear conflict. It leads some doves to reject deterrence altogether, arguing that its possible failure makes nuclear war an eventual inevitability. In between is a vast middle ground whose occupants generally support the concept of deterrence but disagree substantially about how best to achieve stable and lasting deterrence. The evolution of nuclear strategies reflects the ebb and flow of these disagreements, the changing views of strategic thinkers, and the reality imposed by new technologies.

Nuclear Strategic Doctrines: The Soviet Era

After World War II, the United States held a monopoly on nuclear weapons. But scientists realized there was no fundamental nuclear secret and that any industrialized country with the will to do so could build nuclear bombs. Nuclear monopoly wouldn't prevail, although estimates suggested it might take others from 5 to 20 years to duplicate the Manhattan Project's achievement. The United States tried briefly to flaunt its nuclear might in an effort to coerce others to do its bidding. That didn't work; in particular, the Soviet Union ignored the implicit nuclear threat and continued its occupation of eastern Europe with an army that outnumbered U.S. forces in Europe by four to one. At the same time, the Soviet Union began vigorous efforts to develop its own nuclear capability, culminating in its first fission explosion in 1949.

How was the United States to respond to the perceived threat from the Soviet Union's vastly greater conventional forces and its growing nuclear capability? In the early 1950s, the U.S. still enjoyed an overwhelming nuclear advantage, and President Eisenhower's administration decided to base U.S. policy on that advantage. Instead of an expensive effort to match Soviet conventional forces, the United States would counter any Soviet aggression with devastating nuclear retaliation against the Soviet homeland. This policy earned the name **massive retaliation.**

When massive retaliation became official policy in 1953, the United States already had well over 1,000 nuclear weapons and a bomber fleet capable of delivering them to Soviet targets. Since 25 percent of the Soviet Union's population and 50 percent of its industrial capacity were concentrated in 100 cities, 1,000 weapons were far more than sufficient for massive destruction of the Soviet Union. And the American deterrent force was credible: in the early 1950s, the U.S. could have launched a massive nuclear attack without suffering "unacceptable" damage from the then-puny Soviet nuclear arsenal.

The mid-1950s saw significant changes in the nuclear balance. Both the United States and the Soviet Union exploded their first megaton-range fusion weapons, and the Soviets developed intercontinental bombers capable of striking the United States. A U.S. nuclear strike in response to Soviet conventional aggression would no longer go unpunished: now the United States could expect massive Soviet nuclear retaliation. Under those circumstances, would the United States really be willing to launch a nuclear attack on the Soviet Union in order to counter aggression in Europe or elsewhere? That question itself undermined the credibility of massive retaliation as a deterrent strategy.

The eroding credibility of massive retaliation led to an emphasis on smaller, short-range nuclear weapons deployed in what was then a divided Europe. The notion of limited nuclear war developed, the idea being to restore deterrence with the more credible threat of using smaller nuclear weapons in a localized conflict that need not spread to the superpowers' homelands. Another nuclear dilemma: Is it really possible to deter nuclear war by making it more acceptable (i.e., limited)? And would a limited war remain limited? Strategists envisioned a graduated sequence of nuclear war scenarios ranging from a local battlefield conflict to all-out strategic bombing; with many possibilities in between, they argued, escalation to all-out war was unlikely. Nuclear doves rejected the assumption that rational decision making would prevail in the heat of conflict and saw the risk of all-out war increasing with each new scenario.

By the early 1960s, the development of intercontinental ballistic missiles made the threat of massive retaliation seem even more hollow; now a swift half-hour exchange would leave both sides in near total ruin. Although the United States' nuclear forces remained numerically superior at that time, it had become clear that the United States and the Soviet Union were already in a nuclear stalemate. Numerical advantage no longer translated into military advantage, and the notion of nuclear superiority was fast becoming obsolete.

A strategy called **flexible response** evolved in light of the changing nuclear picture. Pioneered by the Kennedy administration in the early 1960s, flexible response called for deterrence of each possible form of aggression with the threat of a counterattack *in kind*—that is, a conventional attack would be countered with conventional weapons, a tactical nuclear strike with tactical nuclear weapons, and a strategic attack with a comparable level of strategic nuclear megatonnage. Only if deterrence failed at one level would we up the ante; thus the United States would use nuclear weapons against a Soviet conventional attack in Europe only if Western conventional forces failed to halt the attack. Again, the idea was to make deterrence more credible. Although the United States might not be willing to attack the Soviet homeland with thermonuclear weapons in response to an invasion of western Europe, it would certainly have fewer qualms about unleashing conventional or even tactical nuclear forces.

Would flexible response work? Again, the same dilemma: flexible response lowers the threshold for using nuclear weapons, making that use more likely. But it does so in the name of deterrence—that is, preventing the use of those weapons. And flexible response requires vastly expanded military capabilities, both conventional and nuclear. No longer are a few hundred strategic nuclear warheads a sufficient deterrent; now it's necessary to have, in kind and in quantity, all the weapons possessed (or thought to be possessed, or possibly under development) by one's potential adversaries. Flexible response may or may not be a suitable deterrent, but it's surely a prescription for an escalating arms race.

Nuclear Targets

Where should we aim our nuclear weapons? At an opponent's population centers? Industries? Military facilities? Questions of targeting go hand in hand with decisions about nuclear strategy.

Most Americans imagine that our nuclear weapons are targeted on potential adversaries' major cities, and they envision nuclear war as an all-out assault on population centers. That impression is only partly correct. Although today's nuclear targets remain classified, most of our weapons are probably aimed at specifically military targets. Some of those targets are, to be sure, in or near cities. Many, though, are isolated missile silos, submarine bases, and remote military installations. On the other hand, even the relatively few weapons targeted against cities are sufficient to kill hundreds of millions of civilians.

In the cold parlance of nuclear strategy, targeting an adversary's population and industry is known as a **countervalue** strategy. The name arises because people and industry are considered to have value, something a country wouldn't want to lose. Targeting military facilities, on the other hand, is a **counterforce** strategy, aimed at destruction of an opponent's armed forces. The original massive-retaliation policy of the 1950s was a

Targets Declassified

In 2015, the United States declassified nuclear target lists from 1956, the height of the Cold War bomber era and predating intercontinental ballistic missiles. The lists put some 1,100 Soviet-bloc airfields as top priorities, but a chillingly large number of targets were major population centers that appeared on a 306-page "urban-industrial target list." Figure 15.1 reproduces one of those pages, which includes the target list for Omsk, a Russian city in southwestern Siberia with a population around 1 million.[4] Targets are coded by category, with several categories identified on the figure. Category 275, which appears at the end of every city's target list, is termed, simply, "population"—indicating that killing of civilians was very much part of the United States' early targeting strategy. Cities on the target list are primarily in the Soviet Union. However, they also include much of eastern Europe as well as locations in China, Vietnam, and North Korea.

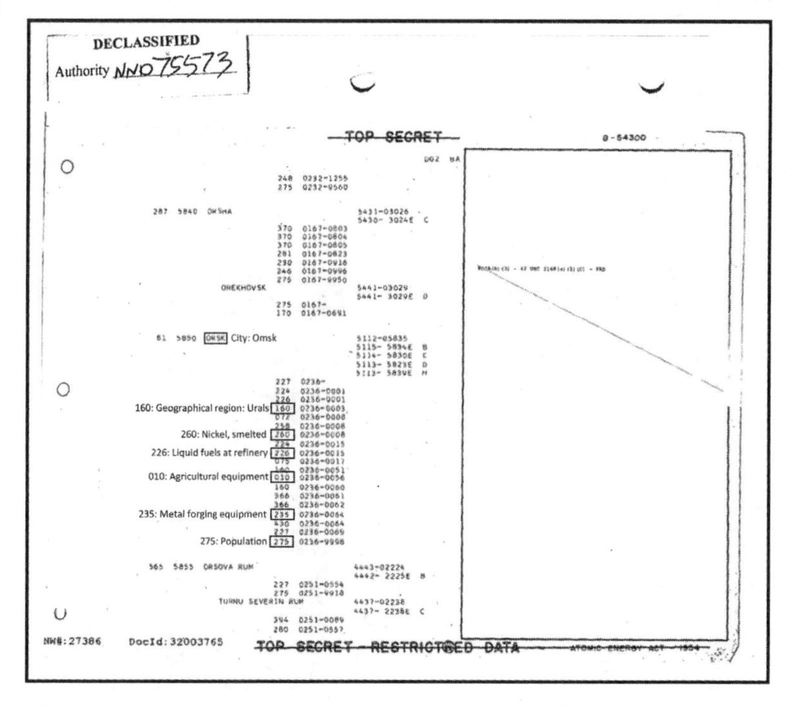

DECLASSIFIED
Authority NND 75573

TOP SECRET

DGZ BA

248 0232-1255
275 0232-9560

287 5840 OKSHA 5431-03026
5430- 3024E C

370 0167-0803
370 0167-0804
370 0167-0805
281 0167-0823
230 0167-0916
246 0167-0996
275 0167-9950

ORKHOVSK 5441-03029
5441- 3029E D

275 0167-
170 0167-0691

81 5850 OMSK City: Omsk 5112-05835
5115- 5834E B
5114- 5830E C
5113- 5823E D
5113- 5839E H

227 0236-
224 0236-0001
226 0236-0001
160: Geographical region: Urals 160 0236-0003
072 0236-0008
258 0236-0008
260: Nickel, smelted 260 0236-0008
224 0236-0015
226: Liquid fuels at refinery 226 0236-0015
075 0236-0017
160 0236-0051
010: Agricultural equipment 010 0236-0056
160 0236-0060
366 0236-0061
366 0236-0062
235: Metal forging equipment 235 0236-0064
430 0236-0064
227 0236-0069
275: Population 275 0236-9998

565 5855 ORSOVA RUM 4443-02224
4442- 2225E B

227 0251-0554
275 0251-9918

TURNU SEVERIN RUM 4437-02238
4437- 2238E C

394 0251-0089
280 0251-0557

NW#: 27386 DocId: 32003765

TOP SECRET – RESTRICTED DATA

Figure 15.1
Page from a 1956 U.S. nuclear target list, showing the then-Soviet city of Omsk in southwestern Siberia. We've added annotations to some categories to show the sorts of materials and activities that made Omsk a strategic target. The comparable listing for Moscow runs four pages. This target listing was declassified in 2015 (see note 4).

countervalue strategy, threatening destruction of Soviet population centers. The flexible-response strategy of the 1960s put more emphasis on counterforce capability, and in the 1970s and 1980s counterforce emerged as the dominant theme in strategic nuclear planning. A counterforce strategy may involve not only the destruction of an adversary's military forces but also **decapitation**—the elimination of civilian and military leadership through precise nuclear targeting.

Which strategy is most likely to prevent nuclear war? Proponents of countervalue argue that the horrible threat of population annihilation remains the most effective deterrent. Their argument rests on an adaptation of the massive-retaliation doctrine to an age in which the superpowers' nuclear arsenals are essentially equivalent. That doctrine goes by the acronym MAD, for **mutual assured destruction**. MAD holds nuclear adversaries' populations hostage against the threat of nuclear attack, maintaining the "balance of terror" which is often alleged to be responsible for keeping the nuclear peace. Under MAD, a nation contemplating nuclear attack is deterred by the knowledge that its own destruction in a nuclear counterattack is certain (figure 15.2).

By assuring nuclear destruction, MAD advocates claim, their strategy guarantees that no rational leader would order a nuclear attack.

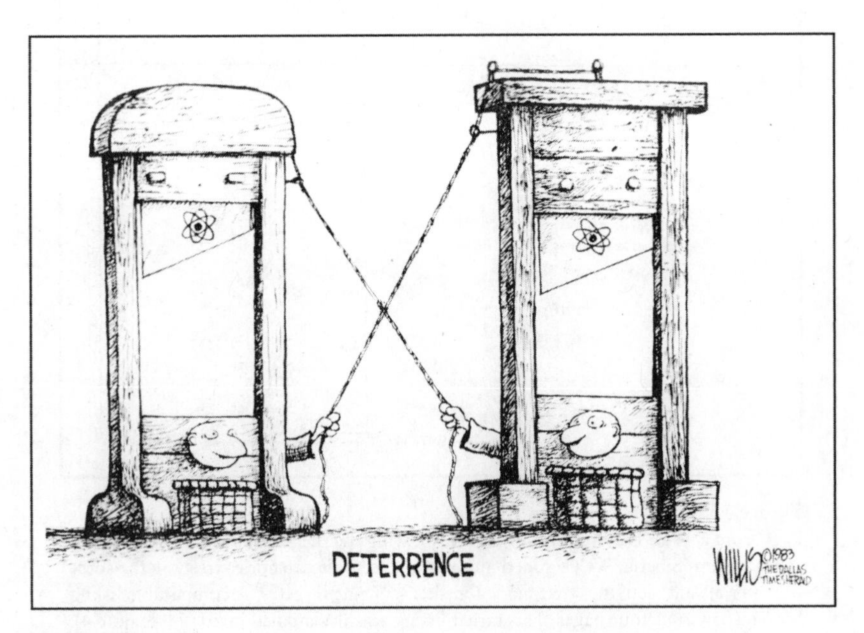

Figure 15.2
A cartoonist's view of MAD. (Scott Willis, ©1983 *Dallas Morning News*)

Counterforce attacks, in contrast, might be viewed as acceptable, rational means to achieve national goals. MAD advocates have claimed that view is illusory for two reasons. First, the presence of military targets in or near cities implies substantial collateral damage and civilian casualties even with counterforce targeting. Second, a nuclear war initially involving only military targets could quickly escalate to an all-out conflict consuming population centers as well.

Proponents of counterforce strategy have countered with the same argument that derailed massive retaliation: the threat of an all-out attack against population centers is too unbelievable to deter any aggression short of an initial all-out attack, so MAD cannot prevent more limited aggression. They argue instead that aggressive behavior is best deterred if an adversary knows that the other side can respond with a specific nuclear strike appropriate to the situation at hand, possibly including destruction of the adversary's own nuclear missiles. Counterforce proponents claim, too, that MAD is militarily ineffective, killing harmless civilians instead of destroying harmful weapons. Some add an ethical twist, arguing that it's immoral to hold civilians hostage to the threat of nuclear annihilation. Better, they say, to target weapons than human beings.

U.S. Nuclear Strategies in the Post–Cold War Era

Our discussion so far has focused on nuclear strategies in the pre-1990s world dominated by the Cold War between the United States and the Soviet Union. The collapse of the Soviet Union in 1991 ended the Cold War and decreased the danger of a direct nuclear confrontation between the U.S. and the new Russian Federation. It also helped enable arms-control treaties that produced significant declines in the two superpowers' nuclear arsenals. We'll discuss those treaties further in chapter 18. In other ways, however, the dissolution of the Soviet Union complicated nuclear strategic policymaking not only for the United States but for other Cold War era nuclear weapon states as well. Among the complicating factors are the development of nuclear weapons by North Korea, the nuclear confrontation between India and Pakistan, nuclear ambitions in Iran, the rise of international terrorism, and an increasingly belligerent Russia. Russian belligerence has many causes, but in part it's a response to the spread of the North Atlantic Treaty Organization (NATO) into formerly Soviet-bloc states, whose protection now falls under the U.S. nuclear umbrella.

Nuclear Posture Reviews

Since the end of the Cold War, the United States has published regular, congressionally mandated reviews of U.S. nuclear strategy, including recommendations for future weapons and policy developments on a roughly decadal timescale. To date there have been four of these **Nuclear Posture Reviews** (NPRs). All four reviews recommend continuing the policy of deterrence through the secure nuclear triad that we introduced in the preceding chapter. However, they differ in subtle details and in whether they move U.S. policy toward or away from reliance on nuclear weapons for the nation's security. The first two of these reviews were classified and so we don't have access to the full documents; the remaining two reviews are available to the public.

The first NPR was issued in 1994 under the administration of President Bill Clinton, at a time when the United States and Russia were enjoying relatively cordial relations in the wake of the Soviet collapse. The 1994 NPR emphasized a declining role for nuclear weapons in U.S. security, ongoing reductions in nuclear arsenals, and a stand-down in readiness of nuclear forces. The report, in the words of the Federation of American Scientists, "struck a prudent balance between leading the way to a safer world and hedging against the unexpected."[5]

The second NPR, issued by the George W. Bush administration in late 2001, continued to recommend reductions in nuclear arsenals but saw the role of nuclear deterrence expanded to include not only nuclear but also biological and chemical attacks, as well as unexpected new threats. The report also called for development of new nuclear weapons, including bunker busters that would burrow underground before detonating, to take out hardened subterranean military facilities. The 2001 NPR was criticized by groups such as the Union of Concerned Scientists both for proposing new weapons and for blurring the distinction between nuclear and conventional conflict.

The third NPR, and first to be unclassified, was issued in 2010 by President Barack Obama's administration. Broader in scope than previous NPRs, this review emphasized the need to prevent both nuclear weapons proliferation and nuclear terrorism. It also clarified previously ambiguous statements about nuclear threats or actual weapons use, declaring that "The United States will not use or threaten to use nuclear weapons against non-nuclear weapons states that are party to the Nuclear Non-Proliferation Treaty and in compliance with their nuclear non-proliferation obligations." The document also emphasized that the sole purpose of nuclear weapons should be deterrence of *nuclear* attacks only—thus reining in the wider-ranging deterrence envisioned in the second NPR. And the third NPR continued the first NPR's emphasis on reducing the role of nuclear weapons in national security. This third report drew praise from the Federation of American Scientists.[6]

The administration of President Donald Trump released the fourth NPR in 2018. This one reverses the 2010 review's clear distinction between nuclear and conventional warfare, by calling for a tighter integration of conventional and nuclear forces and an expansion of the situations under which the United States would use nuclear weapons. The NPR does not spell out such situations in detail but does state that they include "significant non-nuclear strategic attacks."[7] It also calls for deployment of new, low-yield nuclear weapons and a new nuclear-capable submarine-launched cruise missile, both of which could make the transition from conventional to nuclear war more likely. The 2018 NPR also states explicitly that "the United States will not seek Senate ratification of the Comprehensive Nuclear Test Ban Treaty."[8] And although it expresses support for the Non-Proliferation Treaty (NPT), it neglects the treaty's obligation that the U.S. (and other nuclear-weapon states) actively pursue nuclear disarmament. The 2018 NPR has been roundly criticized by the Union of Concerned Scientists, the Federation of American Scientists, and the *Bulletin of the Atomic Scientists*. Particularly worrisome is the statement, quoted above, about a possible nuclear response to "significant non-nuclear strategic attacks." Would this include cyberattacks—for example, an attack that uses neither conventional nor nuclear weapons but manages, through computer hacking, to shut down the U.S. electric power system? Or attacks by non-nuclear antisatellite weapons on space-based communications, navigation, or surveillance assets?

The Trump administration's new nuclear policy saw its first hardware implementation in 2020, with the deployment of a new low-yield warhead (the W76–2, about 6.5 kilotons) on some U.S. ballistic missile submarines. The argument for the new warhead is that it will deter limited nuclear attacks by Russia because the Russians will realize that the U.S. can launch a counterattack without having to use high-yield weapons that could lead to all-out war. Opponents—including former secretaries of state and defense—criticized the move as lowering the threshold for nuclear combat, and pointed out that an attack with the new low-yield weapons couldn't be distinguished from an all-out attack with larger warheads until the weapons had actually detonated.

Others' Nuclear Strategies

Nuclear strategies are sensitive elements of national policy. Often they aren't made public—as was the case with the United States' first two Nuclear Posture Reviews. The U.S. is nevertheless the most transparent of the nuclear-armed nations, although France also has a long-standing policy of nuclear transparency. Here we summarize briefly what's known about other nations' nuclear strategies, based on public statements as well as expert studies. You'll find some of these sources in the Further Reading at the end of the chapter.

Russia

Late in the Soviet era and in the early post-Soviet years, the nuclear strategy of the Soviet Union and then the Russian Federation was predicated on the assumption that a nuclear conflict would be an all-out global affair. The purpose of nuclear weapons was to deter such a conflict. Although the USSR and then Russia initially adopted a no-first-use policy, the Soviet/Russian strategy included inflicting unacceptable damage through a strike launched on warning of an incipient attack.

Russian nuclear strategic doctrine underwent significant changes around the turn of the twenty-first century and beyond. In addition to deterrence of global war, nuclear weapons were now envisioned to have a role in regional conflicts. In particular, the threat of nuclear weapons or their actual but limited use could, the Russians hoped, help deescalate a regional war. That limited use would be against military targets, but it included the possible first use of nuclear weapons. No longer would a threatened Russian nuclear strike necessarily bring unacceptable damage; rather, nuclear damage could be limited to the context of a regional war.

In 2010, Russia tightened the threshold for its use of nuclear weapons. Previously such weapons would be used in situations "critical for national security"; now they would be used only when "the very existence" of the country was threatened. However, Russia soon reclassified NATO from a "challenge" to a "threat," and continued to envision limited use of nuclear weapons to deescalate an initially conventional conflict with NATO. This possibility, according to some analysts, today makes the chances of a U.S.-Russian nuclear conflict greater than it's been since the height of the Cold War.

China

China first tested a nuclear weapon in 1964 and quickly developed a modest nuclear arsenal probably numbering several hundred warheads. The goal of Chinese nuclear strategy is a "lean and effective" deterrent, aimed primarily at preventing a nuclear attack by the United States. China has made no attempt to match the much larger arsenals of the U.S. and Russia. At the same time there's been no reduction in the Chinese arsenal even as U.S. and Russian counts of deployed strategic warheads dropped from tens of thousands to less than 2,000 each. China's nuclear arsenal and the strategy behind it have remained quite consistent over the decades since China became a nuclear power. That includes a clear statement that China's nuclear policy entails no first use of nuclear weapons. In recent years, however, China has begun expanding and modernizing its nuclear capability. This trend, while consistent with China's philosophy of maintaining a modest nuclear deterrent, is likely driven by concern about changes in the United States' nuclear posture. In particular, China seems increasingly intent on being able to inflict unacceptable second-strike damage on the United States. This does not necessarily require a much larger Chinese arsenal but can be achieved through more flexible delivery systems, including road-mobile missiles, fixed-base multiple-warhead missiles, advanced nuclear-missile submarines, multiple-warhead mobile missiles, and hypersonic delivery vehicles. All but the last two of these improvements have been deployed in recent years, and the other two are under active development.

China also faces increasing threats from other Asian nuclear powers. India, in particular, has a nuclear arsenal about half the size of China's, although its delivery systems are less sophisticated. While China appears unfazed by U.S. and Russian strategic arsenals that total nearly 10 times China's, it's less clear that China would tolerate India achieving even nuclear parity. And although China remains North Korea's strongest ally, uncertainties about North Korea and its future—including its growing nuclear arsenal—are surely worrisome to China. Furthermore, the North Korean nuclear program prompts changes in the United States' military posture in Asia that could threaten China as well as North Korea. For example, missile defenses aimed at protecting South Korea and Japan from North Korean nuclear missiles could have the side effect of reducing the effectiveness of China's nuclear deterrence.

All these considerations suggest that the coming decade may see considerable enhancement of China's nuclear capability but without it approaching those of the United States and Russia.

India and Pakistan

India and Pakistan have been archrivals since the founding of Pakistan in 1947. India first detonated a nuclear device in 1974, declaring it a test of peaceful uses for nuclear explosives. No further testing occurred until 1998, when India exploded five nuclear devices over a period of two days. Just weeks later, Pakistan responded with its first nuclear tests—also a series of five explosions. Since then there have been no further nuclear tests by either country, but each has built an arsenal of around 150 nuclear weapons. Delivery systems for each country include aircraft, short- and intermediate-range missiles, and ballistic or cruise missiles on surface ships. India is actively developing submarine-launched missile capability. Neither country is a party to the Non-Proliferation Treaty.

Pakistan's nuclear strategy has not been made public, but Pakistani nuclear weapons are clearly intended primarily to deter aggression by India, including the latter's superior conventional forces. With the deployment of longer-range missiles in the mid-2010s, Pakistan appears to be intent on a second-strike capability that might be perceived as enhancing its deterrent. At the same time, Pakistan has developed low-yield tactical nuclear weapons that could lower the threshold for nuclear conflict. It appears that Pakistan's nuclear strategy is evolving away from simple massive retaliation to a more flexible approach with differing levels of nuclear response. Whether this is stabilizing or destabilizing is a matter of debate.

India, too, sees its nuclear forces as primarily a deterrent to Pakistan—although it also has its eyes on China, with which it has had several military confrontations over disputed border territory. India believes in maintaining a credible minimum deterrent, including a retaliatory force that could survive an adversary's first strike. India has pledged itself to no first use of nuclear weapons, although, unlike China's no-first-use pledge, India's would permit a nuclear response to chemical or biological weapons. And India's policy may be changing as we describe below.

Nuclear weapons are often seen as preventing even conventional wars among nuclear-armed states. But a disturbing aspect of the India-Pakistan nuclear standoff is that military conflicts between the two have actually increased since they emerged as nuclear powers in 1998. Several of these conflicts have led to real fears of escalation to a nuclear exchange, including a 2019 incident that involved airstrikes by each on the other's territory. The recent rise of Hindu nationalism in India and revocation of special autonomous status for Kashmir have contributed further to tensions between India and Pakistan. Those tensions, in turn, raised doubts about India's no-first-use policy, when, in 2019, the Indian defense minister announced, "Till today, our nuclear policy is 'no first use'. What happens in future depends on circumstances."[9] Stay tuned!

France

France, which chose nuclear power for domestic electricity to avoid dependence on imported fossil fuels, is also independent in its military nuclear posture. France's arsenal of some 300 nuclear weapons is distributed among land-based aircraft, carrier-based aircraft, and, predominately, submarine-launched ballistic missiles. Aircraft-deployed weapons are mounted on air-launched cruise missiles, while the submarine-launched missiles are MIRVed, with some five warheads on each. Today's French arsenal is about half its size at the height of the Cold War.

Since its first nuclear test in 1960—making France the fourth member of the nuclear club—the French nuclear arsenal and the programs and personnel that developed it have been very much homegrown. A 2010 treaty with the United Kingdom initiates French-British cooperation in the stewardship of their nuclear arsenals but emphasizes that the two countries' arsenals remain independent. France is so independent in its nuclear posture that, although it's a member of NATO, France's nuclear forces are not integrated with NATO's military command.

France is very clear that its nuclear weapons are "strictly defensive" and would be used only in "extreme circumstances of legitimate self-defense." Nevertheless, France does not subscribe to a no-first-use policy and would conceivably use nuclear weapons to prevent a conventional attack on its territory.

United Kingdom

The U.K. became the third country to develop nuclear arms, testing its first fission weapon in 1952. And although the British bomb was homegrown, the U.K.'s nuclear program soon became intertwined with the United States, as the U.K. acquired some complete U.S. weapons, and also began building its bombs using U.S. designs. Later the U.K. bought Trident submarine-launched ballistic missiles from the United States. Today Britain has some 120 deployed nuclear warheads, about one-fourth of its arsenal at the peak of the Cold War. Since 1988, all of the U.K.'s nuclear weapons have been deployed on nuclear-missile submarines, of which the country currently has four.

The U.K. sees its own nuclear force as a complement to the nuclear umbrella provided by the United States through NATO, and as a hedge should the United States prove unwilling to risk its own homeland in a nuclear defense of Britain. The U.K.'s nuclear strategy is therefore intentionally minimalist, offering "nuclear deterrence at the lowest possible level of conflict." To that end, the British government points to that fact that it has been reducing the number of deployed nuclear weapons, and that those weapons aren't specifically targeted and would take several days to target and launch. The British government also ensures its citizens and the world that its nuclear forces, despite being developed in cooperation with the United States, are under independent control by the U.K. only. The U.K. does not subscribe to a no-first-use doctrine and is intentionally vague about circumstances under which it might use its nuclear weapons.

The future of the U.K.'s nuclear forces is somewhat clouded by Brexit—the nation's 2020 exit from the European Union. In its wake, Scotland, whose citizens largely opposed Brexit, might choose independence from the U.K. That's a nuclear choice as well because Scotland is home to the U.K.'s only nuclear forces, namely its nuclear missile submarines.

Israel

Israel is widely acknowledged to have a nuclear arsenal of at least 80 warheads, deployed primarily as submarine-launched cruise missiles and medium- to long-range ballistic missiles. Israeli aircraft are also capable of delivering nuclear weapons. Israel probably built its first nuclear weapon in 1966, but there's no firm evidence that it has ever conducted a nuclear test—although recently declassified documents lend credence to the view that a bright flash detected by satellite over the South Atlantic in 1979 was an Israeli nuclear test. But an Israeli test wasn't essential: in the 1960s, close collaboration between Israel and France may have given Israel data from French nuclear tests.

However, Israel maintains a deliberate policy of "nuclear opacity," neither denying nor admitting that it has nuclear weapons, so much of our understanding of the Israeli nuclear program is necessarily speculative (as it is to some extent for all nuclear-armed countries). Furthermore, Israel is not a party to the Non-Proliferation Treaty, so its nuclear activities aren't subject to international inspection. However, it's no secret that Israel's greatest worry is hostile neighbors in the Middle East, especially Iran. Although none of those neighbors currently possess nuclear weapons, Iran and possibly Saudi Arabia have nuclear ambitions. In any event, Israel's nuclear arsenal is an ample deterrent to conventional attack by its Middle Eastern adversaries and is probably sufficient to deter a nuclear attack should that threat emerge. And Israel's submarine-based cruise missiles and underground ballistic missiles assure a second-strike capability. Finally, Israel's deployment of its Jericho III ICBM in 2011 gave it the capability to strike well inside Russia—suggesting a role for Israeli nuclear forces in deterring Russian aggression.

Although Israel doesn't publish a formal nuclear strategy for use of the weapons it claims it may or may not possess, as early as 1966 it had declared that a nuclear response would be justified in the event of military incursions into Israel proper, a strike that destroyed the Israeli Air Force, chemical or biological attack, or an adversary's use of nuclear weapons against Israel. Perhaps Israel's deterrent strategy is best summed up in a quote by former Israeli defense minister Moshe Dayan: "Israel must be seen as a mad dog, too dangerous to bother."

North Korea

Reclusive, impoverished, and isolated, North Korea has nevertheless a robust nuclear weapons program. Estimates of North Korea's nuclear arsenal suggest it may have as many as 60 warheads, with fissile material on hand for several dozen more. North Korea's weapons may include fusion-boosted fission weapons or even true thermonuclear (fusion) devices. There is some evidence for miniaturized warheads capable of delivery by missile. And North Korea's ambitious ballistic missile program has already developed a missile that's likely capable of reaching much of the continental United States.

North Korea maintains that the purpose of its nuclear weapons is to deter aggression by the United States. In a 2017 speech at the United Nations, the North Korean foreign minister declared that his country's nuclear force is "a war deterrent for putting an end to the nuclear threat of the U.S. and for preventing its military invasion" and that the ultimate goal was a "balance of power with the U.S." While North Korean leader Kim Jong-Un has stated that his country "will not use a nuclear weapon as long as the aggressive hostile forces do not infringe upon our country's sovereignty and interests," the North Korean government has also declared U.S. economic sanctions to be "an act of war." Some analysts see the North Korean nuclear program, and its threats to expand nuclear and missile tests, as a "blackmail strategy" aimed at extracting concessions from others, especially the United States and South Korea. The breakneck pace of North Korea's nuclear activities—from its first test in 2006 to what may be nuclear-capable intercontinental missiles by 2017—makes it difficult to be confident about the reclusive nation's nuclear strategy.

Technology and Strategy

Counterforce and countervalue—the two main strategies we've been discussing—have distinctly different technological implications. Destruction of population centers and industrial infrastructure doesn't require particularly accurate delivery, although it does demand penetration of an adversary's defenses. Nor does countervalue require speed, since cities don't move. Counterforce strategies, in contrast, make quite different demands on weapons technology. Many military targets do move: submarines submerge, missiles get launched, aircraft take off. Speed and surprise are essential in an effective counterforce attack. Destroying specific objects, especially hardened targets such as missile silos, calls for high-yield, accurate weapons and multiple warheads. Accuracy, speed, and multiple-warhead capability are thus hallmarks of a counterforce arsenal.

Unfortunately, those features are precisely what's needed for first-strike capability. Opponents of counterforce claim the strategy is dangerous for that very reason: deploying weapons that *could* be used effectively in a first strike makes such a strike more likely. Even if weapons are intended only for counterforce deterrence, the fact that they double as first-strike weapons makes them so threatening to a potential adversary that they may invite a preemptive strike. Those policy considerations have a direct bearing on decisions about nuclear weapons technology.

Minimum Deterrence and No First Use

What's the minimum size for a nuclear arsenal that will assure deterrence of potential nuclear aggression? Whatever that **minimum deterrence** figure is, today's larger nuclear arsenals very likely exceed it.

In the early 1960s, U.S. defense secretary Robert McNamara took a serious look at minimum deterrence. By analyzing destructive effects on the Soviet Union, he concluded that 400 one-megaton warheads would be a more than adequate deterrent. The "unacceptable damage" inflicted by those 400 warheads included destruction of nearly 80 percent of the Soviet Union's industrial capacity and half its population. McNamara's 400-warhead minimum was quickly trampled in the proliferation of counterforce strategies that called for more and varied nuclear weapons systems. Later, Strategic Arms Reduction Treaties (START) that followed the collapse of the Soviet Union led to significant reductions in the two superpowers' nuclear arms, but they remained well above any credible minimum deterrence level. And in the twenty-first century's deteriorating climate of U.S.-Russia relations, the futures of those and other arms-control agreements are in doubt. Meanwhile, nuclear newcomers—especially North Korea—continue to grow their arsenals. Nevertheless, recent studies support McNamara's conclusion that a few hundred weapons could constitute a credible deterrent,[10] including work by U.S. Air Force personnel suggesting some 300 weapons as an effective minimum.[11]

Minimum deterrence could be an element of a country's nuclear strategy, one that might lower the danger of a nuclear war starting and limit the damage if it did. But today only two nations—China and India—have incorporated the idea of minimum deterrence into their strategies.

Minimum deterrence is closely related to the policy of **no first use**—the declaration that a country will not be the first to use nuclear weapons in a confrontation. Like minimum deterrence, wide adoption of a no-first-use policy could reduce the risk of nuclear war. Today, China and India are the countries with no-first-use policies, and India's, as discussed above, is conditional. The United States and NATO, in particular, reject no first use so they can credibly threaten a first-use nuclear response to a massive attack with conventional weapons—and, for the U.S. as of its 2018 Nuclear Posture Review, possibly even cyberattacks.

Nuclear Ethics

We've been considering nuclear strategies as means to further national goals, to enhance national and global security, and above all to avoid nuclear war. But what's the *right* way to "use" nuclear weapons? Is morality in nuclear matters synonymous with avoiding nuclear war, or are there conditions under which wartime use of nuclear weapons would be morally justified? And what about nuclear strategies that succeed in avoiding war? Are they above moral reproach? These have been troubling questions since the beginning of the nuclear age.

Long before there were nuclear weapons, philosophers and religious leaders grappling with the morality of human aggression conceived the notion of **just war**. Not all agree with that concept, but it has helped shape moral thinking, particularly in the Judeo-Christian cultures of the West. Central to the just-war concept are the requirements that the damaging effects of war must be *proportionate* to the good expected from the war and that the use of force must be *discriminating*, harming not innocent civilians but only those who actually commit aggression.

How do these moral criteria apply to nuclear war and to the strategies for its prevention? Nuclear weapons are unique in the enormity of their destructive effects. From the millions of dead in the war-fighting strategists' most precise nuclear exchange to the extinction of humanity in the most extreme "nuclear winter" scenarios, conceivable nuclear wars fail miserably on the moral criterion of proportionate damage. And the indiscriminate nature of nuclear destruction makes their use only against aggressors essentially impossible. Innocent civilians within a target country would be slaughtered in even a limited attack, and fallout would spread nuclear death beyond the combatants' borders.

Suppose we agree that nuclear war would be immoral. Then what about nuclear deterrence, based on the threat of mutual assured destruction? MAD-based deterrence can't be an empty threat but requires the military capability and political readiness to use nuclear weapons. Some would argue that threatening the immoral act of nuclear war is enough to make MAD an immoral posture. Others agree that MAD's threat is immoral but argue that it's necessary to prevent the greater immorality of nuclear war itself. Even questioning the morality of nuclear war raises another moral dilemma: if we really believe nuclear war is immoral, doesn't that weaken deterrence by making us less willing to use our deterrent forces? That, in turn, could increase the likelihood of war itself.

Nuclear weapons pose deep and complex moral issues. Those issues aren't just abstractions for philosophical debate; as we've seen, morality even enters the controversy between advocates of counterforce and countervalue strategies. How are we to weigh moral considerations in the face of a technology that radically expands our ability to harm our fellow humans and indeed our entire planet? This nuclear question engages our deepest moral and religious convictions.

Summary

Nuclear weapons have irrevocably altered the role of military forces in international relations. No longer can a nuclear-armed nation expect to win a major war; instead, its nuclear weapons must serve to prevent war. But how best to achieve that goal of nuclear deterrence? For some, the destructive potential assured by the world's huge nuclear arsenals is enough to deter their use. Mutual assured destruction, or MAD, has kept the nuclear peace for three-quarters of a century, and, insane as its acronym implies, MAD nevertheless may be our best hope. Others argue that MAD-based deterrence will inevitably fail at some point and that only drastic reductions in nuclear forces, leading eventually to total nuclear disarmament, can save us in the long run. Still others feel the threat of massive destruction lacks credibility as a deterrent. They argue that counterforce strategies and the ability to fight and prevail in a variety of nuclear war scenarios will enhance deterrence. Opponents counter that preparation for nuclear war can only make such war more likely.

Questions of nuclear strategy have been the focus of political debate since the middle of the twentieth century. Although the world has moved toward a consensus that nuclear war is unacceptable, we're far from agreement on how to make the unacceptable also improbable. Debate on that issue will continue, and it's no exaggeration to say that our very survival hinges on the outcome.

Glossary

counterforce A nuclear strategy that targets an opponent's military forces.

countervalue A nuclear strategy that targets an opponent's population centers.

decapitation Elimination of an adversary's civilian and military leadership, usually with precisely targeted nuclear weapons.

deterrence The use of weapons to deter an adversary's attack by threatening retaliation.

flexible response A nuclear strategy calling for response in kind to an aggressor's actions: conventional attack with conventional counterattack, tactical nuclear strike with tactical nuclear weapons, strategic attack with strategic nuclear weapons.

just war A concept that gives moral justification for war provided it is both discriminating in its victims and proportionate in its response to aggression.

launch-on-warning A strategy that calls for land-based missiles to be launched upon receipt of a warning that an enemy attack is underway.

massive retaliation A nuclear strategy of the 1950s, declaring that the United States would respond to *any* Soviet aggression with a massive nuclear attack on the Soviet homeland.

minimum deterrence A strategic deterrent force of the minimum size and capability necessary to deter an adversary's attack.

mutual assured destruction (MAD) An international nuclear balance in which nuclear-armed adversaries are capable of inflicting certain and devastating destruction on each other.

no first use A pledge by a nation possessing nuclear weapons that it won't be the first to use them in a conflict.

Nuclear Posture Review (NPR) A congressionally mandated review of nuclear strategy and proposed future developments carried out regularly by the United States government's executive branch.

Notes

1. Bernard Brodie, ed., *The Absolute Weapon* (New York: Harcourt, Brace, 1946), 76.

2. Robert McNamara, "The Military Role of Nuclear Weapons: Perceptions and Misperceptions," *Foreign Affairs* 62 (Fall 1983): 79.

3. The White House, Nuclear Weapons Employment Strategy of the United States, fact sheet available at https://obamawhitehouse.archives.gov/the-press-office/2013/06/19/fact-sheet-nuclear-weapons-employment-strategy-united-states.

4. Document posted online by William Burr, National Security Archive, George Washington University, "U.S. Cold War Nuclear Target Lists Declassified for First Time," December 22, 2015, https://nsarchive2.gwu.edu/nukevault/ebb538-Cold-War-Nuclear-Target-List-Declassified-First-Ever/documents/1st%20city%20list%20complete.pdf.

5. Federation of American Scientists, extract from *1995 Annual Defense Report*, available at https://fas.org/nuke/guide/usa/doctrine/dod/95_npr.htm.

6. Hans M. Kristensen for the Federation of American Scientists, *The Nuclear Posture Review*, April 8, 2010, https://fas.org/blogs/security/2010/04/npr2010/.

7. U.S. Department of Defense, *Nuclear Posture Review 2018*, 21.

8. Ibid., 72.

9. Defense Minister Rajnath Singh, quoted in Toby Dalton, "Much Ado About India's No-First-Use Nuke Policy," September 2019, Carnegie Endowment for International Peace, https://carnegieendowment.org/2019/09/26/much-ado-about-india-s-no-first-use-nuke-policy-pub-79952.

10. Sidney Drell and James Goodby, "What Are Nuclear Weapons For? Recommendations for Restructuring U.S. Strategic Nuclear Forces," Arms Control Association, 2007. Available at https://www.armscontrol.org/system/files/20071104_Drell_Goodby_07_new.pdf.

11. Keith B. Payne, "Why US Nuclear Force Numbers Matter"; James Forsyth et al., "Remembrance of Things Past: The Enduring Value of Nuclear Weapons"; and James Forsyth, "The Common Sense of Small Nuclear Arsenals," *Strategic Studies Quarterly* 10, no. 5 (2016). *Strategic Studies Quarterly* is an Air Force sponsored forum on national and international security published by Air University Press. This special issue is devoted to deterrence in the twenty-first century. Available at https://www.airuniversity.af.edu/Portals/10/SSQ/documents/Volume-10_Issue-5/USSTRATCOM.pdf.

Further Reading

Abbasi, Rizwana, and Zafar Kahn. *Nuclear Deterrence in South Asia: New Technologies and Challenges to Sustainable Peace*. London: Routledge/Taylor & Francis, 2020. Two Pakistani scholars give their perspectives on the nuclear balance between India and Pakistan.

Bulletin of the Atomic Scientists 76, no. 1 (January 2020). This special issue is devoted to nuclear weapons policy in the context of the 2020 U.S. presidential election. Particularly relevant to this chapter are articles by John Holdren and James Miller on no-first-use policy, and Brad Roberts on Nuclear Posture Reviews.

Cohen, Avner. *Israel and the Bomb*. New York: Columbia University Press, 1999. The most authoritative public account of the Israeli nuclear weapons program through the twentieth century, including statements of Israeli nuclear strategy cited in this chapter.

Cunningham, Fiona, and M. Taylor Fravel. "Dangerous Confidence? Chinese Views on Nuclear Escalation." *International Security* 44, no. 2 (Fall 2019). A lengthy analysis of Chinese nuclear thinking and strategy, especially on the prospect of escalation from conventional to nuclear conflict and within nuclear conflict.

Ellsberg, Daniel. *The Doomsday Machine: Confessions of a Nuclear War Planner* New York: Bloomsbury, 2017. Ellsberg, famous for leaking the Pentagon Papers, was, in the early 1960s, a consultant to the U.S. Secretary of Defense on nuclear strategies. In *The Doomsday Machine*, Ellsberg unveils the mad logic behind nuclear war planning.

Freedman, Lawrence, and Jeffrey Michaels. *The Evolution of Nuclear Strategy*, 4th ed. London: Palgrave Macmillan, 2019. The new edition of this voluminous study by two British scholars is a complete rewrite of a classic work.

Heginbotham, Eric, et al. *China's Evolving Nuclear Deterrent: Major Drivers and Issues for the United States*. Santa Monica, CA: RAND Corporation, 2017. An up-to-date look at the consistency and evolution of Chinese nuclear strategy and China's nuclear arsenal. Available for download at https://www.rand.org/pubs/research_reports/RR1628.html.

Kristensen, Hans, Robert S. Norris, and Ivan Oelrich. *From Counterforce to Minimal Deterrence: A New Nuclear Policy on the Path Toward Eliminating Nuclear Weapons*. Washington, DC: Federation of American Scientists, 2009. Published as FAS Occasional Paper No.7. Available at https://fas.org/pubs/_docs/occasionalpaper7.pdf.

Kroenig, Matthew. *A Strategy for Deterring Russian Nuclear De-Escalation Strikes*. Washington, DC: Atlantic Council, April, 2018. Available at https://www.atlanticcouncil.org/wp-content/uploads/2018/04/Nuclear_Strategy_WEB.pdf. Kroenig, a former military analyst and now a professor at Georgetown University as well as a senior Fellow in the Atlantic Council's Scrowcroft Center for Strategy and Security, presents worrisome scenarios for nuclear conflict resulting from current Russian nuclear strategy.

Lewis, Jeffrey. *The 2020 Commission Report on the North Korean Attacks Against the United States: A Speculative Novel*. Boston: Mariner Books, 2018. A leading expert in nuclear nonproliferation and especially on North Korea's nuclear program presents a chilling tale of how deterrence might fail. Although fiction, there's plenty of real-world fact and expertise that makes Lewis's scenario all too believable.

The Monist 70, no 3. July 1987. The entire issue of this prestigious philosophy journal is devoted to the ethics of nuclear war. Although dated, it provides provocative discussions of nuclear strategies in a moral context.

Office of the Secretary of Defense. *Nuclear Posture Review February 2018*. Available at https://media.defense.gov/2018/feb/02/2001872886/-1/-1/1/2018-nuclear-posture-review-final-report.pdf. The fourth of the United States Nuclear Posture Reviews emphasizes modernization and flexibility in the U.S. nuclear arsenal and strategies both for its use and its role in maintaining deterrence.

Panda, Ankit. "'No First Use' and Nuclear Weapons." Washington, DC: Council on Foreign Relations, July 2018. Available at https://www.cfr.org/backgrounder/no-first-use-and-nuclear-weapons. Reviews the meaning and implications of no first use and gives an account of where each nuclear-armed country stands on first use.

Quinlan, Michael. *Thinking About Nuclear Weapons: Principles, Problems, Prospects*. Oxford: Oxford University Press, 2009. A comprehensive look at nuclear policy from a career policymaker with NATO and the British government. Includes a chapter entitled "The Ethics of Nuclear Weapons."

Rudolf, Peter. "US Nuclear Deterrence Policy and Its Problems." Berlin: German Institute for International and Security Affairs. SWP Research Paper 10, November 2018. A contemporary look at the United States' nuclear policies by a senior fellow at a leading European think tank.

U.S. Department of Defense. *Nuclear Posture Review Report April 2010*. Available at https://dod.defense.gov/Portals/1/features/defenseReviews/NPR/2010_Nuclear_Posture_Review_Report.pdf. The third of the United States Nuclear Posture Reviews emphasizes arms control and the prevention of nuclear terrorism, while acknowledging the need for modernization to maintain a safe and effective nuclear deterrent.

16

Nuclear Terrorism

We live, unfortunately, in an age of global terrorism. Terrorists with a wide range of motivations and methods strike fear in the hearts of civilian populations throughout the world. How has the threat of terrorism evolved in the nuclear age?

Goals of Terrorism

Our perception of terrorism and terrorists has changed over the years. As recently as the 1990s, terrorists were seen as small actors, hardly strategic, intent on causing harm but with the explicit goal of convincing the public of their ideology or cause. Examples include the Irish Republican Army and Basque separatist groups in Spain. In 1975, the prominent terrorism expert Brian Jenkins saw terrorists' goal as wanting "a lot of people watching, not a lot of people dead."[1] That is, earlier terrorists needed to limit casualties to avoid hurting their cause in the eyes of the public. Nevertheless, horrendous acts did occur, sometimes with casualties in the many hundreds.

When the Cold War ended, the threat of nuclear war between superpowers seemed greatly diminished. However, nuclear terrorism emerged as a serious threat. This was in part because the former Soviet Union was awash in fissile materials and even nuclear weapons that could conceivably fall into the hands of terrorist groups. In addition, new terrorist groups like Al Qaeda had publicly stated their intention to use nuclear weapons if they got their hands on them—a goal they justified under a twisted interpretation of Islam. Western terrorism analysts worried that Al Qaeda could either steal or purchase nuclear weapons or fissile materials, a serious possibility given the organization's considerable financial resources.

Thinking about terrorism, and particularly nuclear terrorism, shifted radically with the September 11, 2001, Al Qaeda attacks on the United States. In a brazen and highly coordinated operation, 19 Al Qaeda terrorists hijacked four airplanes with the aim of hitting valuable targets in the United States. Two airplanes attacked the World Trade Center in New York and one attacked the Pentagon. The fourth plane targeted a building in Washington, D.C., but crashed in a Pennsylvania field due to the actions of brave passengers.

Many readers of this book will be too young to appreciate the fear of nuclear weapons instilled in the global psyche during the Cold War, with the idea that the superpowers might end civilization in mere minutes, but many will be fully aware how the world changed after the September 11 attacks. Essentially, fear of a nuclear exchange between superpowers was supplanted by fear of an attack using weapons of mass destruction (WMD) by terrorists rather than states. This hasn't dissipated in recent years even though no attacks using either nuclear explosives or radiological weapons have occurred. This is often credited to the success of the United States in foiling terrorist plots and of homeland defense both in the U.S. and abroad. However, it might also be that the world has just been lucky or that terrorist groups are biding their time.

In a landmark speech in 2009 in Prague, President Obama declared that "the most immediate and extreme threat" to global security is nuclear terrorism and that aspiring terrorist groups like Al Qaeda would have no qualms about buying or stealing nuclear weapons or nuclear materials with the aim of actually using them. Concern over nuclear terrorism remained a primary security preoccupation of the U.S. government under President Obama and appears to have continued under the Trump administration—at least on paper.

A Focus on Radiological Weapons

The purpose of **radiological weapons** is to expose the public to nuclear radiation, either promptly or over a period of time, and to incite fear and panic. **Radiological dispersal devices** (RDDs) are designed to disperse radioactive materials over large areas, denying access and likely causing massive financial harm. Radiological weapons are different from nuclear weapons since, unlike nuclear weapons, RDDs don't involve nuclear explosions and aren't designed to cause mass casualties.[2]

The most often cited radiological dispersal device is the *explosive RDD* (known in popular parlance as a **dirty bomb**), in which conventional chemical explosives disperse radioactive materials (figure 16.1). **Non-explosive RDDs**, in contrast, use crop-dusting aircraft or other active means to disperse radioactive material in powdered or liquid form.

A related weapon is the **radiological exposure device**, where unwitting victims are exposed externally to radiation from a nearby radioactive

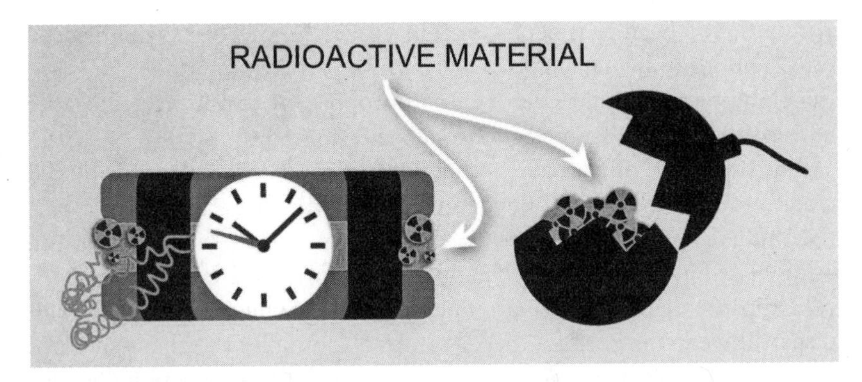

Figure 16.1
Image from a U.S. Centers for Disease Control and Prevention brochure on radiological dispersal devices and how the public should respond to them. The diagram suggests how simple RDDs might be made. Circles with light and dark triangles are the universal warning signs for radioactivity, and here represent radioactive material in the RDDs.

source. An example is a radioactive source placed beneath the seat on a train, where it could be in close vicinity to a person for a prolonged period of time and could thus produce a substantial radiation dose.

How Radiological Weapons Cause Harm

Chapter 4 showed how radiation can do enormous good—for example, diagnosing disease, curing cancer, preserving foods, or eradicating insects that spread disease or destroy crops. But the same properties that make radiation helpful can also wreak harm, especially at the hands of people or groups with nefarious objectives. Chapter 4 discussed these harmful effects along with radiation's benefits. Radiological weapons expose their victims to radiation through three pathways: ingestion, inhalation, and external exposure. With ingestion, someone unwittingly swallows radioactive material in tiny quantities undetectable by taste, with possibly fatal effect. An example is the 2006 poisoning of the Russian spy Alexandr Litvenenko when his tea was laced with the alpha-emitting isotope polonium-210. He was poisoned on November 1, deteriorated quickly, and died November 23. Inhalation exposure is similar to ingestion, except that here radioactive material in the form of airborne particulates enters the lungs through breathing. In both cases, the radioactive materials lodge within the body, where they emit high-energy particles that damage living cells.

It's useful to think of the particles released from radioactive materials as tiny bullets. The greater the number of disintegrations per second (the *activity*, as introduced in chapter 3) the more bullets are "fired" and the more damage they do. The alpha particle is the most dangerous bullet within the body since it deposits all its energy locally, destroying nearby cells. The irony is that while alpha particles are deadly inside the body, the alpha-particle bullets can be easily stopped by just a piece of paper and so external exposure isn't particularly dangerous.

The third type of exposure occurs when a radioactive source is placed close to the body. It's still dangerous because the beta and gamma particle bullets are fired at close range and have strong penetrating power, as discussed in chapter 3. Such external exposure is the goal of radiological exposure devices. These are much less challenging to construct than dispersal weapons because there's no need to modify the shape or form of the radioactive material, nor to intersperse it with conventional explosives. In addition, because of their legitimate use in medicine, industry, and science, strong radioactive sources remain widely available. That's despite a concerted global effort over the past decade to minimize the use of radioactive sources by replacing them with alternatives such as X-rays or particle beams. This has limited terrorists' access to radioactive sources in many places, but unfortunately they're still widely available, especially in the developing world.

We emphasize that the purpose of radiological terrorism is to stoke fear of being irradiated. Thus radiological terrorism can be effective even if its impact is largely psychological rather than causing physical harm. After all, who would want to live in an irradiated area where there is, say, an extra 0.1 percent chance of developing cancer? Radiological terrorism could cause people to flee affected areas, might overwhelm the medical system, and would require expensive and disruptive cleanup—if cleanup is even possible. It could also incite mistrust in government, if, for example, standards for radiation exposure were relaxed following a radiological attack. This happened when Japan raised the permissible exposure from 1 mSv to 20 mSv per year in areas contaminated by the Fukushima disaster. The more lax standard exceeded international recommendations, but the Japanese government wanted to keep contaminated schools open as well as prevent mass evacuations—which themselves cause casualties and hardship. Japan also sought to minimize cleanup costs. Similar harrowing decisions made after a radiological attack could have destabilizing effects not only on the economy but also on trust in government and on psychological health.

That hypothetical 0.1 percent increase in lifetime cancer risk might be typical of exposure during a radiological attack. That actually isn't much, given that roughly 40 percent of all people develop cancer at some point. But radiation often elicits an understandably illogical response. It differs from other poisons. You can't see it, smell it, or taste it, and you may not know whether you've been exposed until it's too late. And if you get cancer, you don't really know if it resulted from radiation exposure at all, let alone from a particular terrorist attack. It's this normal fear of the unknown that terrorists seek to exploit with radiological weapons and why those weapons are expected to be so effective.

The Threat from Radiological Weapons Is Real

A radiological attack by a terrorist organization has not occurred at the time of this writing, but it's not prudent to assume that the threat is merely theoretical. We've had major incidents where strong radioactive sources have slipped out of regulatory control—becoming so-called **orphan sources**. When they've been found they've sometimes inadvertently irradiated people and have caused mass panic. It's through lessons learned from these incidents that we understand the substantial economic disruption that radiological weapons can produce.

An example of an orphan source causing harm and panic occurred in 1987 in the Brazilian city of Goiânia. Scavengers going through an abandoned cancer clinic separated a canister containing a cesium-137 source from a teletherapy machine used to irradiate tumors. The canister was broken up and given to a junk dealer who didn't know what the material was and inadvertently further dispersed the radioactive material. About 200 people were exposed and four died. But that wasn't all. There was a massive problem when the **worried well**—people who worried that they had been exposed—overwhelmed the medical system. In the end, 10 percent of the city's population sought medical attention—vastly more than the few hundred who were actually exposed. In addition, the massive cleanup cost $20 million while lingering fears of contamination destroyed the area's tourism industry.

Similar incidents occurred more recently in Mayapuri, India (2010) and in Tijuana, Mexico (2013). Both cases involved cobalt-60 irradiation units. In Mayapuri, a chemistry department sold a unit to scrap dealers without regard to its dangers. The dealers cut a Co-60 source into pieces, resulting in significant radiation exposure. One person died, and many more were injured in a situation that blatantly exposed callous disregard of India's radiation protection laws (figure 16.2). In the Mexican incident, a truck was legitimately taking an obsolete Co-60 radiation therapy unit for disposal when thieves stole the truck at gunpoint. Apparently after the truck itself, the thieves probably had no inkling about its cargo. Eventually they abandoned the radioactive unit in a cornfield (figure 16.3). Fortunately for

Figure 16.2
Reporters discussing the Mayapuri radiation incident on Indian television. (Courtesy of NDTV, India.)

Figure 16.3
A highly radioactive cobalt-60 radiation therapy unit lies by a Mexican roadside in 2013 after thieves abandoned it. (AP Photo/Marco Ugarte)

them, only one of the thieves showed any evidence of radiation exposure. Of greater worry is that such a theft could have resulted in material being sold to terrorists for use in radiological weapons.

In these cases, radiation exposure to the public and the fear it produced weren't deliberate. The radioactive material hadn't been weaponized, but the effect was nonetheless dramatic. Imagine if terrorists had used the same materials in actual radiological weapons. The consequences could have been much more severe. And incidents like these could happen anywhere, in the business district of a megacity like Lagos in Nigeria or Seoul in South Korea or Frankfurt in Germany or Chicago in the United States. Powerful engines of the world economy could be severely disrupted—even though it's unlikely that more than a few people would die. So it's prudent to take radiological weapons seriously and do everything possible to prevent their use and to have plans in place should prevention fail.

Challenges of Constructing an Effective Radiological Dispersal Device

The principal aim of radiological weapons is to cause mass panic and widespread contamination. For such weapons to be credible, or in cases where terrorists really want to harm people, they have to be effective. At first thought, an explosive RDD might seem easy to construct—as figure 16.1 hints, it could be as simple as strapping explosives to radioactive material. But, in fact, it's more challenging. Consider the choice of radioisotope: it needs to have a long enough half-life that the weapon doesn't lose effectiveness between construction and use. But, as you saw in chapter 3, the longer the half-life, the lower the level of radioactivity. Given that an explosive RDD dilutes its radioactive material by spreading it over a wide area, giving a significant dose to the target population requires a high level of activity. In addition, the material itself should ideally be a powder or readily aerosolize into minute particles so it's easy to disperse and inhale. Another desirable trait would be for the material to be insoluble in water so it's not easily removed in postdisaster cleanup. For example, cesium-137 binds well to concrete, thus hindering decontamination. These requirements limit the isotopes that can be used to make effective radiological dispersal devices. Cobalt-60, which has a five-year half-life and is insoluble in metallic form, might seem a good candidate. However, only 1 percent of cobalt metal aerosolizes in an explosion—meaning that a weapon would require 100 times more Co-60 than if the material would aerosolize completely.

The highly radioactive material needed for an effective RDD presents additional challenges to the terrorists constructing and delivering their weapon. Heavy lead containers would be needed for transportation, and assembly of a weapon will inevitably expose the terrorists to significant radiation doses.

Constructing a radiological exposure device present far fewer challenges. The radioactive material doesn't need to aerosolize or to be made into a powder, but can be kept, unshielded, in the form in which it was acquired.

Obtaining Radioactive Sources

Where can terrorists get their radioactive material? Most likely through theft or illicit purchase, since a terrorist group is unlikely to have access to a nuclear reactor, let alone the means to extract its highly radioactive spent fuel. Fortunately, it's becoming more difficult to steal radioactive sources for two reasons. First, security has improved steadily over the past decade and, second, many applications where radioactive sources would have been used in the past no longer need actual radioactive material.

Yet radioactive sources remain ubiquitous because, as you saw in chapter 4, they're useful in a wide range of applications. These include hospitals, factories, industrial laboratories, universities, and other everyday settings. One estimate suggests that more than 13,000 buildings, in some 100 countries, use radioactive sources strong enough to pose serious threats.[3] On top of that are millions of weaker sources that are less hazardous. You're probably close to one now, namely the americium-241 used in smoke detectors.

The International Atomic Energy Agency (IAEA) characterizes sources according to risk.[4] Category 1 describes *extremely dangerous* sources, category 2 describes *very dangerous* sources, and category 3 sources are merely *dangerous*. These qualitative designations have quantitative meanings in terms of radiation dose and the appropriate physical protections. However, the physical security protocols applied to radioactive sources vary for different countries, regions, and even institutions. Moreover, medical and industrial facilities tend to be ill-equipped to secure their sources, which leads to a worldwide problem of sources slipping outside of regulatory control—that is, they become orphan sources. Even in countries like the United States, with strict regulatory guidelines, sources do go missing. In fact, one estimate is that one source is lost every day in the United States! In low- and middle-income countries, the situation is considerably worse, especially in Central Asian nations that inherited radioactive material after the collapse of the Soviet Union.

The most common use of strong radioactive sources is medical: teletherapy devices for cancer treatment and for blood irradiation to prevent blood from being tainted prior to transfusion. These are life-giving applications, but they use cobalt-60 and cesium-137 sources often in forms that they can be easily dispersed, and they're usually considered category 1 sources: *extremely dangerous*. Another application of strong sources is in the oil industry, using radiation-based well-logging devices described in chapter 4 (recall figure 4.5). They're found wherever oil exploration and drilling take occur—including places like Nigeria where known terrorist groups are active. Most well-logging sources are category 3, *dangerous*, but some are category 2, *very dangerous*. Finally, strong radioactive sources, often category 1, *extremely dangerous*, are used for food irradiation and sterilization of medical equipment.

Sabotaging Nuclear Facilities

Terrorists could target nuclear power plants with the aim of causing mass harm via a core meltdown or station blackout (see chapter 8), or to steal radiological materials to use in radiological weapons. Again, this threat is not theoretical. All nuclear power plants require off-site power for normal operation. Thus links to the external power grid as well as emergency power systems are points of vulnerability—as was made obvious in the Fukushima disaster.

Internal sabotage is also possible. In 2014, an employee at the Doel-4 reactor in Belgium intentionally opened a locked valve, causing the reactor to overheat. The result was a shutdown lasting more than five months and causing over $200 million in damage. Motives remain unclear, and the perpetrator was never caught. After the 2015 Paris bombings that killed scores of people, it turned out that one of the suspects had surveilled a key nuclear scientist connected to a reactor in Belgium. The terrorist's intentions have not been disclosed, but it's reasonable to assume that some sort of extortion was the goal, perhaps by threatening to damage the reactor from within. This **insider threat** is a grave concern for nuclear facilities. The idea that an employee may be motivated either through blackmail or other reasons to sabotage nuclear facilities is chillingly worrisome.

In another incident in 2014, Hamas militants fired rockets towards Israel's Negev Nuclear Research Facility with the express purpose of targeting the facility's reactor. One rocket was intercepted by Israel's Iron Dome missile defense system, while the other two landed harmlessly on open ground. But most nuclear reactors aren't protected by antimissile systems, and not all are hardened enough to withstand an attack by terrorist missiles. In South Korea, for example, the outer walls surrounding nuclear power plants aren't designed to withstand a missile attack.[5] This is an acute vulnerability, perhaps not as much for terrorist action but for an attack from North Korea—and it continues despite IAEA recommendations to implement precautions commensurate with the threat.

Reactors are also vulnerable to commando-style attacks. In 2007, two groups of armed men infiltrated a South African nuclear plant at Pelindaba where large amounts of highly enriched uranium (HEU) were stored. One group exchanged gunfire with a guard and left. Another group successfully disabled the electrified security fence without raising any alarms, disabled a security camera, and then shot a guard. This event ended without nuclear material being stolen, but it served to underscore the vulnerability of nuclear plants to armed attacks.

Finally, cyberattacks have proven effective against power plants and other facilities. In 2010, a cyberattack—possibly mounted by the United States and Israel—seriously damaged gas centrifuges at the Natanz uranium enrichment facility in Iran. In the 1990s, three attempted cyberattacks on the Ignalia nuclear power plant in Lithuania, and an associated extortion attempt, led to temporary shutdown of two reactors while authorities searched for possible sabotage; none was found. Cyberweapons were also used by North Korea to hack South Korean nuclear power plant computer systems. And in 2019, a utility in the western United States revealed that it had been the victim of a cyberattack that probably blocked access to power grid data, although it's not clear whether any nuclear facilities were affected. All these incidents raise the specter that a sabotaged nuclear facility itself could become a radiological weapon.

It isn't just civilian power reactors that are vulnerable. In 2012, three peace activists, including a nun in her eighties, staged a predawn intrusion into a nuclear weapons plant at Oak Ridge, Tennessee. They cut fences and managed to get to a building storing 400 tons of highly enriched uranium—enough to make 10,000 nuclear bombs. The intruders were apprehended only after defacing the building with blood and spray paint. They were convicted of intending to harm national security, and all were given jail sentences. The intruders' goals were entirely peaceful and even praised by the sentencing judge, but nevertheless the incident highlighted lax security at a key nuclear weapons facility. The intrusion also resulted in a two-week shutdown and the loss of a $23 billion contract to Babcock & Wilcox, the company operating the facility under contract to the U.S. Department of Energy.

Improvised Nuclear Devices

Radiological weapons are bad enough, but far more horrific would be a terrorist attack with actual nuclear explosives. Primitive weapons known as **improvised nuclear devices** (INDs) could conceivably be constructed by nonstate actors like terrorist organizations. We classify such weapons differently from state-developed nuclear weapons on the assumption that they will be crude devices with relatively low yields. Be aware, however, that much of the know-how needed to design nuclear weapons, especially simple gun-type weapons, can be found in open-source literature. Furthermore, competent scientists and engineers may be just as motivated by terrorist causes as others. The barrier to construction and ultimately use of such terrorist weapons is access to fissile material such as highly enriched uranium (HEU). That's why it's so critical to convert or shutdown research reactors that use HEU and to monitor institutions and individuals that work with this material. Research reactors are often on university campuses with only modest security precautions, making nuclear materials vulnerable to theft. U.S. government studies indicate that a crude bomb, and even an implosion-type weapon, could be constructed by a well-funded terrorist organization.[6] In fact, even a "fizzle" with a plutonium core could still produce an explosion well in excess of 1 kiloton—enough to produce devastating effects.

Theft or Purchase of Complete Nuclear Weapons

Another possibility is that a subnational actor could steal, or even purchase, a complete, ready-made nuclear weapon. Theft would require a tactical operation, overwhelming force, and meticulous planning but is not as far-fetched as it may seem. In fact, there have been many incidents at military bases overseas that call into question the security of nuclear weapons. These include highly publicized infiltrations by peace activists, such as at the Kleine Brogel base in Belgium, known to store at least 20 B-61 nuclear bombs. In a first attack, in February 2010, activists from the group Vredesactie (*Peace Action* in Dutch) were able to access the area near hardened facilities that house nuclear bombs. In a second attack, the same group infiltrated the facility again, this time recording the entire event and sharing it on YouTube. The activists spent more than an hour inside the facility before being confronted by an unarmed security officer.

NATO bases aren't alone in being targeted. Far more sinister events have occurred in Pakistan, such as a 2007 attack on the Mushaf Air Base, where nuclear weapons are likely stored, or at the Minhas Air Base, where eight Taliban militants infiltrated and a long battle ensued. Fortunately, the militants weren't able to access the interior of the base (or perhaps they had no interest). Pakistan claims to store nuclear components away from air bases themselves, but they can't be very far if they need to be assembled and delivered quickly.

However, possessing a nuclear weapon doesn't mean it can be detonated. Modern U.S. weapons employ safety and control devices to prevent inadvertent or unauthorized detonation. These **permissive action links** (PALs) act as mechanical combination locks that need deciphering to open electrical circuits that arm nuclear warheads. The arming system is inside a physical barrier forming an exclusion zone isolating the system from lightning or other spurious electrical signals. The exclusion zone is composed of *strong links* and *weak links*, and all links must pass tests before the warhead can detonate. Strong links are switches that pass only specific electrical signals that act as unique authorization codes. Furthermore, during the delivery flight they must have experienced trajectory characteristics commensurate with the mission. Weak links, in contrast, are deliberate vulnerabilities in the components of the arming system such that under abnormal conditions these components will become inoperable. If any links don't pass, the warhead won't detonate. The importance of such safety and security measures became abundantly clear in 2016, when the very base in Turkey that's believed to house 50 U.S. nuclear weapons was the origin of attacks on the Turkish parliament in an attempted coup d'état.

Security of other countries' nuclear weapons is unclear, but it is known that terrorists have surveilled Russian nuclear weapon storage sites and that as a result security was significantly upgraded.

Regardless of your view on the necessity of nuclear weapons, it's in everyone's interest to keep nuclear weapons secured to prevent the possibility of terrorists stealing them. A related concern is that if locations of nuclear weapons are known they could be targeted with radiological weapons to spread panic and thus weaken security.

A Market for Nuclear Weapons?

Another concern is that nonstate actors or non-nuclear states might purchase nuclear weapons from states that do possess them. It could be that the seller state shares the same ideology as the prospective buyer, or the motivation may be simply economic. Is this an easy route to a bomb? Perhaps, but most nuclear weapons likely have security measures to prevent unauthorized detonation unless they've been specifically modified before the sale—somewhat like selling an unlocked cellphone.

Concern about possible weapons purchase is particularly serious in the case of state actors like Saudi Arabia, which has openly expressed interest in purchasing nuclear weapons from Pakistan if Iran were to produce its own weapons. This isn't surprising given that Saudi Arabia bankrolled the Pakistani nuclear program. Pakistan denies any expectation of a quid pro quo, but the concern remains real.

It's also very worrisome that Saudi Arabia has purchased short- and medium-range missiles from China—reportedly with the help of the U.S. Central Intelligence Agency. And analysts from the Middlebury Institute of International Studies recently revealed that Saudi Arabia may manufacture solid propellant missiles, ostensibly to deliver nuclear weapons. These moves point to Saudi Arabia's interest in deliverable nuclear weaponry—even though they themselves don't yet have the capability to manufacture their own weapons.

For years it's been reported that Al Qaeda had an interest in and possibly even purchased a nuclear weapon from Pakistan. However, there is little hard evidence for these claims. There was also a statement in the Islamic State's propaganda magazine *Dabiq* that ISIS could purchase a nuclear weapon from corrupt officials in Pakistan. However, the article itself professed that this was "hypothetical" and "far-fetched." Still, the ideological ties between groups like Al Qaeda and ISIS to extremists within the government and military establishments of nuclear weapons states, especially Pakistan, remain worrisome.

Could North Korea sell nuclear weapons to enemies of its enemies? This is certainly plausible and underscores the importance of bringing North Korean nuclear weapons under an agreement that at least prevents transfer of those weapons out of North Korea. Since North Korea at the time of this writing is under punishing economic sanctions, the reclusive nation certainly has financial motivation for considering its nuclear weapons as saleable commodities.

If the Unthinkable Were to Happen

Surprisingly, there are simple ways to improve survivability from the terrible effects of a terrorist attack using an improvised nuclear device (IND). If it's a ground detonation—likely, since terrorists may not have aerial delivery—the fireball itself will destroy most everything within a certain radius. Let's assume that the device yields 10 kilotons, somewhat

Figure 16.4
Recommendations from the U.S. National Nuclear Security Administration for responding to a terrorist attack with an improvised nuclear device.

less than the Hiroshima bomb. Several kilometers away the effects of the blast wave, heat, and prompt radiation will be diminished, and people sheltered inside sturdy buildings will likely survive. Recent computer simulations have shown that a building of formidable construction will protect people in an IND attack just as it would in a hurricane, tornado, or earthquake. Staying inside sturdy buildings for more than a day also protects you from dangerous radioactive fallout since the most intensely radioactive isotopes will have decayed during that time. If you're outside and survived the initial blast, then you have about 15 minutes to get deep inside a building to escape radioactive fallout. If you are exposed to fallout, it's critical that you immediately remove your clothing, bag it, and place it as far from yourself as possible. Stay inside and await further instruction by radio since all cellphone service, Internet, and power will be down for many days. As long as you and your family *get inside, stay inside, and stay tuned*, the odds of surviving an IND attack are higher (figure 16.4). While the possibility of terrorists constructing, delivering, and detonating an IND is small, we should be prepared for such an eventuality.

The Good News

Alternatives to Radioactive Sources Do Exist

For most legitimate strong-source radiation uses, alternatives exist that can meet or exceed the effectiveness of radioisotope sources. Following a 2008 report by the U.S. National Academy of Sciences on radiation sources and replacements,[7] many facilities have converted to alternative technologies that require less strict security protocols and in many cases are even more effective than radioisotope sources. One example: teletherapy devices that use radioactive cobalt-60 sources similar to the ones that caused the Goiânia incident are still employed in many developing countries. These sources can't destroy tumors deep inside the body because the beam doesn't penetrate that far. Furthermore, they tend to burn the skin, produce an uncertain radiation dose, and require a very strict security protocol. A hospital that uses such sources is potentially liable if the source is used in an RDD. Thankfully, alternatives exist. Called *medical linear accelerators*, they produce particle beams electronically rather than using radioactive sources. The beam energy and hence penetrability can be customized to specific tumors. However, these devices are expensive and require steady supplies of water and electricity, making them less suitable for developing countries. For this reason, a network of doctors, physicists, and engineers met in 2017 at the European Nuclear Research Laboratory (CERN) to explore development of an alternative linear accelerator that's affordable and effective for a range of uses.

Another application where radioactive sources are being supplanted by alternate technologies is in blood sterilization, which commonly uses cesium-137 in the form of cesium chloride. That chemical is easily dispersed and poses a risk if nonstate actors acquire it for radiological weapons. Alternatives include linear accelerators, X-ray machines, or ultraviolet sterilization. In Europe, cesium-137 blood sterilizers have been significantly curtailed, with Denmark banning them altogether and Norway requiring "compelling justification" for their use. Sweden and Finland both strongly encourage the use of X-ray irradiators. However, the United States has been slow to respond to the threat posed by cesium-chloride blood irradiators, with the exception of hospitals in New York City and within the University of California system.

Efforts Around the World to Minimize Use of HEU

Consensus holds that terrorists would likely construct an improvised nuclear device using a gun-type design rather than an implosion bomb. The fissile material of choice for this is highly enriched uranium with more than 80 percent U-235. Such HEU is only weakly radioactive, making it easy to slip across borders undetected. It's clear that preventing terrorist INDs requires limiting access to HEU. The problem is that HEU is widely used around the world as a fuel for research reactors or as a target to produce medical isotopes. It's also used in reactors for ship propulsion, including in submarines, aircraft carriers, and icebreakers. And of course, there's HEU in some nuclear weapons.

In order to counter the threat of HEU ending up in the wrong hands, the United States and Russia introduced programs decades ago to either convert research reactors from HEU fuel to low-enriched uranium (LEU) fuel or to close and decommission such reactors and repatriate the fuel. Principles of conversion require that reactor operations be minimally impacted, meaning that the reactor infrastructure and footprint shouldn't be enlarged nor its productivity decreased. Many research reactor designs were unique and already optimized to get the highest productivity using compact cores, making it challenging to modify them to use LEU fuel without violating the principles of conversion. As a result, each of the dozens of reactor conversion projects around the world required detailed calculations and testing, including scientists as well as diplomacy, to make it happen. In the age of tensions between the United States and China, scientists from the United States, China, and Ghana nevertheless collaborated to complete the successful conversion of a Chinese origin miniature neutron source reactor in Ghana and repatriate a kilogram of weapons-grade HEU back to China (see box and figure 16.5).

Reactor conversion went into high gear after the September 11, 2001, terrorist attacks but became supercharged during the Obama administration when many reactors were converted or decommissioned. Since the conversion program started, well over 100 reactors have been successfully converted and HEU fuels removed to secure storage locations. The fuel takeback programs have removed more than 3,000 kilograms (3 tons) of HEU. As a result, more than 33 countries have been completely cleared of highly enriched uranium.[8]

While this effort is impressive, there are still research reactors, mainly in the former Soviet Union, the United States, and Europe, that use HEU but could be converted to LEU. A particular problem has been high-performance reactors that require special fuels and that are still in their testing phase.

Summary

In this chapter, you've seen the risk that radioactive sources pose when they're weaponized as radiological weapons. You learned how radiological weapons differ from improvised nuclear devices that are full-fledged nuclear explosives capable of multikiloton yields. The purpose of radiological weapons is to produce panic rather than mass casualties as true nuclear weapons do. However, this doesn't diminish the economic devastation resulting from widespread contamination nor the deleterious psychological effect that a radiological weapon could have on a population. We can consider the Goiânia incident or the Fukushima accident as foreboding inklings of what could happen in a modern city should a

An HEU Conversion

In 2009, one of us (FDV) attended a research reactor conference in Beijing. Little did we know that in side meetings, behind closed doors, a deal was struck between the United States and China to convert the Chinese origin miniature neutron source reactors. These reactors have long been on the U.S. government's wish list for conversion since each contains about one kilogram of weapons-grade HEU. Furthermore, China exported these reactors to areas of concern such as Iran, Pakistan, and Syria as well as Ghana and Nigeria. A series of follow-up meetings resulted in a plan where the United States provides a special facility known as the Zero Power Test Facility to determine the suitability of reactor cores for conversion, while China would pay for the costs of the actual conversion. Over the next few years, specialists from the United States, China, and the two African countries worked closely to ensure that the performance and safety of these reactors wouldn't be adversely affected by conversion.

In 2016, an international team successfully converted a Chinese prototype reactor, followed soon after by conversion of the Ghana reactor, and in late 2018, the Nigerian research reactor. The Nigerian conversion is important because the Boko Haram terrorist group is active in the region. These Chinese neutron source reactor conversions are among many other successful conversions around the world, from Kazakhstan to Nigeria, carried out for decades by diplomatic, scientific, safety, and legal specialists from the United States and Russia.[9]

Figure 16.5
Members of the Chinese technical team make adjustments after the LEU core is inserted in Ghana's GHARR-1 research reactor. (U.S. National Nuclear Security Administration)

radiological weapon incident occur. We might consider radiological dispersal devices as *weapons of mass disruption* rather than weapons of mass destruction that nuclear weapons are. They share the same acronym, but their effects are very different.

Governments around the world take the risks of radiological and nuclear weapons from nonstate actors seriously, through active programs of finding alternatives to using radioactive sources and of converting or decommissioning research reactors that use highly enriched uranium for fuel or as targets for radioisotope production. You've seen that most medical therapy devices using radioactive sources can be replaced with X-rays or particle beams that are easily turned off and don't leak radiation. Similarly, most research reactors can be converted from highly enriched to low-enriched uranium cores without significantly compromising their performance.

As we head into an uncertain time where the use of a nuclear weapon seems less far-fetched than it did at the end of the Cold War, it's important to note new research on the effects of nuclear weapons—in particular, that if your city is a victim of a nuclear detonation, there are nevertheless ways you can protect yourself and your family to improve your chances of survival.

Glossary

dirty bomb A radiological dispersal device that disperses radioactive material using conventional explosives. Also called an *explosive RDD*.

improvised nuclear device (IND) A crudely made nuclear weapon, such as a terrorist group might construct.

insider threat The concern that employees of a nuclear facility, even if well vetted, may be coopted or may willingly commit sabotage or steal nuclear materials.

non-explosive RDD An RDD that does not use explosives to disperse radioactive materials.

orphan source A radioactive source that is not under regulatory control, meaning it's not tracked, and its location is unknown.

permissive action link (PAL) A security system whose purpose is to prevent unauthorized arming or detonation of nuclear weapons.

radiological dispersal device (RDD) A type of radiological weapon that disperses radioactive material over a large area.

radiological exposure device A radiological weapon that exposes people to radioactive material without the material being ingested or inhaled.

radiological weapon A weapon that uses radioactive material to cause mass panic, render areas uninhabitable, and/or cause economic damage.

worried well The phenomenon of people thinking they have been exposed to radiation and overwhelming the medical system, impeding response for people who actually are affected.

Notes

1. Brian Michael Jenkins, "Will Terrorists Go Nuclear?" (Santa Monica, CA: Rand Corporation, Rand Paper Series, 1975), 4. Available at https://www.rand.org/pubs/papers/P5541.html.

2. See https://emergency.cdc.gov/radiation/pdf/Infographic_Radiological_Dispersal_Device.pdf for the entire CDC brochure on RDDs.

3. Matthew Bunn, Martin B. Malin, Nickolas Roth, and William H. Tobey, "Preventing Nuclear Terrorism: Continuous Improvement or Dangerous Decline?" Project on Managing the Atom, Belfer Center for Science and International Affairs (Cambridge, MA: Harvard University Kennedy School of Government, 2016), 98. Available at https://www.belfercenter.org/sites/default/files/legacy/files/PreventingNuclearTerrorism-Web.pdf.

4. A full description of IAEA radioactive source categories is available at https://www-pub.iaea.org/MTCD/Publications/PDF/Pub1227_web.pdf.

5. "S. Korea's Nuclear Power Reactors Not Designed to Deal with Military Attacks," Yonhap News Agency, 2016. Available at http://english.yonhapnews.co.kr/news/2017/04/16/0200000000AEN20170416002800320.html.

6. Matthew Bunn and Anthony Wier, "Terrorist Nuclear Weapon Construction: How Difficult?," *Annals of the American Academy of Political and Social Science* 607, no. 1 (2006): 133–149.

7. National Research Council 2008, *Radiation Source Use and Replacement: Abbreviated Version* (Washington, DC: National Academies Press), https://doi.org/10.17226/11976.

8. National Nuclear Security Administration, "NNSA Removes All Highly Enriched Uranium from Nigeria," December 7, 2018, https://www.energy.gov/nnsa/articles/nnsa-removes-all-highly-enriched-uranium-nigeria.

9. Miles Pomper and Ferenc Dalnoki-Veress, "The Little Known Success Story of U.S.-China Nuclear Security Cooperation," Nuclear Threat Initiative, June 10, 2020, https://www.nti.org/analysis/articles/little-known-success-story-us-china-nuclear-security-cooperation/.

Further Reading

Baylon, Carolyn, Roger Brunt, and David Livingstone. *Cyber Security at Civil Nuclear Facilities: Understanding the Risks*. London: Chatham House, 2015. This report focuses on the vulnerability of nuclear facilities to cyberterrorism. While the emphasis is on the civilian nuclear industry, the findings in the report are relevant to other sectors as well.

Buddemeier, Brooke, and Jessica S. Wieder. "Can You Survive Nuclear Fallout?" YouTube video, January 8, 2019. https://www.youtube.com/watch?v=GHBb25lzNVM. This video provides a quick description of what measures should be taken if a nuclear incident occurs.

Bunn, Matthew. "Insider Threats: A Worst Practices Guide to Preventing Leaks, Attacks, Theft, and Sabotage." YouTube video, May 24, 2017. https://www.youtube.com/watch?v=tkB4FLEEq74. This presentation discusses recent research by Harvard professor Matthew Bunn on the risk of insider threats to nuclear facilities.

Ferguson, Charles D., Tahseen Kazi, and Judith Perera. *Commercial Radioactive Sources: Surveying the Security Risks*. Monterey, CA: Monterey Institute of International Studies, Center for Nonproliferation Studies, 2003. http://www.nonproliferation.org/wp-content/uploads/2016/09/op11.pdf. This authoritative study examines specific uses of radioactive sources and the threats they pose.

Ferguson, Charles D., and William C. Potter. *The Four Faces of Nuclear Terrorism*. New York: Routledge, 2005. This important book alerted policy makers and the public to the risks of nuclear terrorism. Unfortunately, the 2005 book's findings are still all too relevant today.

Hafemeister, David. *Nuclear Proliferation and Terrorism in the Post-9/11 World*. Cham, Switzerland: Springer International, 2016. See especially chapters 12 and 13 for an excellent discussion on the motivations of terrorists, radiological weapons, and INDs.

National Research Council. *Radiation Source Use and Replacement: Abbreviated Version*. Washington, DC: National Academies Press, 2008. https://doi.org/10.17226/11976. This is a landmark study by the U.S. National Research Council that for the first time considered how radioactive sources can be replaced with other technologies.

Pomper, Miles A., and Gabrielle Tarini. "Nuclear Terrorism–Threat or Not?" In *American Institute of Physics Conference Proceedings 1898*, no. 050001. AIP Publishing, 2017. https://doi.org/10.1063/1.5009230. This well-researched and up-to-date summary of the threat of nuclear terrorism is highly recommended.

Pomper, M. A., F. Dalnoki-Veress, and G. M. Moore. "Treatment, Not Terror: Strategies to Enhance External Beam Therapy in Developing Countries While Permanently Reducing the Risk of Radiological Terrorism." Stanley Foundation and James Martin Center for Nonproliferation Studies, 2016. Available at https://www.stanleyfoundation.org/publications/report/TreatmentNotTerror212.pdf. This report describes the findings of a study focused on replacing external beam cancer therapy sources in low- to middle-income countries (with a focus on Africa) with linear accelerators and outlines a strategy to improve cancer therapy while reducing the risk of terrorism.

United States Government Accountability Office. "Nuclear Terrorism Response Plans: Major Cities Could Benefit from Federal Guidance on Responding to Nuclear and Radiological Attacks." 2013. https://www.gao.gov/assets/660/658336.pdf. A look at U.S. cities' plans for response to nuclear terrorism.

17

Defense in the Nuclear Age

Despite decades of arms control agreements, public concern about nuclear weapons, and the fall of the Soviet Union, there remain thousands of missiles with nuclear warheads aimed at targets around the world. We seldom use the term *mutually assured destruction* these days, but cloak that concept in prettier words like *deterrence*—the idea that if you do something bad to me I'll do it right back to you so you'd better not do it in the first place. However, the basic MAD dilemma still exists, and not only between the U.S. and Russia but also between India and Pakistan. Is there a way out of MAD, a way to attain security that doesn't rest on threatened annihilation? President Ronald Reagan thought so, as he described in a historic 1983 speech:

> After careful consultation with my advisers, including the Joint Chiefs of Staff, I believe there is a way. Let me share with you a vision of the future which offers hope. It is that we embark on a program to counter the awesome Soviet missile threat with measures that are defensive. Let us turn to the very strengths in technology that spawned our great industrial base and that have given us the quality of life we enjoy today.
>
> What if a free people could live secure in the knowledge that their security did not rest upon the threat of instant U.S. retaliation to deter a Soviet attack, that we could intercept and destroy strategic ballistic missiles before they reached our own soil or that of our allies?[1]

Reagan's speech launched the **Strategic Defense Initiative** (SDI), euphemistically called "Star Wars," a program whose ambitious aim was complete defense against ballistic missiles. (Among Reagan's advisors and a vigorous SDI proponent was Edward Teller, who had helped make fusion weapons practical 30 years earlier.) A successful SDI would have circumvented the 1972 Anti-Ballistic Missile Treaty, which among other items banned space-based **ballistic missile defense** (BMD) technology. In the end, SDI was a failure and, after hundreds of billions of dollars, we're no safer. Today we have minimal ground-based ballistic missile defenses, but they're far from Reagan's impenetrable shield. Although the United States' withdrawal from the ABM Treaty in 2002 gave the U.S. free rein to develop new missile defense systems, so far there's been only modest development. The Trump administration announced plans to greatly expand missile defense programs, ostensibly to protect the U.S. homeland from North Korean ICBMs. U.S. adversaries have responded: Russia, as we discussed in chapter 14, has enlarged its portfolio of offensive weapons specifically to circumvent U.S. missile defenses, and China is increasing its number of ICBM warheads as well as developing new missiles that can get around defense systems. There's a clear danger that continuing unabated development of ballistic missile defense could lead to a new arms race.

This chapter discusses developments in BMD, largely from a United States perspective, because we have the clearest information about U.S. missile defense programs. We'll describe systems for defense against intercontinental missiles as well as regional and local threats from nuclear-capable missiles.

A New Shield against a New Arrow

It wasn't until 2017 that the United States first destroyed an ICBM-like target in a missile-defense test, but such defense is not a new concept. As soon as humans invented the bow and arrow they developed protective shields. The same defensive instinct was triggered when German V-2 missiles struck London during World War II. British General Frederick Alfred Pile suggested that the supersonic V-2s could be stopped by firing defensive shells in front of incoming missiles. He calculated that to destroy a V-2 would require 90 tons of ammunition, or 320,000 shells. The barrage of falling shells would do far more damage than the V-2 itself! So even in World War II it was clear that defense against incoming ballistic missiles was a daunting challenge. A 1945 U.S. Army report recommended that "High velocity guided missiles, preferably capable of intercepting and destroying aircraft flying at speeds up to 1,000 miles per hour at altitudes up to 60,000 feet or destroying missiles of the V-2 type, should be developed at earliest practicable date."[2] It was clear that to destroy a missile you need another missile.

In 1957, the Soviet Union launched Sputnik, the first artificial satellite, and that act of civilian space exploration spurred the development both of ballistic missiles and missile defense systems. During the 1960s, BMD development proceeded despite debate over the utility and desirability of missile defense. U.S. technology converged on a two-phase system consisting of

Sprint low-altitude interceptor rockets and high-altitude *Spartan* rockets for interception above the atmosphere. Originally the Spartans were to defend the entire United States, with low-altitude Sprints grouped for additional defense. The combined missile defense system was termed *Sentinel.*

But the American ballistic missile defense effort faced a problem: its effectiveness against current and future Soviet ICBMs was far from certain. By 1967, the U.S. government conceded that a BMD system couldn't counter a full-scale Soviet attack. Instead, the system would serve as insurance against an accidental missile launch or a future Chinese ICBM threat. Note the pattern here: initial optimism for full-scale missile defense followed quickly by relegation of the proposed system to less significant roles.

Missile-defense advocates in the late 1960s faced mounting public opposition to nuclear-armed BMD sites near cities. Spartan missiles themselves carried 5-Mt nuclear warheads, which did not spur enthusiasm in people living nearby. An influential 1968 article by prominent physicists Richard Garwin and Hans Bethe—both of whom had played important roles in nuclear weapons development—described how adversaries could deploy countermeasures to defeat missile defenses. As you'll see again and again throughout this chapter, arguments against antimissile defense just won't go away.

In 1969, the Nixon administration reaffirmed Sentinel's impotence in the face of an all-out attack, and reshaped the U.S. antiballistic missile program toward protection of ICBM missile silos only. The new system, called *Safeguard,* barely won approval in Congress. Construction of a Safeguard system to protect ICBMs at Grand Forks, North Dakota, soon followed (see figure 17.1). However, Safeguard was never completed and the system was abandoned in 1976 after being operational for less than a year.

The ABM Treaty

Even as ballistic missile defense development continued through the 1960s, the United States and the Soviet Union moved slowly toward the first negotiations aimed at limiting strategic nuclear arms. Both sides had come to recognize their essential nuclear stalemate, and both realized that maintaining the MAD balance was in their mutual interest. The Strategic Arms Limitation Talks (**SALT**) began formally in 1969 and led in 1972 to the SALT I agreements.

SALT negotiators recognized that missile defenses could have several adverse consequences. A truly effective system would give a defended country the ability to strike first with impunity, knowing its defenses would

Figure 17.1
Missile control building at the Stanley R. Mickelsen Safeguard Complex in North Dakota, one component of the Safeguard ABM project. Safeguard became active in 1975 but was shut down less than a year later. It is now owned by a traditionally pacifist Hutterite colony. (Library of Congress, Prints and Photographs Division, Historic American Engineering Record, HAER ND-9-B-8)

fend off a retaliatory strike. Even a partially effective BMD system might encourage a first strike by limiting damage from weakened retaliatory forces to "acceptable" levels. And in any event, antimissile technology would surely lead to an expensive and potentially destabilizing offensive arms race, as each side sought to maintain the nuclear balance with offensive weapons capable of overwhelming its adversary's defenses.

The SALT I agreements acknowledged the danger of defensive systems with the **ABM Treaty** (for Anti-Ballistic Missile Treaty), limiting each superpower to only two ABM sites, one to defend its capital city and the other at a selected ICBM field. A 1974 protocol further restricted deployment to only one site. The Soviet Union chose BMD for Moscow, while the United States went with its Safeguard site in North Dakota. Safeguard was abandoned in 1976 for being ineffective and costly, leaving the Soviet *Galosh* defenses ringing Moscow as the world's only operational antimissile defense system. The Galosh interceptors were nuclear tipped with megaton-range warheads.

The ABM Treaty embodied an essential paradox of nuclear arms: that effective defense may be more likely to cause war than to prevent it. And the treaty formalizes hostage roles for the superpowers' populations. By denying them defense against nuclear attack, it upholds MAD and makes that attack less probable. Another dilemma for missile defense is that a defense system is designed for a particular threat, but the adversary can always counter by fielding more offensive missiles—and at lower cost than it takes for the adversary to enhance its defenses.

What Makes Ballistic Missile Defense So Difficult?

As I write, it was just a few minutes ago that a Twitter message announced another North Korean missile test. This follows multiple tests the previous month, and the introduction of possibly new missiles and missile launchers this week. A missile test is just that—a test of a missile's operation—but tests are also used politically to signal displeasure with an adversary and to affirm intention to actually use missiles if provoked. So why not just knock these missiles out of the air, especially because it's not always possible to distinguish a test from a belligerent launch? In fact, President Trump voiced just that view in 2017. When a North Korean test missile flew over the Japanese island of Hokkaido, Trump asked why a nation of samurai warriors would not shoot down missiles overhead. Setting aside the fact that the very act of shooting down the missile could provoke North Korea into a war, did President Trump have a point? Why not swat a missile away as you would a mosquito? In fact, it's notoriously difficult to shoot down missiles.

To demonstrate the challenge of missile defense, let's return to the analogy in chapter 14 where we compared firing a missile to throwing a ball. Ask a friend to throw you a ball on a slow arc, then throw another ball to try to hit the first one so as to knock it off course. Even if you're a top athlete you'll find this challenging. Now imagine that the first ball can curve in midflight or execute abrupt unexpected movements like a knuckleball, or deploy many other balls in midflight where only a few actually carry the warheads that you're trying to hit. Or imagine that they fly so fast you don't even have time to calculate where and when to throw your defensive ball. All of these issues are at play in missile-defense planning. Ballistic missile defense is partly a political issue, but unlike many other geopolitical challenges it's fundamentally a technical problem. Understanding the problem of ballistic missile defense requires a detailed look at the flight of an intercontinental ballistic missile. As figure 17.2 shows, the trajectory of a typical ICBM comprises four distinct phases. Each phase offers separate challenges and opportunities for a BMD system.

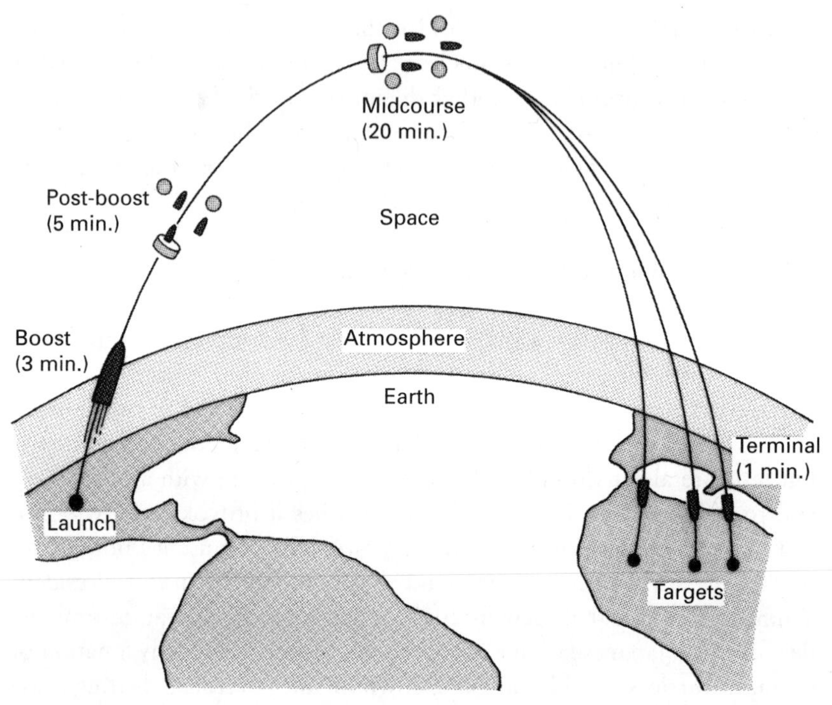

Figure 17.2
Phases of an ICBM's trajectory. By midcourse, up to 10 warheads and 100 or more decoys may have been released. Only the actual warheads survive atmospheric reentry. Maximum altitude and thickness of the atmosphere are exaggerated for clarity.

Boost Phase Intercept

During the **boost phase**, rocket motors lift the missile to an altitude outside Earth's atmosphere. For solid-fueled missiles the boost phase typically lasts three to four minutes. Attacking missiles in the boost phase offers significant advantages to the defense. First, flame from the rocket boosters is tremendously bright, making it easy for space-based sensors to detect and track the missile during the boost phase. Tracking satellites relay information to a battle management system, which calculates the trajectory needed to intercept the missile. Also, the boost-phase missile constitutes a single relatively large target, with possibly multiple warheads riding on one vehicle. Destruction of a single missile during the boost phase could put many warheads out of commission. Finally, the missile in the boost phase is starting to accelerate from rest to its burnout velocity, so at the beginning it's relatively slow and thus easier to hit.

Boost-phase defense is decidedly advantageous, but it's not without difficulties. The boost phase is relatively short, giving the defense limited time for the decision to attack. In an attack involving many missiles, those are formidable tasks to accomplish in a few minutes. Also, a missile's boost-phase trajectory is nearly vertical, requiring an interceptor to be very fast or to be based relatively close to the offensive missile's launch site. This is difficult, although antimissile missiles on the United States' newly deployed F-35 fighter jet may be able to achieve boost-phase interception.

The rocket motors that boost a missile into space fall away as their fuel is exhausted. The boost phase ends as the last rocket stage shuts down and separates, leaving the bus on which the warheads are mounted. There follows a **post-boost phase** of about five minutes, during which the bus uses small rockets to orient itself in different directions, releasing individual warheads on trajectories that will take them to their targets.

It's unfortunate for the defense that what was one threatening object may now have become multiple individual warheads on separate trajectories. Think of the previous analogy: you're trying to throw a ball (the interceptor) at another ball (the incoming missile) in midflight, but suddenly the single ball becomes many (the warheads). However, things may get even worse: The bus can also release **decoys**, typically lightweight metalized balloons that superficially resemble warheads. In the vacuum of space, the absence of air resistance means that the decoys travel on the same ballistic paths as the much heavier warheads. Since decoys are light, inexpensive, and compact until they inflate, a typical missile could carry 10 or more decoys for every warhead. A single missile thus becomes a **threat cloud** comprising 100 or more objects that all look like nuclear warheads. In addition to decoys, the bus might release clouds of aluminum-foil strips to confuse defensive radars and aerosol mists to shield warheads from the prying eyes of infrared sensors. A nuclear strike involving thousands of missiles could easily put a million objects on ballistic trajectories toward the target country. An effective defense must destroy all this junk or somehow single out the several thousand actual warheads.

Midcourse Phase

The **midcourse phase** begins once the bus has released all its stuff. A system for midcourse defense faces stiff challenges. You've seen that it may have to deal with a million potentially threatening objects. And with no more rockets firing, those small, cold objects are very hard to detect—all the more so because they're moving several miles each second. Despite formidable problems involving detection and discrimination of warheads from decoys, there are two advantages to midcourse defense. First, midcourse is the longest phase of missile flight, about 20 minutes for an ICBM. Second, the entire threat cloud spends midcourse following ballistic trajectories, moving in accordance with simple physical laws under the influence of gravity alone. If an object can be tracked to determine its position and velocity, then its future trajectory can be predicted accurately. That predictability, coupled with the relatively long time available, might allow a midcourse defense to pick off its many targets one by one. The United States' current **Ground-based Midcourse Defense** (GMD) system uses ground-based interceptor rockets (GBIs) to attack ICBMs in their midcourse phase, while the Aegis system can intercept intermediate-range ballistic missiles in midcourse.

Terminal Phase

As it nears the target country, a missile's threat cloud reenters the atmosphere. Reentry marks the beginning of the **terminal phase**, which lasts only about a minute before the weapons detonate on or above their targets. Terminal-phase defense has little time to act, and there's no room for failure. Nuclear warheads, still traveling at several miles per second, *must* be destroyed or they will wreak their unimaginable damage. Terminal defense is a last-ditch effort, but it does have one advantage: drop a rock and a piece of paper and you'll see how air resistance dramatically alters the motion of the paper. The same thing happens to the lightweight decoys—they're stripped away by air resistance high in the atmosphere, leaving only the true warheads still plummeting at high speed toward their targets.

The U.S. Terminal High Altitude Area Defense (THAAD) system uses mobile launchers and has been deployed in many locations, including Guam, South Korea, the United Arab Emirates, Israel, and Romania. It can intercept short- and medium-range missiles in the terminal phase but not ICBMs, because they move too fast.

Whether terminal defense can work depends on many variables, including interceptor specifications and the threat missile's trajectory. A missile that's intercepted too low could still cause significant damage. Success or failure of terminal defense also depends on the attacking missile's target. A missile silo would be saved if an attacking warhead detonated a few miles up instead of in a direct-hit ground burst. Under the same circumstances, a city would still be destroyed.

Submarine-launched ballistic missiles also go through the four phases of flight described above. But SLBMs would typically be launched from just off the target country's coast, making the overall flight time much shorter. The constraints on a defense system are correspondingly tighter, and many experts believe that systems designed for ICBM defense could not function effectively against submarine-launched missiles.

Layered Defense

Ronald Reagan's Strategic Defense Initiative was intended to produce a 100 percent effective shield against ballistic missile attack on the United States and thus render Soviet nuclear weapons "impotent and obsolete." SDI favored a **layered defense**, with separate weapons engineered for effectiveness against boost, midcourse, and terminal phases of the attacking missiles' flight. Ideally, boost-phase defenses would destroy most of the attackers. The few that got through would be largely eliminated in midcourse. If any survived midcourse, terminal defenses would take care of them. The formidable problems faced by midcourse and terminal defense would be reduced by elimination of missiles in earlier phases. However, as costs mounted and technical problems persisted, SDI was scaled back so that instead of stopping all incoming missiles, the goal was to destroy enough missiles in flight so as to convince the enemy that an effective attack was impossible. So again we see that same pattern: first great optimism and later a scaling back of the system to less ambitious goals.

In the 1990s, the defense effort reoriented once again from a national missile defense to a regional or theater defense. After the fall of the Soviet Union, the ballistic missile threat was seen to be from shorter-range missiles launched by regional actors such as North Korea or Iran. During this time the Patriot and THAAD antimissile systems were developed. However, this wasn't the end of the homeland missile defense program; instead, it evolved to focus on ground-based interceptors designed to destroy warheads in midcourse. Those interceptors form the backbone of the current Ground-based Midcourse Defense system. We'll discuss the technology of ground-based interceptors later in this chapter.

Modern-Day Ballistic Missile Defense

In the aftermath of the September 11 terrorist attacks on the U.S. homeland, the George W. Bush administration in 2002 withdrew from the ABM treaty, believing that it prevented the U.S. from protecting itself against missiles launched by rogue states or nonstate actors. The administration said at the time that it would offer a series of amendments to the original treaty or withdraw if Russia didn't comply with the changes.

In fact, the Bush administration never did offer those amendments and unilaterally withdrew from the treaty.

In 2002, the Bush administration instituted Presidential Security Directive NSPD-23, actively funding ballistic missile defense initiatives. The directive mandated a layered defense to deploy 10 ground-based interceptors (GBIs) in Poland to counter possible ICBM threats from Iran or North Korea. A GBI consists of a solid fuel booster and a device known as a **kill vehicle**, whose purpose is to hone in on the target missile and physically destroy it by colliding at very high speed.

The U.S. fielded multiple BMD systems that, initially, were inadequately tested and riddled with technical problems. The presidential directive had anticipated this: "The United States plans to begin deployment of a set of missile defense capabilities in 2004. These capabilities will serve as a starting point for fielding improved and expanded capabilities later."[3] The Bush administration presented a wish list of layered, boost-phase, midcourse, and terminal-phase ballistic missile defense components, as well as ground- and sea-based interceptors called Aegis, for defense against short- and medium-range missiles. Also included was initial deployment of the Terminal High Altitude Area Defense. The result, shown in table 17.1, is a hodge-podge of U.S. missile defense systems currently deployed.

Table 17.1
Capabilities and characteristics of U.S. missile defense systems

Name	Layer	Threat (range)	Components	Approximate defended footprint (radius)
GMD	Midcourse	ICBM (>5,500 km)	GBI, EKV	Homeland/ National
Aegis	Midcourse	IRBM (3,000–5,500 km) MRBM (1,000–3,000 km) SRBM (<1,000 km)	SM-3 series	Theater/Regional 1,000s of km
THAAD	Terminal	SRBM/MRBM	Kinetic warhead	200 km
Patriot	Terminal	SRBM	PAC-3	15–20 km

Acronyms: GMD: ground-based Midcourse Defense; SRBM: short-range ballistic missile; MRBM: medium-range ballistic missile; IRBM (intermediate-range ballistic missile); ICBM (intercontinental ballistic missile); THAAD: Terminal High Altitude Area Defense; GBI: ground-based Interceptor; EKV: exoatmospheric kill vehicle; SM: Standard Missile, PAC: Patriot Advanced Capability interceptor.

The full U.S. missile defense system includes GMD for homeland defense, Aegis for theater or regional defense, and THAAD and Patriot terminal-phase systems. Although these systems remain imperfect, advocates of missile defense have argued that "missile defense needed to demonstrate that it could crawl before it could be asked to walk, and it had to walk before it could run" and "Something deployed when the country remained vulnerable is better than nothing deployed, even if that something was a rudimentary, minimal system."[4] However, "something deployed" may also give a dangerous false confidence to leaders who believe they're more protected than they really are. The Russian Foreign Affairs Ministry warned exactly that, stating in 2018 that "the availability of a missile shield may give grounds for a vile feeling of invincibility and impunity."[5]

The Obama administration took office in 2009 with a different view of threats facing the United States. They initially reduced the number of deployed ground-based interceptors from a planned 44 to 30, although these were ramped back up to 44 in 2013. They also scaled back Bush-era plans to deploy anti-ICBM GBIs to Poland and instead introduced the **European Phased Adaptive Approach** (EPAA), which uses the U.S. Navy's Aegis interceptors armed with kinetic warheads. Destroyer-based interceptors (known as Aegis Afloat) are deployed in the Baltic Sea, along with virtually identical land-based systems in Poland and Romania (Aegis Ashore, each with a radar and 24 interceptors). An additional radar is based in Turkey and a command center in Germany. EPAA is supposed to address the threat posed by short- and intermediate-range ballistic missiles that could target European and U.S. assets.

The essence of ballistic missile defense is "hitting a bullet with a bullet." That's an enormously complex technological challenge—one on which the United States has, since 2002, spent $142 billion without rigorous congressional oversight. The result is a set of systems that, despite being partially deployed, is still a work in progress.

Scrutinizing the Flight Test Record

How close are we to a viable ballistic missile defense? The **U.S. Missile Defense Agency** emphasizes an 80 percent "hit to kill" test record, but that's misleading because it's averaged over the first three systems shown in table 17.1. The THAAD terminal-phase system has indeed a strong record of interception in 16 out of 16 tests. However, it hasn't been shown how THAAD would perform under a salvo of incoming missiles. Although it's been deployed in Guam since 2013, in range of North Korean IRBMs, THAAD has only once been tested with an IRBM. So

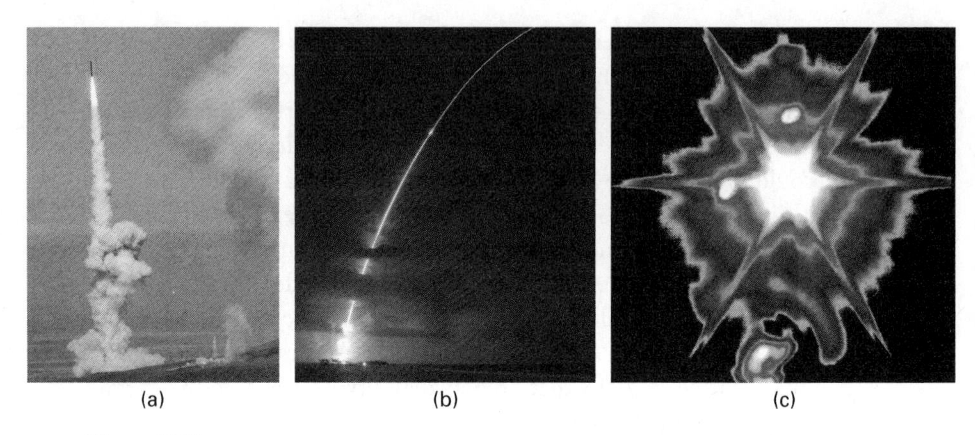

Figure 17.3
The first-ever salvo test of ground-based interceptors against an ICBM target. (a) The second of two interceptors launching from Vandenberg Air Force Base in California on March 25, 2019. Cloud at right is from the first interceptor's launch. (b) The interceptors' ICBM target launching from the Ronald Reagan Ballistic Missile Defense Test Site in the Marshall Islands. (c) Infrared image showing successful interception of the ICBM target. (U.S. Department of Defense, Missile Defense Agency)

one wonders how well it would do if several North Korean missiles were simultaneously to target the U.S. base at Guam.

The Aegis missile defense system has a 78 percent success rate in 42 tests of the SM-3 interceptor. But one-third of tests conducted in the late 2010s failed. The ground-based interceptors for the anti-ICBM mid-course defense system had 11 failures out of 19 tests over more than two decades of development since 1999. In 2017, however, after three decades and hundreds of billions of dollars, there came a clear success: an interception by a single GBI of an ICBM-like missile launched from Kwajalein Atoll in the Marshall Islands. A similar test in 2019 was also successful, this one using a salvo of two GBIs (figure 17.3).

North Korea monitors U.S. BMD tests very carefully, looking for ways to exploit vulnerabilities in the defensive systems. It's probably no coincidence that North Korea carried out multiple short-range missile tests in August 2019, only months after the U.S. antimissile success depicted in figure 17.3. Indeed, the North Korean central press agency stated that one reason for the August tests was the recent U.S. ICBM-interception tests. One of North Korea's newest missiles is the KN-23, whose depressed trajectory keeps it under radar detection, and which undertakes unexpected maneuvers that makes it difficult to track. It's the knuckleball of missiles (except that it flies at seven times the speed of sound)! How would

U.S. missile defense fare under attack by multiple missiles with these capabilities?

How Is Ballistic Missile Defense Supposed to Work?

To fully appreciate the complexity and challenge of ballistic missile defense, we describe briefly how the full system is supposed to work (see figure 17.4). Consider a North Korean ICBM attack on San Francisco—this time for real, unlike Hawaii's false missile alert in 2018. The first detection would occur almost immediately as satellites' infrared sensors would detect the hot boost-phase plume, then send alerts that a threat missile had been launched. In post-boost, the missile would deploy the bus containing single or multiple warheads and perhaps also a cloud of decoys and other countermeasures to confuse the defense. Radars such as the TPY-2—components of the THAAD systems based in Korea and in Guam, as well as ship-based Aegis SPY-1 radars—would track the threat and attempt to determine which objects are warheads and which are countermeasures.

In our hypothetical attack on San Francisco, the powerful Sea-Based X-Band Radar (SBX; see figure 17.5) located near Hawaii and the Cobra

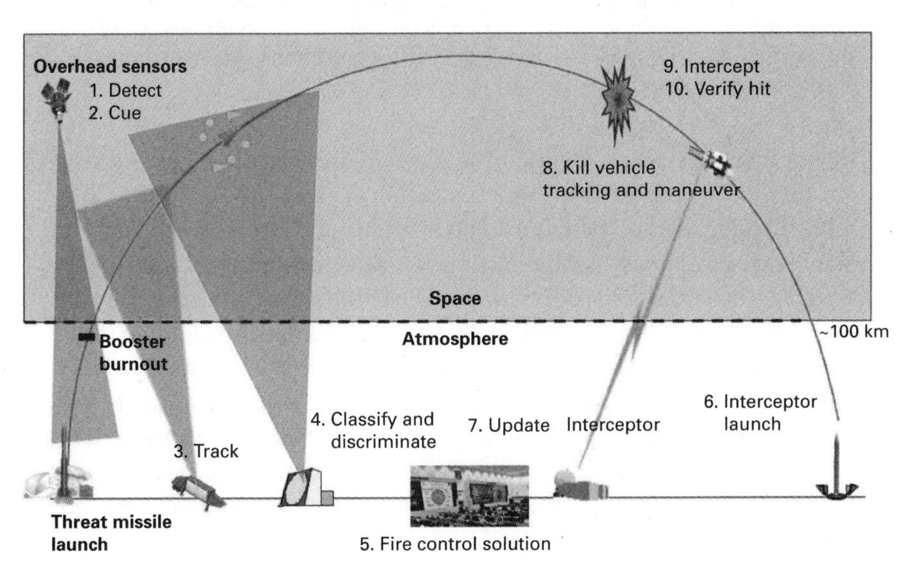

Figure 17.4
The 10 steps in a successful missile intercept. The defense begins with space-based assets detecting the threat in the boost phase and predicting its trajectory, followed by tracking and discrimination by ground- and sea-based radars. The intercept point is calculated and the interceptor is launched. The interceptor then deploys the kill vehicle, which crashes into the threat missile to destroy it. (U.S. Department of Defense, Missile Defense Agency)

Figure 17.5
The Sea-Based X-Band Radar (SBX) detects ballistic missiles, provides precise tracking information, and distinguishes warheads from decoys. Part of the U.S. Ballistic Missile Defense System, SBX supplies data for ground-based interceptors and assesses the results of intercepts. (U.S. Navy)

Dane radar in Alaska's Aleutian Islands would track the threat and send information to control and communication centers in Alaska and Colorado to calculate the best interception point and determine when and from where to launch the interceptor. These centers integrate data from satellite sensors and radars to visualize the battle space. After the interceptor is launched and its kill vehicle deployed, the centers transmit real-time data to the kill vehicle to make any needed corrections to its trajectory. The kill vehicle (here an exoatmospheric kill vehicle or EKV) uses its telescopes and infrared sensors to determine which object in the threat cloud is the warhead. A kill assessment is ideally done after the hit to determine if a second interceptor should be fired or, if there's not enough time left, whether assets should be engaged for terminal phase interception.

Success Is a Risky Game of Chance

A successful interception involves a sequence of steps that each must be successful, like a relay race in which each runner passes a baton to the next runner. If there are two handoffs and each runner has only a 50 percent chance of passing the baton successfully, then there's only a 25 percent (0.5 times 0.5) chance for the athletes to complete the race. Missile defense is similar but with many more steps. Satellites have to spot the launch, radars have to accurately track the attacking missile, the interceptor has to launch successfully, the interceptor has to choose the warhead over decoys, and the interceptor has to hit the warhead. Every step is independent of the others, but if one step fails and there are no backup options, then the whole interception fails. If each step has only a 50 percent chance of success, then with so many steps it's very unlikely that the interception will succeed (multiply 0.5 by itself as many times as there are steps; the result is a tiny number). Also, if just one step has a really low probability of success, then it doesn't matter what the probabilities are for the other steps—the intercept will likely fail. The full chain of events is only as strong as its weakest link. Therefore thorough and rigorous testing is crucial to maximize the probability of a successful ballistic missile defense—something we noted previously has not always happened with the U.S. missile defense program.

The probability that an interceptor will destroy an incoming warhead is known as the **single-shot kill probability** (SSKP). This important parameter allows missile-defense capability to be characterized under specific scenarios, such as a salvo of incoming missiles or a salvo of outgoing interceptors. President Trump made an offhand comment with respect to the U.S. capability to intercept a North Korean ICBM: "We have missiles that can knock out a missile in the air 97 percent of the time." This implies that the single-shot kill probability for U.S. ground-based interceptors is 97 percent, which, based on tests, is not correct. Ascertaining the actual SSKP for ground-based interceptors is complex. But over the past decade we've had three failures and three successes, so it's reasonable to conclude that the likelihood of success is around 50 percent. This implies a 75 percent chance of taking out an incoming threat with two interceptors. The probability obviously decreases precipitously when there's a salvo of incoming missiles.

Acceptable "Leakage"

An important question for society is how much **leakage** of incoming missiles is acceptable in a missile-defense system. No system is 100 percent effective, so it's important to quantify how effective the system should be. If the defense is 95 percent effective (a leakage of 5 percent), then 1 in 20 missiles might make it through the defense. Just a single warhead getting through to a city would be an enormous catastrophe, so 5 percent leakage might be deemed unacceptable. Given our estimate of 50 percent success for a single interceptor, we can calculate that we need 5 interceptors to achieve a 5 percent leakage.[6] Then if North Korea targets San Francisco with 4 ICBM's, we would need 20 interceptors to give San Francisco a 95 percent chance of surviving. But if acceptable leakage is 1 percent, then we'd need 7 interceptors per incoming missile, or 28 total in this scenario. And this doesn't even account for decoys that, if the defense can't distinguish them from warheads, would require even more interceptors.

Missile-Killing Technology: A Closer Look

Early ballistic missile defense systems used nuclear-tipped interceptors with large—often megaton-range—warheads. The idea was that a huge nuclear explosion anywhere near an incoming missile would either destroy it, disable it, or knock it off course. Today, with more accurate tracking, guidance, and sensors, U.S. BMD instead takes a precision approach of destruction by direct impact.

Kinetic Kill Vehicles

Driving fast through rain, you experience loud taps as raindrops impact your car's windshield. Stop the car and it's more a soft dribble. All those drops make tiny collisions, and the higher the relative speed the more violent the collision. This is the same concern we have with space debris, where small particles moving at high speeds can destroy space assets. And it's the basis of today's BMD kill mechanism, dubbed **hit to kill**.

A kill vehicle works by colliding deliberately with a missile or reentry vehicle. There are no explosives involved. The relative speed of the interceptor and its target is immense—some 10 miles per second—so the collision alone destroys the incoming threat.

Types of Kill Vehicles

The interceptor used in the U.S. Ground-based Midcourse Defense program (GMD) carries an **exoatmospheric kill vehicle** (EKV; see figure 17.6), which separates after the three-stage booster rocket exhausts its fuel. At this point the vehicle is deep in space, so its nonaerodynamic shape isn't a problem. The EKV itself has 10 small rocket motors, allowing it to maneuver to intercept one warhead in the threat cloud. It uses a suite of instrumentation to select the warhead from among the threat-cloud objects and then directs the EKV toward the warhead. Kill-vehicle instrumentation includes an optical telescope used to maneuver to the threat cloud as well as for star-based navigation. The EKV uses passive seeking, meaning it doesn't send out infrared or radar but rather uses its telescope

Figure 17.6
The exoatmospheric kill vehicle (EKV) used in the U.S. Ground-based Midcourse Defense system. Note the large telescope for infrared imaging of warheads. (Raytheon Corporation)

to image infrared emitted from the target. As the kill vehicle approaches the threat cloud, the discrimination improves and, with the warhead identified, infrared imaging guides the EKV to its target.

Currently under development is a miniaturized anti-ICBM kill vehicle known as the **MOKV** (multi-object kill vehicle), of which as many as six can be deployed in a single interceptor. It's the MIRV concept in reverse: In contrast to multiple nuclear warheads on one attacking missile, here one interceptor can destroy multiple incoming threats. MOKV isn't a new idea; it was first suggested in 2004 and funded. The program was canceled in 2009 after spending $700 million but revived in 2015. However, it's not clear when, if ever, MOKV will be deployed.

The **Aegis** missile-defense system, for use against shorter-range missiles, also uses exoatmospheric interceptors. Aegis is an evolving technology but has a better test record than the GBI interceptors used against ICBMs. The current most advanced version of the Aegis interceptor is the product of a unique Japanese-U.S. collaboration. Figure 17.7 shows an interceptor's infrared view of its target missile.

The **Terminal High Altitude Area Defense** (THAAD) is a self-contained mobile system including a powerful radar that can discriminate threats

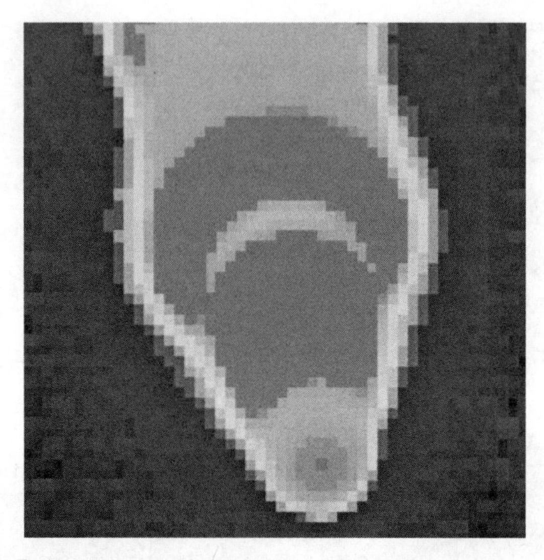

Figure 17.7
Image of a missile nosecone by an infrared seeker on a kinetic warhead several milliseconds before the intercept. (U.S. Department of Defense, Missile Defense Agency)

1,000 km into the sky as well as providing tracking information to interceptors. The THAAD interceptors, too, use hit-to-kill kinetic warheads—although they're not exoatmospheric. THAAD can defend a half-spherical "bubble" some 150 kilometers (about 93 miles) in radius centered on the launcher and destroys shorter-range incoming missiles in their terminal phase. The THAAD test record is remarkably strong (16 tests and all successful), and THAAD can be a powerful component of a layered defense. But there are only eight interceptors per battery, and the system—which has never been tested in combat—can likely be overwhelmed by a swarm of incoming missiles. THAAD might also be destabilizing: a THAAD battery installed in 2017 in South Korea drew Chinese protests because its radar could peer deep inside China, potentially helping track Chinese ICBMs in their early stages. Here's yet another situation where the presence of a defensive system could actually provoke an ICBM launch.

Cruise and Drone Defense
In this chapter, we've primarily emphasized defense against ICBMs and related strategic weapons. We've said little about cruise missiles or their relatives, drones. That's partly because these unmanned aerial vehicles (UAVs) are designed more for battlefield situations—although ship- and,

Patriot Missile Defense: Saudi Case Study

Patriot (Phase array tracking to intercept on target; see figure 17.8) is a terminal-phase, small-area ballistic missile defense system designed to protect specific assets, such as radars, cities, military bases, or airports. Its range is a hemispherical volume with a radius of 15 to 20 kilometers. Patriot, too, uses a hit-to-kill kinetic warhead, although it's also armed with a small explosive that can detonate in proximity to its target. The system saw combat in both Iraq wars, with mixed success. In fact, the U.S. House Subcommittee on National Security stated, "The public and the United States Congress were misled by definitive statements of success issued by administration and Raytheon representatives during and after the war." However, that was 1991 and 2003. The manufacturer, Raytheon, has sold the system to some 17 countries. Raytheon said of a 2018 sale, "Romania is purchasing the most advanced, capable, cutting edge tactical ballistic missile defense system in the world. Patriot has been tested thousands of times in peace, and repeatedly proven itself in combat. Simply put, Patriot saves lives."[7] Yet recent experience says otherwise.

Figure 17.8
A Patriot antimissile missile launching from its mobile battery. (U.S. Army)

(*continued*)

Patriot Missile Defense: Saudi Case Study (*continued*)

The Center for Nonproliferation Studies (CNS), where one of your authors (FDV) is based, did an independent assessment of the Patriot missile-defense system in Saudi Arabia.[8] The Saudi government is a staunch supporter of the United States and purchases a great deal of military equipment from the U.S., including Patriot systems. Recently it used Patriot in its fight against the Iran-backed Houthi rebels in Yemen. In one case, a Burkan-2 missile, a fast, 600-km-range SCUD missile variant, made it to an airport where Saudi military hardware was located. The Saudis stated that the missile was successfully intercepted. However, video seems to show an interceptor that explodes in the boost phase and another interceptor that makes an unusual U-turn in the air and "comes screaming back at Riyadh, where it explodes on the ground." CNS colleagues mapped where the debris fell from the missile and interceptor, and it appears that the attacking missile reached its target, while interceptor debris fell far from the airport. The fact that the offensive missile warhead landed only a few hundred meters from Terminal 5 in Riyadh's King Khalid International Airport showed clearly that the offensive missile was successful. Another missile attack that CNS analyzed destroyed a Honda dealership, despite the Saudis claiming that the missile was intercepted. These are just two examples, but they indicate that the system might not be as successful as advertised. For Raytheon, false claims of success provide an opportunity to advertise and sell more Patriot systems, while for the buyer it gives their public confidence that their government is protecting them. President Trump has promulgated the false impression of Patriot success: one day after the attack on the Riyadh airport, he stated, "Our system knocked the missile out of the air. That's how good we are. Nobody makes what we make, and now we're selling it all over the world." It's critical that the governments purchasing the Patriot and other missile defense systems, along with the public they serve, are fully aware of the limitations and true performance record of these systems.

especially, bomber-based cruise missiles could carry nuclear warheads to strategic targets in an adversary's homeland. Another reason is that it's hard to defend against these UAVs because, despite being slow, they fly at low, radar-evading altitudes and they're maneuverable so their trajectories aren't predictable.

Another issue is that drones and even simple cruise missiles are cheap, while interceptor systems aren't. In chapter 14 we described possible military uses for swarms of perhaps thousands of drones. Even if defense against drone swarms were technologically possible, mounting such a defense could bankrupt the defending nation!

Future Ballistic Missile Defense Systems

The most exotic weapons concepts proposed in the 1980s for the Strategic Defensive Initiative were subsequently discarded, but some are now being reconsidered as **directed-energy weapons** that deliver lethal effects at or near the speed of light. The idea is that, at near light speed, even the fastest missiles can be intercepted.

Laser Weapons

A **laser** is a source of light or other electromagnetic radiation—including infrared radiation and X-rays—capable of producing a high-intensity beam that travels long distances with little spreading. A laser differs from a conventional light source not only in its intensity but also because the individual waves making up a laser beam are all exactly in step. This property, called *coherence*, is what allows a laser to maintain its intensity over long distances. Lasers for ballistic-missile defense would use several kill mechanisms to disable their targets. Visible-light or infrared lasers would shine on the target long enough (typically from seconds to minutes) to cause melting and structural failure. X-ray lasers would deliver energy in a sudden burst, driving a destructive shock wave into the target. Either of these kill mechanisms requires enormous laser power, which presents a huge technological challenge. The Missile Defense Agency plans to deploy lasers on a high-altitude long-endurance unmanned aerial vehicle that could loiter for 36 hours at an altitude exceeding 65,000 feet (about 20 kilometers).

Laser-based BMD systems face major technical and logistical hurdles. Even under laboratory conditions, no laser has yet achieved performance close to what would be needed for long-range missile defense. Challenges include being able to track targets at hundreds of kilometers standoff distance, so that the weapon system that fires the laser stays away from adversary's countermeasures. Then the laser needs to maintain focus on a particular spot on the attacking missile in order to deposit as much energy as possible—not easy when the target is hundreds of kilometers distant. Another challenge is atmospheric distortion of the laser beam. Finally, laser power must scale up at least threefold, to about 100 kilowatts, from the laboratory-demonstrated power of about 30 kilowatts—and increasing power also tends to worsen the beam quality. And for deployment in space or high in the atmosphere, the system must simultaneously exhibit low weight and yet maximum power.

Particle Beams

Particle beam weapons constitute a second class of directed-energy weapons. These devices would use accelerators to bring protons, electrons, or other particles to nearly the speed of light, whence they could penetrate several inches into missile warheads and destroy electronic circuits, guidance systems, and other delicate mechanisms. Technology to achieve the requisite particle energies is widely used in scientific research, but producing intense beams of high-energy particles and maintaining them over thousands of miles is an altogether different matter. One complication comes from Earth's magnetic field, which deflects electrically charged particles out of straight-line paths. Neutralizing the particles as they form the beam can alleviate this issue but adds a further technological complication. Particle beam weapons technology is far from the practical stage, and even if perfected it will require space basing because particle beams don't penetrate the atmosphere.

Countermeasures to Directed Energy Weapons

Just as there are cheap, easily deployed countermeasures that can thwart kinetic missile defense systems, so there are countermeasures against directed energy weapons as well. For example, lasers need to deposit energy at a specific point on the attacking missile and keep it on that point as long as possible. However, if the missile has added protection where electronics or other sensitive components are, then the beam power or its time on target will have to increase. Or the incoming missile can rotate, making it impossible to maintain the beam on a single point. And, just as with kinetic ballistic missile defense, the adversary can always deploy decoys to saturate the defense. Once a laser is trained on one target, it requires a few seconds before it can be targeted on another object, so increasing the number of objects in the threat cloud makes it easier to overwhelm the defense.

BMD as Antisatellite Weapons

There's one offensive task for which antimissile technology is ideal: destroying satellites. Any BMD system designed to find, track, and destroy warheads in space minutes after their unexpected launch can easily eliminate satellites in predictable orbits relatively close to Earth. This doesn't include high-altitude satellites like the Global Positioning System (GPS) constellation orbiting some 12,000 miles up or geostationary communications satellites at 22,000 miles. But it does include many crucial satellites in low orbit at altitudes under about 1,200 miles. These include surveillance satellites, missile-launch warning sensors, military communications satellites, and, eventually, space-based missile defenses. Destruction of such assets could be provocative, leading a nation whose spacecraft were destroyed to conclude that a nuclear attack might be imminent and that they would be without warning or defensive capability. An attack on space assets could also cripple civilian and military control systems that might otherwise contain an escalating crisis. Might the nation with crippled space assets feel compelled to launch a preemptive nuclear strike? And even in terrestrial peacetime, antisatellite attacks could disable the spy satellites that verify treaty compliance and warn of military buildups. Again, crisis stability would suffer.

To date, four countries—Russia, China, the United States, and India—have demonstrated the ability to destroy orbiting satellites. So far these demonstrations have been limited to their own satellites—but nevertheless each creates a mess of space debris that can endanger other spacecraft, including the International Space Station. Expect more development of **anti-satellite weaponry** in the coming years, especially in light of the United States' creation, in 2019, of its new military branch, the Space Force.

Missile Defense in Perspective

The United States' missile defense program has been so expensive that each interceptor is literally the economic equivalent of its weight in gold. Adversaries respond, creating a terrible feedback loop that one prominent missile expert describes as "Challenge begets responses, and responses beget more challenges."[9] The U.S. may be waking up to this conundrum: Major General Rick Evans, the U.S. Strategic Command's deputy commander, recently said, "We cannot 'active defense' our way out of the problem sets. We simply don't have the money, capability and capacity."[10] The general and some military colleagues have suggested more emphasis on destruction of missiles before or right after launch, and on cyber or electromagnetic weapons for missile defense. It seems that decision makers are realizing there's no "silver bullet" to counter missile threats.

Your authors agree that there's no single "silver bullet" that will defend us against nuclear-armed ballistic missiles or nuclear weapons in general. But we go further: we don't see any prospect for anything approaching Ronald Reagan's 100 percent effective shield against ballistic missiles. Perhaps we've made some progress in providing limited defense against rogue states with relatively small nuclear arsenals, like North Korea. But against the thousands of warheads deployed by Russia (or, from Russia's

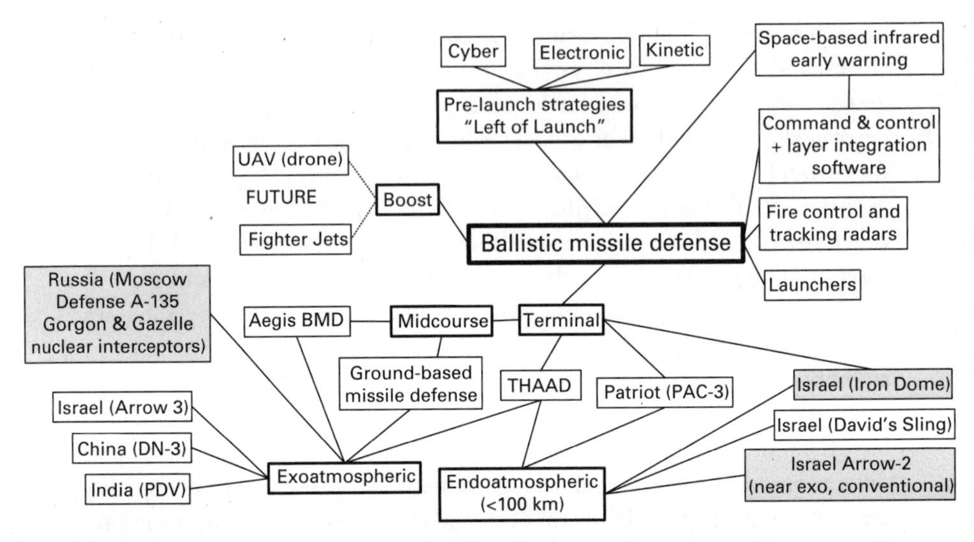

Figure 17.9
Summary diagram relating ballistic-missile defense systems deployed globally.

viewpoint, by the United States), there is simply no defense to prevent unimaginable disaster once missiles start to fly.

Summary

We began this book's section on nuclear weapons with this statement: "Our nation can be destroyed in half an hour, and there's nothing we can do to prevent it." In this chapter, we've considered the challenges of ballistic missile defense. We've described the technology, current approaches, and ideas that may be used in the future (see figure 17.9 for a summary of global BMD systems). And we've covered the modest successes and many failures of what remains an imperfect defense against ballistic missiles.

The history of ballistic missile defense shows a constant cat-and-mouse game where a BMD system is developed only to have adversaries respond by (1) building faster missiles, (2) building more missiles so as to overwhelm the defense, (3) deploying countermeasures that make it difficult for the defense to identify true threats, or (4) developing radar-evading quasi-ballistic missiles with the speeds of ICBMs but the maneuverability of cruise missiles. This vicious cycle leads to geopolitical instability and is economically unsustainable. So we'll say it again: our nation can be destroyed in half an hour, and there's nothing we can do to prevent it.

Glossary

ABM Treaty Part of the SALT I agreements, limiting missile-defense systems. The treaty was in effect from 1972 to 2002, when the United States withdrew.

Aegis The U.S. Navy's theater missile ballistic missile defense system.

anti-satellite weapon (**ASAT**) A weapon designed to seek out and destroy an adversary's militarily significant satellites, including those for warning, communication, intelligence gathering, and missile defense.

ballistic missile defense (**BMD**) The system of radars, sensors, interceptor missiles, and firing units that makeup a missile defense system.

boost phase The first part of a missile's flight, in which rocket motors fire to boost the missile above the atmosphere.

decoys Lightweight objects carried by an ICBM and released along with warheads to confuse defensive systems.

directed-energy weapon A weapon that delivers intense bursts of energy in the form of electromagnetic radiation or subatomic particles.

European Phased Adaptive Approach (**EPAA**) The U.S. contribution to NATO's missile defense system, being deployed in three phases. EPAA's primary purpose is to protect Europe against short- through intermediate-range ballistic missiles launched especially from Iran.

exoatmospheric kill vehicle (**EKV**) An antimissile weapon that destroys its target by direct collision above Earth's atmosphere. This is the kill vehicle used by the United States' ground-based midcourse defense system (GMD).

Ground-based Midcourse Defense (**GMD**) The United States' primary system for defense against intercontinental ballistic missiles.

hit to kill Antimissile strategy that uses so-called kill vehicles to destroy an attacking warhead by direct collision at high speed.

kill vehicle (**KV**) A weapon that destroys incoming warheads by direct collision rather than with explosives. Also called a *kinetic kill vehicle*.

laser A source of intense, coherent light, X-rays, infrared, or other electromagnetic radiation.

layered defense An antimissile strategy using different weapons tailored for use against each phase of a missile's flight.

leakage The percentage of incoming threats that make it through the BMD system.

midcourse phase The longest part of a ballistic missile's flight, during which warheads and other elements of the threat cloud travel on ballistic trajectories under the influence of gravity alone.

MOKV (**multi-object kill vehicle**) Still under development, MOKV is an interceptor that deploys more than one kill vehicle from a single booster.

particle-beam weapon A space-based device using beams of high-energy subatomic particles to destroy its targets.

Patriot Terminal phase system for local area defense; stands for "Phased Array Tracking Radar to Intercept on Target."

post-boost phase The brief second part of an ICBM's flight, during which individual warheads and decoys are released.

SALT Strategic Arms Limitation Talks.

single-shot kill probability (SSKP) The overall probability that an interceptor will destroy an incoming warhead.

Strategic Defense Initiative (SDI) A U.S. program begun in the 1980s with the aim of providing complete defense against ballistic missiles.

Terminal High Altitude Area Defense (THAAD) A mobile system whose interceptors are designed to destroy ballistic missiles either within or outside the atmosphere during the terminal phase of their flight.

terminal phase The final phase of a ballistic missile's flight, as warheads plunge through the atmosphere.

threat cloud The totality of objects—warheads, decoys, chaff, etc.—released once a ballistic missile is above the atmosphere.

U.S. Missile Defense Agency (MDA) A research, development, and acquisition agency within the U.S. Defense Department, whose purpose is to develop BMD systems.

Notes

1. From "Defense Spending and Defensive Technology," Reagan speech, televised on March 23, 1983.

2. Quoted in U.S. Missile Defense Agency, *Missile Defense: The First Seventy Years*, 3. Available at https://www.mda.mil/global/documents/pdf/first70.pdf.

3. National Security Presidential Directive NSPD-23, December 6, 2002. Posted by the Federation of American Scientists at https://fas.org/irp/offdocs/nspd/nspd-23.pdf.

4. Steve Lambakis, *The Future of Homeland Missile Defenses* (Fairfax, VA: National Institutes Press, 2014). Available at https://www.nipp.org/wp-content/uploads/2014/12/Future-of-Homeland-Missile-Defenses.pdf.

5. Russian News Agency TASS, April 24, 2018, https://tass.com/politics/1001627.

6. This can be solved in Wolfram Alpha for an SSKP of P = 0.5 and leakage of 5 percent: "solve (1-P)^n=0.05 for P=0.5" gives n=4.3 interceptors or 5 to be sure none are missed. Changing "=0.05" to "=0.01" gives 7 interceptors for a 1 percent leakage rate.

7. Mike Ellison, Raytheon Country Manager for Romania, quoted at http://raytheon.mediaroom.com/2018-11-01-Romania-to-procure-additional-Patriot-Air-and-Missile-Defense-systems.

8. Lewis, Jeffrey. "Patriot Missiles Are Made in America and Fail Everywhere." *Foreign Policy*, March 28, 2018. https://foreignpolicy.com/2018/03/28/patriot-missiles-are-made-in-america-and-fail-everywhere/.

9. Uzi Rubin, concluding slide at https://rusi.org/sites/default/files/session_4_the_disruptors_-_the_disruptive_impact_of_space_cyberwarfare_and_automation_on_missile_defence_-_uzi_rubin.pdf.

10. Major General Rick Evans, quoted at https://www.defensenews.com/digital-show-dailies/smd/2019/08/06/should-the-dod-shift-focus-toward-passive-missile-defense/.

Further Reading

Grego, Laura, George Nelson Lewis, and David Wright. *Shielded from Oversight: The Disastrous U.S. Approach to Strategic Missile Defense*. Cambridge, MA: Union of Concerned Scientists, 2016. https://www.ucsusa.org/sites/default/files/attach/2016/07/Shielded-from-Oversight-full-report.pdf. An excellent report analyzing the problems and limitations of current missile defense programs and their acquisition procedures.

Hafemeister, David. *Nuclear Proliferation and Terrorism in the Post-9/11 World*. Cham, Switzerland: Springer International, 2016. See chapter 5 for an excellent discussion of the state of ballistic missile defense.

Kelleher, Catherine M., and Peter Dombrowski, eds. *Regional Missile Defense from a Global Perspective*. Stanford, CA: Stanford University Press, 2015. Well-selected writings from writers around the globe, emphasizing missile defense policy in the early 2000s.

Lambakis, Steve. *The Future of Homeland Missile Defenses*. Fairfax, VA: National Institute Press, 2014. A good description of the status and plans for homeland/national missile defense. Available at https://www.nipp.org/wp-content/uploads/2014/12/Future-of-Homeland-Missile-Defenses.pdf.

Lewis, George N. *mostlymissledefense* (Blog). https://mostlymissiledefense.com/. Professor Lewis of Cornell University tracks activities in ballistic missile defense. An excellent resource for keeping up-to-date on BMD.

Mantle, Peter J. *The Missile Defense Equation: Factors for Decision Making*. Reston, VA: American Institute of Aeronautics, 2004. The only textbook on ballistic missile defense, although quite dated. It discusses missiles and interceptors and shows calculations for defended areas. Highly recommended for a deeper study of ballistic missile defense.

Pella, Peter. *The Continuing Quest for Missile Defense: When Lofty Goals Confront Reality*. San Rafael, CA: Morgan & Claypool, 2018; ebook at https://iopscience.iop.org/book/978-1-6817-4942-6.pdf. In this publication of the Institute of Physics' Concise Books series, physicist Pella provides a historical and technical perspective on missile defense.

Union of Concerned Scientists. *All Things Nuclear* (Blog). https://allthingsnuclear.org/. The UCS blog deals with nuclear issues in general and frequently focuses on missile defense.

United States Missile Defense Agency. The MDA website has useful resources describing historical and current BMD systems. https://www.mda.mil/.

Wilkening, Dean A. "A Simple Model for Calculating Ballistic Missile Defense Effectiveness." *Science & Global Security* 8, no. 2 (2000): 183–215. Good discussion on ballistic missile defense from a probabilistic point of view.

18

Arms Control and Nuclear Proliferation

In 1963, President John F. Kennedy voiced his fear that nuclear weapons would proliferate around the globe: "I see the possibility in the 1970s of the President of the United States having to face a world in which 15 or 20 or 25 nations may have these weapons."[1] Kennedy's fear wasn't realized, but nevertheless the number of nuclear-armed states has more than doubled (from four to nine) since Kennedy's time, and there's no reason to believe it won't increase in the future. So how do we manage the risk posed by nuclear weapons and by any technologies—including so-called peaceful ones—that could help spread nuclear weapons or increase the chances of nuclear war? The key is a multipronged, concerted effort focused on (1) preventing states that currently possess nuclear weapons from acquiring more of them and, ideally, reducing their nuclear arsenals; (2) preventing other states and nonstate actors from acquiring nonpeaceful nuclear technology; (3) managing safely and securely civilian use of nuclear technology; (4) minimizing use of materials and products that might support development of nuclear weapons and related technologies, by replacing with less sensitive materials. Here we'll discuss these objectives, focusing especially on controlling nuclear technology itself, the materials and equipment used to produce nuclear weapons or delivery systems, and how to achieve a global consensus on a coordinated approach to accomplishing all four objectives.

So far, the world's approach to curbing the spread of nuclear weapons has been piecemeal and reactionary. In some respects the nuclear-armed states are like toddlers playing with matches. Disaster lurks at every turn, but immediate political goals often overshadow the long-term stabilizing effect of reducing nuclear arsenals. Nevertheless, just as toddlers do, nations interact and invent new rules for living with humankind's relatively new nuclear toys.

This chapter reviews the tools the international community has developed to slow nuclear proliferation and decrease nuclear risk, collectively called the **nuclear nonproliferation regime**. The regime is far from fully successful, and its tools are continually being questioned and recalibrated as world circumstances change. As of this writing, we've just buried one treaty (the Intermediate-Range Nuclear Forces Treaty) and another appears to be on its deathbed (New START), while others, subjects of the next chapter, lie in perpetual limbo. Export control is increasingly used to prevent proliferation of sensitive equipment, and, although largely successful, is at risk from emerging technologies and regional rivalries. The International Atomic Energy Agency (IAEA) continues to promote the peaceful use of nuclear technology and discourage production of undeclared fissile material, but some states' perceived use of nuclear inspections as political tools can cause tensions with the agency.

Developing and maintaining the nonproliferation regime is a complex effort involving government officials, diplomats, scientists, international organizations like the United Nations and the International Atomic Energy Agency, numerous nongovernmental organizations (NGOs), and even industry groups. That's a diverse bunch of people. But is it diverse enough? The box on the next page explores this important question.

How Nuclear-Armed States Obtained Their Weapons

The United States' nuclear monopoly ended in 1949, when the Soviet Union exploded its first fission bomb. Today nine nations have well-established nuclear arsenals and delivery systems. Others, including Iran, Saudi Arabia, and South Korea, may have nuclear ambitions. Table 18.1 lists the nine countries known to have nuclear weapons capability, along with the year each first tested a nuclear device, the estimated size of its strategic nuclear arsenal, and its status with regard to two treaties that we'll discuss shortly.

How did these states acquire their nuclear capabilities? For the United States, the answer is the Manhattan Project, reviewed at length in chapter 12. The Soviet Union began nuclear weapons research in 1939, but only after World War II did the Soviet project accelerate. The Soviets were aided by physicists Theodore Hall and Klaus Fuchs and machinist David Greenglass, who were at Los Alamos and, unbeknownst to their colleagues, passed nuclear secrets to the Soviets for several years. These efforts may have shaved a year or two off the Soviet bomb program, but even without any espionage the Soviets were certainly capable of producing their own bomb—as are all states that are willing to make the investment

Too Much Testosterone?

In 1987, Dr. Carol Cohn, then at Harvard Medical School's Center for Psychological Studies in the Nuclear Age, argued that diversity is seriously lacking among defense intellectuals and others involved with nuclear weapons policy. Cohn, an expert on gender issues in global politics and security, embedded herself in the world of defense intellectuals for a year, learning their language and *how* they use that language. She reported her findings in the essay "Sex and Death in the Rational World of Defense Intellectuals"[2] (figure 18.1). Cohn learned terms that you readers have seen in this book—including ones that we use coldly, matter-of-factly, just as defense analysts do: "limited nuclear war," "counterforce exchanges," "first strike," "countervalue targets," and the like, which Cohn calls **technostrategic discourse**. These terms hide the morality of what they're expressing or, as Cohn said recently: "What struck me was how removed they were from the human realities behind the weapons they discussed." When your authors teach the physics of nuclear weapons, we're aware of how interesting and even empowering such nuclear knowledge can be. Therefore we purposely inject pictures from, for example, Hiroshima, without context, to remind students that physics, like the technostrategic terms, hides the immorality of nuclear weapons.

Figure 18.1
Illustration from Dr. Carol Cohn's article "Slick 'Ems, Glick 'Ems, Christmas Trees, and Cutters: Nuclear Language and How We Learned to Pat the Bomb," *Bulletin of the Atomic Scientists* 43, no. 5 (1987). (Cartoon by Tom Herzberg.)

(*continued*)

Too Much Testosterone? (*continued*)

Cohn recognizes the bizarre way that metaphors are used in the subculture of defense analysis. That language, she found, is inextricably linked with sexual imagery, as when a defense lecturer solemnly announced that "to disarm is to get rid of all your stuff" implying, as Cohn emphasizes, that in this culture to disarm is emasculation, and so "how could a real man consider it?" Other terms you've seen also reek of sexualized imagery: "penetration aids" and "transporter erector launcher" (chapter 14).

But we needn't go back to the 1980s: President Trump's talk of nuclear weapons is the ultimate technostrategic discourse: "North Korean Leader Kim Jong-un just stated that the 'Nuclear Button is on his desk at all times.' Will someone ... please inform him that I too have a Nuclear Button, but it is a much bigger and more powerful one than his, and my Button works." The metaphor, of course, is penis size. Such sexualized technostrategic discourse is everywhere when nuclear weapons are discussed. It's also evident in metaphors related to marriage, virginity, male birth, and God. Of the bombs "Little Boy" and "Fat Man" that destroyed Hiroshima and Nagasaki, Cohn says: "These ultimate destroyers were the progeny of the atomic scientists." Scientists at the first nuclear test "hoped the baby was a boy, not a girl—that is, not a dud." We've used technostrategic terms throughout this book, and for some of you the sexual connotation may have been obvious. Unfortunately, we need these terms because they're still widely used in practice. So we're left with a dilemma: If we continue with this language we'll become desensitized to it, and, in Cohn's words, "invite the transformation, the militarization, of our own thinking." But if we use plain English we won't be taken seriously and will remain outside the relevant "spectrum of opinion." That was Cohn's experience when she tried conversing in plain English versus technostrategic language.

Cohn also emphasizes that rarely do defense intellectuals use the word "peace," supplanting it instead with "strategic stability"—which she emphasizes "refers to a balance of numbers," not all the nuanced, holistic meaning of *peace*. Defense intellectuals use such language to avoid becoming emotional. How could this ever be considered cool-headed objectivity? How can we change our discourse when it comes to nuclear weapons and defense?

Perhaps the answer lies in this box's title: there's been too much testosterone in military discourse, especially around nuclear weapons. The dominant nuclear decision makers in the United States are white males. Expanding the spectrum of intellectuals working in the field is not just good practice that can help change awareness of how language is used, but might also alter the way we view militarization and nuclear weapons in general. This is beginning to happen, with analysts in arms control and nonproliferation pledging to not participate in "manels" (exclusively male panel discussions) and to include more women in the field, but we're only at the tip of the iceberg and need to be much more inclusive. Might our survival in the nuclear age depend on it?

Table 18.1
Nuclear weapons holders and treaty status

	First Test	Deployed warheads*	Stored warheads (available for deployment with modest preparation)	Retired warheads awaiting dismantlement	Total warheads	NPT Signed	CTBT Status
United States	1945	1750	2050	2650	6450	yes	signed
Russia	1949	1600	2750	2500	6850	yes	ratified
United Kingdom	1952	120	95		215	yes	ratified
France	1960	280	10	10	300	yes	ratified
China	1964		280		280	yes	signed
India	1974		130–140		130–140	no	not signed
Pakistan	1998		140–150		140–150	no	not signed
Israel			80		80	no	signed
North Korea	2006	?	?	?	(20–60)	withdrew 2003	not signed

Note: Data from tables and associated footnotes in *SIPRI Yearbook 2018*, chap. 6, "World Nuclear Forces" (Stockholm: Stockholm International Peace Research Institute, 2018); chap. 6 available at https://www.sipri.org/sites/default/files/SIPRIYB18c06.pdf.

* Deployed warheads are defined as those actually mounted on missiles or, in the case of aircraft-delivered bombs, located at airbases with the bombers. That's right: China, India, Pakistan, Israel, and possibly North Korea are thought not to have their warheads actually deployed per this definition.

and can obtain needed technology and materials. Both the United States and the Soviet Union followed their fission devices with successful fusion weapons in the mid-1950s. The United States conducted 1,030 nuclear tests at dozens of sites, whereas the Soviet Union conducted 715 tests at locations in Kazakhstan, Ukraine, and Uzbekistan.

The British were active in nuclear weapons research from the beginning, even before the United States, and were partners with the U.S. in the Manhattan Project through World War II. After the war, the United States withdrew from weapons collaboration, and in 1947 the British resolved to develop their own nuclear weapons. The first British fission explosion was in 1952, by which time Britain had well-developed plutonium-production facilities and other essentials of a nuclear weapons industry. A British fusion test followed in 1957. Today's British nuclear missiles (Trident II D5 SLBMs and their MK4/4A reentry vehicles) are leased from the United States, but their W76-like warheads are British.[3] The British conducted 45 nuclear tests in Australia and the Pacific island Republic of Kiribati.

France, fiercely independent in its military posture, began in 1954 to develop a fully indigenous nuclear force. A French fission test occurred in 1960 and a fusion explosion in 1968. France today sustains its own nuclear force and, while committed to NATO, French nuclear forces do not support the alliance's integrated military command structure. France has conducted 210 tests in Algeria and the South Pacific.

Next to join the nuclear club was China, with its 1964 fission explosion. During the 1950s, the Chinese had received considerable nuclear help from the Soviet Union, including uranium-enrichment facilities and plutonium-production reactors. With the Sino-Soviet split in 1960, China chose to defer plutonium production in favor of its better-developed uranium capability. As a result, China remains one of two nations we know of whose first nuclear test involved a uranium device (the other is Pakistan). A Chinese thermonuclear test followed in 1967. China conducted some 45 tests at its Lop Nor nuclear test site, about half of them above ground. As a result, Lop Nor is believed to be the most contaminated nuclear site on Earth, with estimates of 30 to 35 percent greater cancer incidence in the surrounding region.[4]

India's first nuclear test, in 1974, was ostensibly an experiment in the peaceful use of nuclear explosives. This was a serious shock to the nonproliferation regime because India used plutonium produced in a Canadian-supplied CANDU research reactor (figure 18.2), moderated with heavy water from the United States. This bad behavior—India had promised it would not use the research reactor for weapons purposes—was rewarded

Figure 18.2
This small CANDU-type research reactor at the Bhabba Atomic Research Center in Trombay, India, was used to produce plutonium for India's first nuclear test. (United Nations)

further by the United States and others by continuing to provide fuel for Indian reactors. India even mastered reprocessing with help of the Atoms for Peace program, which spread nuclear technology all over the world. Then in 1998 India conducted five nuclear weapons tests.

Pakistan, India's bitter rival, had vigorously pursued a nuclear deterrent and soon sent another shock through the nonproliferation regime. The tenacity of the Pakistani leadership's interest in nuclear weapons is exemplified by a famous quote by Pakistan's former president Zulfikar

Ali Bhutto saying "If India builds the bomb, we will eat grass or leaves, even go hungry, but we will get one of our own. We have no alternative." Pakistan learned to enrich uranium with centrifuge plans smuggled by A. Q. Khan from the Netherlands. In fact, A. Q. Khan also established an extraordinarily elaborate network of buyers and sellers of nuclear components; more on this later. In the 1980s, thanks to the Khan network, Pakistan had produced enough HEU for a uranium bomb, although at that point it didn't test one. But mere weeks after India's 1998 tests, Pakistan responded with five tests of uranium-fueled implosion devices.

Israel is known to have a nuclear weapons program although details are scant. Israeli whistleblower Mordechai Vanunu gave evidence to the British press in 1986. After his exposé, Vanunu was lured to Italy by a classical honey pot operation (i.e., seduced by a female Israeli agent), where a ship waited offshore. He was drugged and smuggled back to Israel and imprisoned. There's no clear evidence that Israel has ever conducted a nuclear test, although a VELA surveillance satellite detected a mysterious flash over the South Atlantic in 1979, and recently declassified documents strengthen the view that this may have been an Israeli test.

North Korea is the most recent country to acquire nuclear weapons. It's conducted six tests, from 2006 through 2017. Based on the size of the seismic signal, the 2017 test had a yield of more than 250 kilotons and may have been a thermonuclear device.

Although not shown in table 18.1, one more country deserves mention here. Beginning in the 1960s, South Africa developed significant nuclear technology, including an innovative uranium enrichment scheme using an aerodynamic nozzle. Then, in the 1980s, South Africa constructed a half dozen gun-type uranium weapons. South Africa's weapons were probably never tested, in part because of pressure from western nations, who had been alerted by the Soviet Union to South African preparations for a nuclear test. However, it's still not known whether South Africa was involved in the 1979 VELA event (mentioned above in connection with Israel), so it remains possible that the event was a South African nuclear test. South Africa voluntarily ended its nuclear weapons program in 1989, becoming the only nation to give up nuclear weapons that it had itself developed.

The Nuclear Nonproliferation Regime

The nuclear-armed states have been locked in a delicate dance for many decades. The risk of nuclear conflict remains high, and a misstep could affect not just one or two countries but the entire planet.

That we've avoided nuclear conflict is partly a matter of luck, but it's also thanks to communication channels between rival militaries, accepted norms of behavior in diplomacy, cooperation and collaboration on security issues, and carefully crafted arms-control agreements. Arms-control measures—the tools of the nonproliferation regime—include restricting types of delivery systems, limiting numbers of nuclear weapons and delivery systems, and preventing acquisition of nuclear-sensitive materials, all with the goal of increasing international stability and preventing the spread of nuclear weapons. Fears that the Cold War arms race was escalating led to the original establishment of the regime, which is now a patchwork of treaties, agreements, and initiatives to limit proliferation of nuclear weapons. Unfortunately, we seem to be rapidly losing the communication and diplomacy skills that enabled the nonproliferation regime, and we're even abandoning some hard-won nuclear agreements.

The Nuclear Non-Proliferation Treaty (NPT)

The bedrock treaty of the nonproliferation regime is the **Non-Proliferation Treaty** (NPT, formally the Treaty on the Non-Proliferation of Nuclear Weapons), which aims to prevent weapons proliferation while facilitating the peaceful use of nuclear technology. The NPT, which went into force in 1970, divides nations into two camps: **non-nuclear-weapon states** (NNWS) and **nuclear-weapon states** (NWS), the latter comprising the United States, Russia, China, France, and the United Kingdom—the five nuclear-armed countries as of 1967. Under the treaty, each non-nuclear-weapon state has specific obligations to refrain "from developing or acquiring nuclear weapons" and to submit to safeguards agreements with the International Atomic Energy Agency (IAEA). The nuclear-weapon states, in turn, have obligations to not share nuclear weapons technology with non-nuclear-weapon states—but they are expected to help NNWS develop peaceful nuclear technologies. Given our emphasis on possible connections between nuclear power and nuclear weapons (more in the final chapter), there's a definite ambiguity in that last one!

Furthermore, even the obligation "not to share" is problematic when it comes to the NATO alliance (which, of course, includes both NWS and NNWS). This is because of the nuclear sharing agreement among NATO states. The nuclear-weapon states within NATO maintain control over all nuclear weapons but some non-nuclear-weapon states not only host nuclear weapons on their territories but are responsible for procuring delivery systems and for arming and using the weapons in a conflict. The moment that happens the non-weapon states would be in violation of the

Non-Proliferation Treaty. It's as if the United States were saying to NATO allies: "I'll train you to use the gun, but I won't give you the key to the gun until the critical moment; after that, you can use it." In this metaphor the guns are nuclear weapons and the critical moment is war. In 2015, in fact, Russia accused NATO member states of violating the NPT because of nuclear sharing. However, proponents of NATO's nuclear sharing agreement argue that nuclear sharing predates the NPT, which was negotiated in 1966 with full expectation of nuclear sharing among NATO states.

Finally, the NPT requires that *all states* (our emphasis) are obligated to "pursue good faith negotiations toward ending the nuclear arms race and achieving nuclear disarmament." Yet 50 years after the NPT entered into force, nuclear disarmament is still not a high priority for the nuclear-weapon states. All countries are members of the NPT except India, Pakistan, Israel, and, since it withdrew in 2003, North Korea.

Nuclear-Weapon-Free Zones

Article VII of the Non-Proliferation Treaty acknowledges the rights of member states to create **nuclear-weapon-free zones** (NWFZs). These regions (often geographical areas of countries themselves) prohibit the "development, manufacturing, control, possession, testing, stationing, and transportation of nuclear weapons." This provision prohibits the nuclear-weapon states from transporting or stationing nuclear weapons in any nuclear-weapon-free zone. Weapon states also agree never to use nuclear weapons against nations that have declared themselves NWFZs. There are currently five NWFZs, which include most developing countries except for the Middle East and those that already possess nuclear weapons. Separate treaties ban nuclear weapons from Antarctica as well as on and beneath the ocean floor. Figure 18.3 shows that wide regions of the planet are committed as nuclear-weapon-free zones.

Outer Space Treaty (OST)

Nations are prohibited from deploying and testing nuclear weapons in outer space, on the Moon, or on other celestial bodies by the 1967 Treaty on Principles Governing the Activities of States in the Exploration and Use of Outer Space, commonly called the **Outer Space Treaty** (OST). This treaty essentially extends nuclear-weapon-free zones into space.

In 2019, the United States launched a sixth branch of the U.S. armed forces known as the *Space Force*, which could eventually cause conflict with the OST. However, the military has always had a presence in space. The GPS system was originally developed for the military, and other

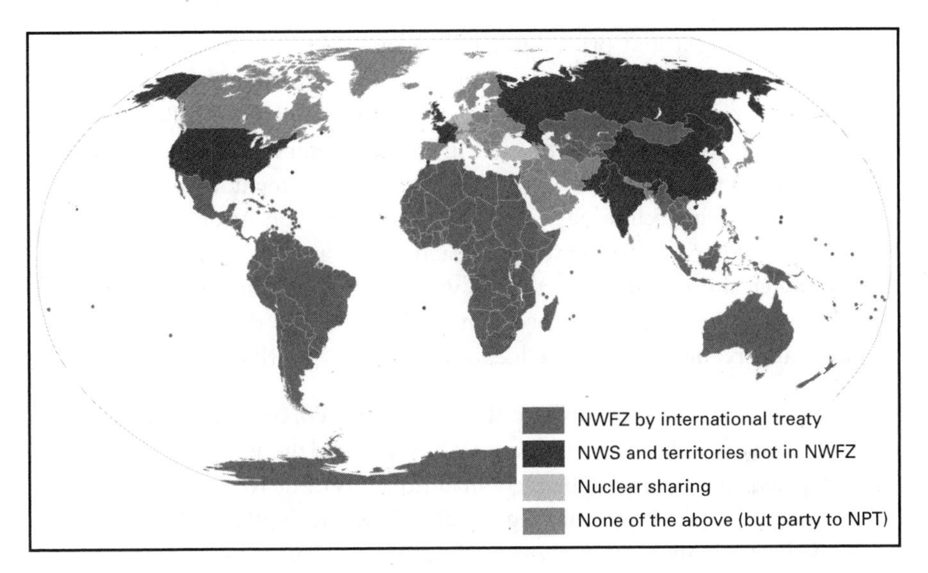

Figure 18.3
Nuclear-weapon-free zones. Only whole countries are shown, even when distinct regions are treated separately. (Wikimedia Commons, CC BY-SA 3.0)

military space assets include satellites for surveillance, communications, and many other functions. The OST bans weapons of mass destruction in space but not others—including anti-satellite weapons and components of ballistic missile defense systems.[5]

Bilateral Treaties to Reduce Nuclear Risks

The United States and Russia/Soviet Union have successfully negotiated treaties to reduce their nuclear weapons from some 65,000 at the height of the Cold War to fewer than 14,000 today. Treaties that have already come and gone are the Strategic Arms Limitation (**SALT**) and Strategic Arms Reduction (**START**) agreements, the INF and ABM treaties, and the Strategic Offensive Reductions Treaty (**SORT**), which was superseded by the **New START** treaty in 2011. New START itself is up for renewal. This treaty caps the number of deployed nuclear weapons for both nations at 1,550 nuclear warheads on some 700 delivery vehicles. President Trump called New START a "one-sided deal like all other deals we make." Although New START isn't dead yet, its prognosis does not look good.

Treaties involving other nations include the Brazilian-Argentine Agency for Accounting and Control of Nuclear Materials (ABACC) and the India-Pakistan Non-Attack Agreement of 1991. ABACC establishes a "set of criteria and procedures for verification and control to ensure that nuclear materials are not diverted for the manufacture of nuclear weapons or other nuclear explosive devices." Argentina and Brazil were once bitter rivals heading toward a nuclear competition, but ABACC allowed them to establish a joint safeguards agreement while pursuing peaceful nuclear technology.[6] The India-Pakistan Non-Attack Agreement prohibits either side from attacking each other's nuclear facilities, including power and research reactors as well as fuel fabrication, enrichment, isotope-separation and reprocessing facilities. Under the agreement, the two states regularly exchange locations of such facilities.

Rest in Peace: The Intermediate-Range Nuclear Forces (INF) Treaty

While START and New START successfully reduced the number of deployed warheads, the **INF Treaty** banned an entire class of delivery systems. Well, not quite: that class remained legal, but only when ship-based.

Military planners once envisioned European conflict as the most likely way for a U.S.-Soviet nuclear war to begin. For 45 years after World War II, Europe bristled with nuclear and conventional weapons as the United States and its NATO allies faced the Soviet Union and Warsaw Pact. Those weapons included short- and medium-range nuclear missiles, and NATO policy called for their first use against aggression by the Warsaw Pact's larger conventional forces. Particularly worrisome to the Soviet Union were medium-range missiles capable of striking well inside Soviet territory from bases in Germany; NATO forces, in turn, were alarmed at Soviet nuclear modernization programs that brought accurate MIRVed missiles to bear on western Europe.

Negotiations on medium-range missiles began in 1981 and led, haltingly at first, to the 1987 U.S.-Soviet Intermediate-Range Nuclear Forces Treaty, which boldly eliminated all ground-launched ballistic and cruise missiles with ranges of 300 to 3,400 miles (500 to 5,500 kilometers).

The INF Treaty did not significantly reduce the number of nuclear warheads in the superpowers' arsenals. But it was nevertheless a major advance in arms control, eliminating an entire class of nuclear weapon systems and requiring physical destruction of existing missiles. (INF did not, however, require destruction of nuclear warheads; in fact, U.S. INF warheads were recycled for use in long-range missiles.) And the treaty established an unprecedented level of international cooperation and openness in the verification process, with American and Soviet observers stationed at each other's missile-production facilities. For arms-control advocates, INF marked the beginning of a serious superpower effort to bring nuclear weapons under control.

In 2014, however, the Obama administration accused Russia of violating its INF obligations to not "possess, produce, or flight-test" a ground-launched cruise missile in the INF range.[7] It was confirmed that the cruise missile, termed SSC-8, was deployed in multiple locations. Russia vehemently denied being in violation of the INF, and in turn accused the U.S. of noncompliance over plans to deploy vertical launch systems for cruise missiles in Romania and Poland. These systems can launch SM-3 interceptors and Tomahawk intermediate-range cruise missiles. The latter would indeed violate the INF Treaty.

On October 20, 2018, President Donald Trump announced his desire to withdraw from the INF Treaty, citing Russian noncompliance and worries over China's mid-range missile arsenal. In February 2019, the administration proclaimed a suspension of U.S. commitments under INF and officially announced withdrawal from the treaty in six months. Shortly thereafter, Russian president Vladimir Putin announced that Russia would suspend its INF obligations. On August 2, 2019, the United States formally withdrew from the INF Treaty. A mere 17 days later, the U.S. tested its first INF-violating cruise missile (figure 18.4), amidst Defense Department hints that intermediate-range missiles would be deployed "sooner rather than later." All indications are that demise of the INF is fueling an accelerated nuclear arms race.

The IAEA and Its Mandate

The International Atomic Energy Agency (IAEA) was established in 1957 as part of the Atoms for Peace program and is now an independent organization that reports to the United Nations. It supports the safe, secure, and peaceful use of nuclear technology by its 152 member states. Its primary nonproliferation tasks are to establish safeguard agreements with member states and to carry out inspections. Other obligations include encouraging safe operation of nuclear technology and supplying member states with technical help. Safeguards have several goals. First is to verify that a state's declarations concerning its nuclear materials and operations are "correct and complete." Second is to ensure the "timely detection of diversion of significant quantities of nuclear materials from peaceful nuclear activities to the manufacture of nuclear weapons or of other nuclear explosive devices or for purposes unknown, and deterrence of such diversion by the risk of early detection."

Figure 18.4
First land-based vertical launch of a U.S. cruise missile in August 2019, just days after the U.S. withdrew from the INF Treaty—which this launch would have violated. (U.S. Department of Defense)

An important safeguard concept is the **significant quantity** (SQ), the amount of material required to make a nuclear bomb, taking into account losses due to machining and manufacture. For uranium the significant quantity is 25 kilograms of U-235 in uranium enriched to more than 20 percent, and for plutonium it's a mere 8 kilograms. Crucially, though, critical masses are substantially lower! Detection techniques must be sensitive to a fraction of a significant quantity so that diversion is discovered well before a full SQ has been diverted. The IAEA accomplishes this through bilateral safeguards agreements with member states. For example, **comprehensive safeguards agreements** apply to non-nuclear-weapon states, where the "IAEA has the right and obligation to ensure that safeguards are applied on all nuclear material in the territory, jurisdiction or control of the State for the exclusive purpose of verifying that such material is not diverted to nuclear weapons or other nuclear explosive devices." Safeguards are also introduced in the few non-NPT nuclear-armed states except for North Korea, based on *item-specific agreements* with the IAEA. These agreements ensure that stipulated items aren't used for military purposes. The five nuclear-weapon states also have a special type of safeguards agreement known as the voluntary offer agreement, where the IAEA safeguards specific facilities that weapon states have voluntarily offered.

The Missile Technology Control Regime

So far we've considered agreements, mostly binding ones, designed to limit the spread of nuclear weapons and materials that can be used to make them. But, as chapter 14 shows, nuclear weapons aren't very useful without the means to deliver them. In today's world, those means are primarily missiles.

The **Missile Technology Control Regime** (MTCR) is a voluntary agreement among some 35 nations, including most of those that manufacture missiles. Its goal is to limit exports of missiles and associated technologies. Originally it was intended to prevent the spread of missiles capable of delivering payloads greater than 500 kilograms (1,100 pounds) or more to distances in excess of 300 kilometers (about 200 miles). Those criteria were later broadened to include any missile technologies with the capability to deliver weapons of mass destruction, including chemical, biological, and nuclear weapons. In addition to ballistic missiles, the Missile Technology Control Regime is concerned with cruise missiles, rockets for space launches, and even drones.

MTCR participants establish export controls on so-called category I items, which include complete missiles and facilities for manufacturing them, and on category II items, which include special materials, rocket propellants, and missile subcomponents. The Control Regime has been successful in halting a joint ballistic missile program by Argentina, Egypt, and Iraq; in encouraging Poland and the Czech Republic to destroy existing missiles; and in getting Brazil, South Africa, South Korea, and Taiwan to shelve or eliminate missile and rocket programs. However, the MTCR hasn't prevented the development of ballistic missiles on the Indian subcontinent, by North Korea, or in the Middle East. In addition to equipping their own militaries, these countries, especially cash-strapped North Korea, vigorously peddle missile technologies to willing buyers.

As a voluntary agreement, the MTCR is "softer" than arms-control treaties, with more reliance on individual nations establishing their own export policies and on voluntary compliance. Furthermore, member nations can ask for exceptions to MTCR requirements. For example, MTCR member South Korea is permitted to have missiles with ranges up to 800 kilometers (500 miles), which allows it to target all of North Korea.

Although the MTCR is not as widely known as other arms-control agreements, and doesn't have the legally binding authority of treaties, it continues to work in a mostly quiet way to limit the spread of missiles and related technologies. However, some MTCR critics worry that the regime isn't adapting to the burgeoning growth in drone technology, to the blurring of distinctions between missiles and drones, and to the widespread development of armed drones.

A Fragile Agreement: Iran and the Joint Comprehensive Plan of Action

Iran's involvement in nuclear technology goes back to the 1950s, when the Shah of Iran got technical help under the U.S. Atoms for Peace program. U.S. support ended with the 1979 Iranian Revolution, but Iran remained dedicated to nuclear technology and developed a comprehensive nuclear fuel cycle, including enrichment plants at Fordow and Natanz. Much of the evidence that Iran had a covert nuclear weapons program comes from the United States and other IAEA members. The IAEA has proof that Iran obtained nuclear knowledge and associated equipment outside ordinary procurement lines from the 1980s to the mid-2000s, including through the A. Q. Khan network. Iranian authorities argue that they've been compelled to pursue nuclear items on the black market because the U.S. and other Western nations have prevented them from acquiring legitimate nuclear technology. What makes the Iranian nuclear program suspicious is the fact that procurement efforts and other initiatives were operated by Iranian military organizations, including the defense ministry. Although there's strong evidence for past Iranian interest in pursuing nuclear weapons, there's no evidence it has ever developed such weapons.[8] The IAEA is sure that Iran continued to develop covert nuclear installations for uranium processing and recycling during the 1990s and 2000s. One concern is that uranium enrichment sites at Natanz and Fordow were only declared after Western intelligence or external sources uncovered them.

The Iranian nuclear fuel program, especially enrichment plants, fissile material stocks, and breakout capability became the subject of intense negotiations as well as sanctions that took place between 2002 and 2015. By *breakout* capability we mean producing a nuclear weapon using known, declared facilities—in contrast to *sneakout*, which uses clandestine facilities. Negotiations concluded in July 2015 with the **Joint Comprehensive Plan of Action** (JCPOA), a 25-year nuclear deal between Iran and the so-called P5 + 1 (China, France, Germany, Russia, the U.K., and the U.S.) that limits Iran's nuclear ability in return for sanctions relief.

The Iranian nuclear deal, as it's called, satisfied two major objectives. First, it limited the possibility of diverting materials from a peaceful nuclear energy program for nuclear weapon production, and it enabled the means to detect Iranian cheating with *reasonable certainty*. The emphasis on *reasonable certainty* is deliberate, because no verification program can be 100 percent effective. It's also important to remember that the NPT grants full rights to peaceful nuclear energy programs for non-nuclear-weapon states like Iran as long as they don't seek nuclear weapons. Unfortunately, the NPT isn't completely clear about what is and is not legal, and the problem is compounded by the fact that peaceful and military nuclear programs intersect at various points in the nuclear fuel cycle, as in uranium enrichment.

Iran has constructed several centrifuge enrichment facilities to produce fuel for its reactors. The problem, as discussed in chapter 9, is that exactly the same facilities can also enrich uranium to weapons grade. The JCPOA essentially closed such cheating scenarios by first cutting Iran's uranium enrichment capability by two-thirds and then dramatically slashing its stockpile of low-enriched uranium, increasing the time required to produce enough enriched uranium for a bomb from two to three months to upwards of a year.

Other measures in the deal include limiting research, development, and testing of more advanced centrifuges, and having continuous surveillance at facilities that manufacture centrifuges. All these measures were time-limited, with some lasting 10 years, while others, such as continuous access to uranium mines, in force for as long as 25 years.

Another concern was Iran's construction of its Arak research reactor, which was to use unenriched uranium and could produce enough weapons-grade plutonium for more than one bomb per year. This was a legitimate concern because, as you've seen, India constructed its first nuclear explosive from plutonium produced in a similar reactor operated under the pretext of a civilian nuclear program. Under the joint agreement, Iran redesigned Arak to operate at lower power and to use low-enriched uranium so as to reduce its ability to produce weapons-grade plutonium.

Iran also agreed to ship spent fuel from all its reactors out of the country for reprocessing and not to engage in domestic reprocessing for 15 years.

The last possible route to an Iranian bomb would be to use clandestine facilities to enrich uranium or extract plutonium. Iran agreed to abide by enhanced safeguards measures through the IAEA's Additional Protocol, which would give the IAEA increased access to sites that it deems relevant.

All these measures still allowed Iran continue peaceful nuclear research, produce isotopes for medicine and industry, and expand nuclear power. However, the agreement limited and delayed more proliferation-sensitive steps in the fuel cycle like indigenous enrichment and spent-fuel reprocessing. As you saw in chapter 9, most countries that produce nuclear energy don't enrich uranium but purchase enriched uranium from trusted outside suppliers. The IAEA isn't alone in verifying Iranian compliance; rather, it's supported by powerful national technical means of all stakeholders (e.g., surveillance satellites) and by societal groups that monitor Iran's compliance.

Opponents of the Iran deal often claimed that there's no guarantee of its prevent cheating. But a more realistic assessment of the agreement is whether cheating can be detected with reasonable certainty. A compromise is always made among what's possible, reasonable, and affordable.

United States Violation of the JCPOA and Its Consequences

Even before his inauguration, President Donald Trump criticized the joint agreement, calling it "one of the most incompetently drawn deals I've ever seen." In 2017, Trump accused Iran of violating the spirit of the agreement and refused to certify compliance; in 2018, his administration announced that the United States would unilaterally leave the JCPOA. The U.S. then reinstituted nuclear-related sanctions on Iran and promised to increase those sanctions, contrary to JCPOA. The U.S. decision disappointed not only Iran but also the other members of the P5 + 1, who stated that they would maintain the deal without the United States. Shortly after the official U.S. withdrawal, in 2019, Iran began enriching uranium to levels beyond those allowed by the agreement, as well as installing more advanced, JCPOA-prohibited centrifuges.

Then, in early 2020, Trump ordered a drone attack that killed Qassim Soleimani, a top commander in Iran's Revolutionary Guard, who was at the time in Iraq. Iran responded with a missile strike on U.S. military bases in Iraq, and the two nations briefly tottered on the brink of war. Eventually, calm prevailed. But Iran announced that it would no longer abide by the joint agreement, and it accelerated uranium enrichment to a pace even greater than what it was before the JCPOA. Britain, France, and Germany remained formally with JCPOA but issued an official complaint under a provision of the agreement. With the joint agreement coming unraveled, it's entirely possible that Iran will have a nuclear weapon long before the minimum 15-year delay that would result from adherence by the joint agreement.

Shocks to the Nonproliferation Regime

Several incidents have shocked the nonproliferation regime and caused the international community to reconsider the strength of existing rules to prevent nuclear weapons proliferation.

The First Shock: India's Nuclear Test

The first shock was India's 1974 nuclear explosion, enabled by technology from the United States and Canada that India had expressly promised would only be used for peaceful purposes. The 1974 nuclear test was a clear violation of the Canada-India Columbo Plan, where India agreed to "ensure that the reactor and any products resulting from its use will be employed for peaceful purposes only." After the 1974 test, Canada and the United States largely ceased nuclear cooperation with India until 2008. A group of 20 nuclear-technology-exporting nations known as the Zangger Committee responded by tightening export restrictions for materials and items that could be used for nuclear weapons.[9] While the group was established before India's nuclear test, the test galvanized the committee to finalize a list of sensitive items that would trigger IAEA safeguards. A larger group of nations, the **Nuclear Suppliers Group**, formed in response to the Indian nuclear test. The Suppliers Group adopted two sets of guidelines for nuclear exports and nuclear-related exports. The Zangger Committee's sensitive-materials list became a technical annex to the first set, while the second focused on dual-use components that could be used for the civilian nuclear fuel cycle as well as weapons purposes. The second set was a response to 1992 revelations surrounding Iraq's covert nuclear weapons program.

In a decision widely criticized by nonproliferation analysts, the United States under George W. Bush accepted India into the Nuclear Suppliers Group in 2008 and normalized nuclear trade with India. This is despite the fact that some of India's reactors and all of its enrichment and reprocessing facilities aren't under safeguards. Consequently, India was able to import uranium for reactor fuel, freeing up its own stocks to enrich for nuclear weapons production. India's 1974 nuclear test was so disruptive to the nonproliferation regime precisely because it was a textbook case of a country that developed nuclear explosives under the guise of a peaceful nuclear program.

The Second Shock: Iraq's Nuclear Weapons Program

In the 1990s, the IAEA learned that Iraq had a well-funded but clandestine nuclear weapons program. After the first Gulf War in 1991, the IAEA found that Iraq had secretly enriched uranium and conducted reprocessing experiments at locations that weren't covered in its agreements with the IAEA or were off-limits to inspectors. Unrelated to Iraq, the IAEA also became aware that North Korea and Romania had secretly reprocessed spent fuel. It was clear that normal comprehensive safeguards agreements weren't rigorous enough, as all these operations happened directly under the IAEA's eyes. No checks, inspections, or protocols had uncovered them.

In reaction to these transgressions, IAEA member states approved an *Additional Protocol* to complement standard safeguards agreements and enhance the IAEA's capabilities for detecting concealed nuclear programs. While the goal of the comprehensive safeguards agreement was to "verify the accuracy of a state's declared nuclear activities," the Additional Protocol's goal was to "verify the completeness of its declaration." The protocol comprises two components: an *expanded declaration* that details information not usually indicated on comprehensive safeguards agreements, and increased IAEA access to undeclared facilities "to assure the absence of undeclared nuclear material and activities." As of 2020, 136 states had the Additional Protocol in force and another 15 had signed but didn't yet have it in force.

The Third Shock: The A. Q. Khan Network

The world learned in 2004 about an elaborate scheme by Pakistani scientist Abdul Quadeer (A. Q.) Khan to sell nuclear technology on the black market to North Korea, Iran, and Libya. The U.S. had suspected Khan of proliferating nuclear materials but didn't grasp the magnitude of his network until 2003, when U.S. officials captured the *BBC China*, a German cargo vessel, on its way to Libya. The ship contained parts for 1,000 gas centrifuges at a supposed price of $140 million. Khan's nuclear network thrived throughout the 1980s and 1990s and was connected to intermediaries and companies in over 20 nations. It all started when Khan worked in the Netherlands for the nuclear fuel company Urenco. After the 1974 Indian nuclear test, Khan offered his help to Pakistani prime minister Zulfikar Ali Bhutto for a Pakistani nuclear program. In 1975, he returned to Pakistan with stolen centrifuge blueprints, which he used to start a uranium enrichment program. Fueled by his disdain for the West, Kahn's program grew to include nuclear exports to many countries. Khan wrote a letter to the German magazine *Der Spiegel* expressing what many were thinking: "I want to question the bloody holier-than-thou attitudes of the Americans and the British. Are these bastards God-appointed guardians of the world to stockpile hundreds of thousands of nuclear warheads and have they God-given authority to carry out explosions every month? If we start a modest program, we are the satans, the devils."[10] The A. Q. Khan network was disruptive to the proliferation regime because it demonstrated that illicit networks can work to support **horizontal proliferation**—the spread of weapons to states that haven't previously possessed them. Figure 18.5 shows the Khan network, including both suppliers and customers.

The international community's response to the Khan network and other potential illicit nuclear supply rings was to enact U.N. Security Council Resolution 1540, which seeks to counter all proliferation pathways for weapons of mass destruction (WMDs), including nuclear, chemical, and biological weapons. In particular, Resolution 1540 was designed to lower the danger that terrorist organizations could acquire WMDs.[11] Analysts realized that the principal reason for Khan's success was the lack of national export controls, so the U.N. directive requires that all countries implement "appropriate effective laws which prohibit any non-State actor to manufacture, acquire, possess, develop, transport, transfer or use nuclear, chemical or biological weapons and their means of delivery," as well as establish robust export controls and physical protection.

Summary

Horrifyingly destructive nuclear weapons have been with us for three-quarters of a century. During that time, they've increased in explosive yield, accuracy, and means of delivery. Born in the United States, they've spread to at least eight other sovereign nations. Although weapons

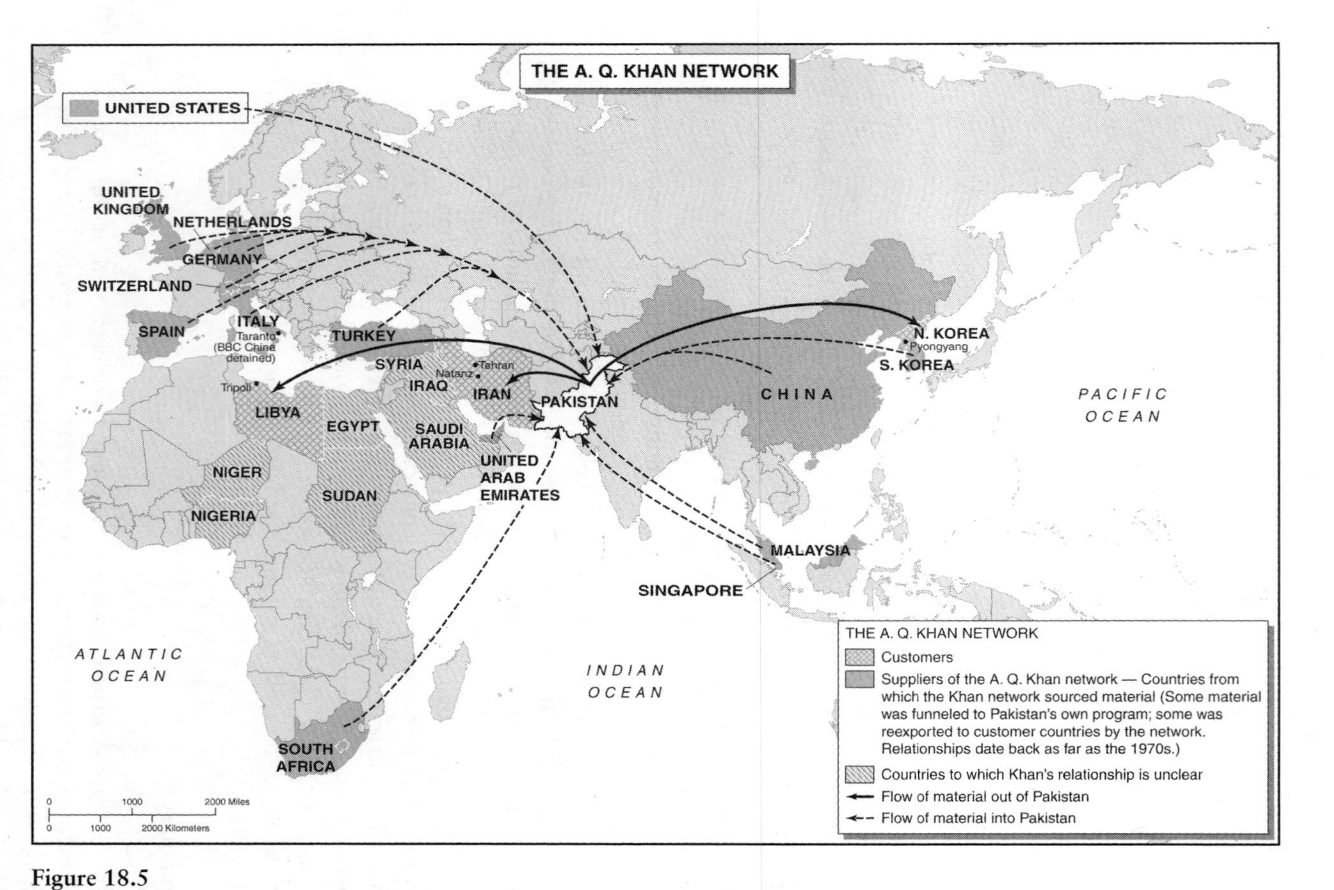

Figure 18.5
The A. Q. Khan network, with arrows depicting flows of nuclear-sensitive material to Pakistan and from Pakistan to Khan's customers. (Reproduced with permission from Gordon Corera, *Shopping for Bombs: Nuclear Proliferation, Global Insecurity, and the Rise and Fall of the A.Q. Khan Network* [New York: Oxford University Press, 2009].)

numbers have declined in recent decades, thanks to arms-control treaties between the U.S. and Russia, nuclear arsenals are still far more than ample to destroy human civilization. Can we bring nuclear weapons under control, or are we doomed to a future in which nuclear weaponry is commonplace and increasingly likely to be used? Will we be commemorating the centenary of the bombings of Hiroshima and Nagasaki while nuclear weapons are still deployed?

You've seen in this chapter that international controls on nuclear weapons are possible, and that some have been in place for decades. What we call the nonproliferation regime is a broad international framework of agreements and organizations aimed at preventing the spread of nuclear weapons and encouraging progress in arms control and disarmament. Those controls are imperfect, as difficult international negotiations and domestic politics lead often to compromises that weaken arms-control agreements. Those agreements are also sensitive to political environments, as you've seen with the demise of the Intermediate-Range Nuclear Forces Treaty and the Iranian nuclear deal JCPOA. But the incentive to control nuclear weapons is increasingly shared throughout the world, and that incentive comes from the nearly universal recognition that nuclear war is unacceptable, inhumane, and immoral. That the 2017 Nobel Peace prize was awarded to the International Campaign to Abolish Nuclear Weapons for their role in drawing "attention to the catastrophic humanitarian consequences of any use of nuclear weapons" is a testament to that recognition. Our next chapter will further examine the goal of abolishing nuclear weapons that's implicit in the name "Campaign to Abolish."

The best-known of the international arms-control agreements are bilateral and multilateral treaties, such as the Non-Proliferation Treaty, the INF Treaty, and the START treaties. Lesser agreements help, too, in reducing the threat of nuclear war: The 1963 hot line communication agreement, for example, was the first of many U.S.-Soviet measures aimed at improving communications in time of crisis.

Have arms-control agreements moved the world away from nuclear war? The Non-Proliferation Treaty may have helped curb the spread of nuclear weapons; on the other hand, it may have encouraged potential nuclear nations to go as far they can without actually producing completed weapons. The INF Treaty, while it was in effect, eliminated an entire class of nuclear weapons but only for land-based missiles. Even New START, with its wholesale reduction in strategic warheads, still leaves the world with thousands of warheads' worth of nuclear destructive power—enough to bomb all of us into the Stone Age many times over.

Perhaps what's important about these treaties is not just the technical limits they place on nuclear weapons, but the international cooperation and trust they engender. If we can negotiate with our global neighbors toward the goal of mutual security, then maybe in the process we'll achieve enough understanding to prevent the nuclear threat from becoming reality.

Glossary

comprehensive safeguards agreement A legally binding agreement between the IAEA and a non-nuclear-weapon state party to the Non-Proliferation Treaty, concluded as a condition of membership in the NPT.

horizontal proliferation The spread of nuclear and other weapons of mass destruction to states that have not previously possessed them.

INF Treaty The Intermediate-Range Nuclear Forces Treaty, a bilateral agreement between the United States and the Soviet Union/Russia. It was the first treaty to require the elimination of weapons. The INF Treaty was in effect from 1988 through 2019, when the U.S. and then Russia withdrew.

Joint Comprehensive Plan of Action (JCPOA) A 2015 agreement between Iran and the P5 + 1 on the Iranian nuclear program, also called the Iran nuclear deal. The U.S. withdrew in 2019, and in 2020 Iran announced that it would no longer abide by the agreement.

Missile Technology Control Regime (MTCR) A voluntary agreement among nations, especially those that build missiles or related items, to reduce the spread of missile technology.

New START Successor to START and SORT, this treaty took effect in 2011 and reduced U.S. and Russian strategic arsenals to 1,550 warheads each.

non-nuclear-weapon states (NNWS) Under the Non-Proliferation Treaty, NNWS are states that had not detonated a nuclear device prior to January 1, 1967, and who agree in joining the Non-Proliferation Treaty to refrain from pursuing nuclear weapons.

Non-Proliferation Treaty (NPT) The bedrock treaty of the nonproliferation regime, whose goal is to prevent the spread of nuclear weapons while encouraging sharing of peaceful nuclear technologies.

nuclear nonproliferation regime The broad international framework of agreements and organizations aimed at preventing the spread of nuclear weapons and contributing to progress in arms control and disarmament. Fears that the Cold War arms race was spiraling out of control led to the initial establishment of the regime, intended to promote stability and reduce the likelihood of nuclear weapons use.

Nuclear Suppliers Group Established in 1975, the Suppliers Group and its members commit to exporting sensitive nuclear technologies only to countries that adhere to strict nonproliferation standards.

nuclear-weapon-free zone (NWFZ) A geographical area in which nuclear weapons may not legally be built, possessed, transferred, deployed, or tested.

nuclear-weapon state (NWS) As defined in the Non-Proliferation Treaty, NWS are the five states that had detonated a nuclear device prior to January 1, 1967 (China, France, the Soviet Union/Russia, the United Kingdom, and the United States). Coincidentally, these five states are also permanent members of the U.N. Security Council. States that acquired and/or tested nuclear weapons subsequently are not formally recognized as nuclear-weapon states.

Outer Space Treaty (OST) Short for the Treaty on Principles Governing the Activities of States in the Exploration and Use of Outer Space, which prohibits weapons of mass destruction in Earth orbit, on the Moon or other celestial body, or otherwise in outer space.

SALT Acronym for Strategic Arms Limitations Talks (SALT I & II), a series of discussions between the Soviet Union and the United States aimed at limiting missile systems and other strategic armaments.

significant quantity (SQ) An IAEA designation indicating the minimum amount of material with which the construction of a nuclear weapon may be possible.

SORT The 2002 Strategic Offensive Reductions Treaty, also known as the Moscow Treaty, committed the United States and Russia to reduce their deployed strategic nuclear forces to 1,700–2,200 warheads apiece. It was superseded in 2011 by New START.

START The Strategic Arms Reduction Treaty between the U.S. and the USSR/Russia, which took effect in 1994 and produced the first significant drop in those countries' nuclear arsenals.

technostrategic discourse "Specialized language ... [that] both reflects and shapes the nature of the American nuclear strategic project and functions in a deeply gendered way." Using technostrategic terms can remove the "emotional reality behind the consequences of nuclear weaponry through euphemism and metaphor to engage in a culture of masculinity through sexual subtext." Carol Cohn, "Sex and Death in the Rational World of Defense Intellectuals," *Signs: Journal of Women in Culture and Society* 12, no. 4 (1987).

Notes

1. President John F. Kennedy, News Conference 52, March 21, 1963, John F. Kennedy Library and Museum, Boston.

2. Carol Cohn, "Sex and Death in the Rational World of Defense Intellectuals," *Signs: Journal of Women in Culture and Society* 12, no. 4 (1987): 687–718.

3. Hans M. Kristensen and Robert S. Norris, "British Nuclear Forces," *Bulletin of the Atomic Scientists* 67, no. 5 (2011): 89–97.

4. Zeeya Merali, "Did China's Nuclear Tests Kill Thousands and Doom Future Generations?," *Scientific American*, July 1, 2009, https://www.scientificamerican.com/article/did-chinas-nuclear-tests/.

5. For a discussion of the Outer Space Treaty, see Union of Concerned Scientists, "International Legal Agreements Relevant to Space Weapons," February 11, 2004, https://www.ucsusa.org/resources/legal-agreements-space-weapons.

6. Mariana Oliveira do Nascimento Plum and Carlos Augusto Rollemberg de Resende, "The ABACC Experience: Continuity and Credibility in the Nuclear Programs of Brazil and Argentina," *Nonproliferation Review* 23, no. 5–6 (2016): 575–593.

7. For details on the Obama administration's concerns about Russia and the INF, see https://www.armscontrol.org/factsheets/INFtreaty.

8. IAEA Report, "Implementation of the NPT Safeguards Agreement and Relevant Provisions of Security Council Resolutions in the Islamic Republic of Iran," November 8, 2011, http://www.iaea.org/Publications/Documents/Board/2011/gov2011-65.pdf.

9. George Perkovich, *India's Nuclear Bomb: The Impact on Global Proliferation* (Berkeley: University of California Press, 2001).

10. A. Q. Khan, quoted in William Langewiesche, "The Wrath of Khan," *The Atlantic*, November 2005, https://www.theatlantic.com/magazine/archive/2005/11/the-wrath-of-khan/304333/.

11. For more on Resolution 1540 and its implementation, see the Nuclear Threat Initiative's report at https://www.nti.org/analysis/articles/unscr-1540/.

Further Reading

Berry, Ken, Patricia Lewis, Benoît Pélopidas, Nikolai Sokov, and Ward Wilson. *Delegitimizing Nuclear Weapons: Examining the Validity of Nuclear Deterrence*. Monterey, CA: James Martin Center for Nonproliferation Studies, 2010. Available at https://www.fdfa.admin.ch/dam/eda/de/documents/aussenpolitik/sicherheitspolitik/Delegitimizing_Nuclear_Weapons_May_2010.pdf. This report, by experts in the field, examines evidence for nuclear deterrence that we generally take for granted as being valid. If that evidence turns out to be weak, then that also weakens arguments for the existence of nuclear arsenals.

Braut-Hegghammer, Malfrid. *Unclear Physics: Why Iraq and Libya Failed to Build Nuclear Weapons*. Ithaca, NY: Cornell University Press, 2016. Braut-Hegghammer is a Norwegian political scientist. In *Unclear Physics* she provides a detailed account of two nations' failed attempts to develop nuclear weapons Despite *Physics* in the title, the book is as much about the role of authoritarian leaders and of economics.

Cohn, Carol. "Sex and Death in the Rational World of Defense Intellectuals." *Signs: Journal of Women in Culture and Society* 12, no. 4 (1987): 687–718. This is Carol Cohn's important work on technostrategic discourse used by defense intellectuals.

Corera, Gordon. *Shopping for Bombs: Nuclear Proliferation, Global Insecurity, and the Rise and Fall of the A. Q. Khan Network* (Oxford: Oxford University

Press, 2009). A nuanced account of the Khan network and why it flourished, by a BBC security correspondent.

Doyle, James. *Nuclear Safeguards, Security and Nonproliferation: Achieving Security with Technology and Policy*, 2nd edition. Oxford: Butterworth-Heinemann, 2019. Focus on safeguards and nonproliferation as well as the CTBT (discussed in the next chapter). Excellent case studies as well. Highly recommended resource.

IAEA information on international safeguards: https://www.iaea.org/topics/safe guards-agreements.

Niemeyer, Irmgard, Mona Dreicer, and Gotthard Stein, eds. *Nuclear Nonproliferation and Arms Control Verification: Innovative Systems Concepts*. Charn, Switzerland: Springer, 2020. Using a systems approach, the writings assembled here explore achievements and challenges in nuclear nonproliferation, including approaches that don't require intergovernmental agreementshttps://tutorials.nti .org/nonproliferation-regime-tutorial/

Nuclear Threat Initiative (NTI) tutorial on the nonproliferation regime: https:// tutorials.nti.org/nonproliferation-regime-tutorial/https://tutorials.nti.org /nonproliferation-regime-tutorial/.

Potter, William C., and Gaukhar Mukhatzhanova, eds. *Forecasting Nuclear Proliferation in the 21st Century*. Vol. 2. Stanford Security Studies, 2010. This book discusses 12 case studies of countries attempting to get or getting nuclear weapons. A must-read if one is interested in *why* countries acquire nuclear weapons.

Solingen, Etel. *Nuclear Logics: Contrasting Paths in East Asia and the Middle East*. Princeton, NJ: Princeton University Press, 2007. A scholarly analysis of why some nations seek nuclear weapons while others don't. The author develops her own theory that seems to fit the case studies she examines in detail.

19

A World without Nuclear Weapons?

In the preceding chapter, you saw how arms-control agreements have reduced global nuclear arsenals and helped slow the spread of nuclear weapons. But after decades of such agreements, we've still got enough nuclear weapons to destroy civilization. Even much greater reductions—down to hundreds of warheads—would still leave nuclear war the greatest catastrophe ever to befall humankind.

Can we do better, ridding ourselves of these doomsday weapons by reducing the world's nuclear arsenals to zero? In a historic speech in Hiroshima, where he was the first sitting U.S. president to visit, Barack Obama echoed this goal, stating, "Among those nations like my own that hold nuclear stockpiles, we must have the courage to escape the logic of fear and pursue a world without them."

A World without Nukes: Realistic? Desirable?

A world without nuclear weapons wouldn't have to live with the threat of immediate annihilation. Many argue that the ultimate goal of our nuclear negotiations should be the complete elimination of nuclear weapons. Others disagree. Here we consider whether elimination of nuclear weapons is realistic and whether it's even desirable. We'll then look at steps that have been or are now being taken toward an international ban on nuclear weapons.

Keeping the Peace?

In 2016, U.S. Army Chief of Staff General Mark Milley testified before Congress in support of more spending on the U.S. nuclear triad. Milley said, "I can tell you that in my view ... the nuclear triad has kept the peace since nuclear weapons were introduced and has sustained the test of time." Later, in 2019 confirmation hearings for his appointment as chairman of the Joint Chiefs of Staff, Milley argued forcefully for modernization of U.S. nuclear forces, development of new low-yield nuclear weapons, and deployment of intermediate-range missiles that had formerly been banned by the INF Treaty. General Milley's view is a widely held one: that nuclear weapons have kept the peace for the many decades they've been in existence. Ironically, of course, if it's true that nuclear weapons *have* kept the peace, then they've done so by virtue of their vast destructive power. But is it true? Absent nuclear weapons, would we have already seen a conventional World War III, perhaps with more deaths than the approximately 80 million who perished in World War II?

But the nuclear era hasn't been an era of peace. Armed conflicts rage around the globe, some of them involving the superpowers either directly, indirectly, or by proxy. Nuclear weapons didn't prevent the Vietnam War, with some 3 million deaths, nor the 1990–1991 Gulf War, which killed nearly half a million, nor the Iraq war that began in 2003 and took a similar toll, nor the ongoing conflict in Afghanistan whose deaths exceed 100,000. And those are just the obvious conflicts involving the United States. Civil wars have raged in Sudan, Rwanda, Syria. Some 100,000 died in the 1992–1995 Bosnian conflict. Russia battled break-away Chechnya in the 1990s, leaving some 50,000 dead. The list goes on and on. So nuclear weapons didn't prevent war, even war involving the superpowers. They may, however, have prevented direct conflict between the superpowers. So there could be just a grain of truth to General Milley's assertion.

Nuclear War: A Nonzero Probability

As long as nuclear weapons exist, there's a nonzero chance they'll be used. And given their vast destructive power, we risk annihilating human civilization. It may, in fact, be pure luck that's kept nuclear war at bay so far—recall chapter 14's box on Soviet naval officer Vasili Arkhipov, who may well have prevented the 1962 Cuban Missile Crisis from turning into an all-out nuclear war. And there have been other close calls, either at times of international tension or because of failures in warning systems. Furthermore, the likelihood of nuclear war is probably rising with the demise of agreements like the INF Treaty. A strong argument for elimination of nuclear weapons is that we can't survive forever with a nonzero probability of nuclear annihilation.

However, one thing we can't eliminate is the knowledge of how to make nuclear weapons. So would it be safe to ban them, knowing that a rogue state or terrorist group might produce them clandestinely? That's a tough question for advocates of a complete nuclear weapons ban. On the other hand, an international community that could, in good faith, negotiate a nuclear weapons ban should also have the wherewithal to establish a verification regime that would catch cheaters long before they could establish civilization-annihilating arsenals. But would it be enough to prevent nuclear blackmail by a party that managed to build a few weapons without the rest of the world finding out?

However realistic or desirable a nuclear weapons ban might be, such a ban is surely a long way off in the present international climate. But that doesn't mean we can't move in the direction of a world without nuclear weapons. Indeed, the arms-reduction agreements of the preceding chapter are a step in that direction. So are other nuclear agreements already implemented, or now in negotiation, or being discussed as future steps.

Nuclear Weapons Testing

Important to nuclear weapons development is testing one's nuclear explosives. Testing helps in designing new weapons, in verifying reliability of existing weapons, in analyzing weapons effects, and in making weapons more secure. If nuclear testing were banned, it would be more difficult for non-nuclear nations to acquire nuclear weapons. And it would be harder—but not impossible—for nuclear-armed states to modernize their arsenals. So a ban on testing could help move the world toward a long-term goal of banning nuclear weapons themselves.

As of 2020, a total of 2,063 nuclear tests had been conducted by the nuclear-armed states (see figure 19.1). These were conducted to verify weapons design, safety, yield, and effects, as well as for the political purpose of demonstrating nuclear capability, and even for so-called peaceful nuclear explosions.

Nuclear tests have been conducted atop towers, on barges, from balloons, underwater, underground, and in space. In some cases they've clearly resulted in harm to humans, often indigenous people living near test sites or military personnel witnessing tests.[1] And fallout from atmospheric tests has affected everyone who was alive in the 1950s and early 1960s, with measurable quantities of fission products in our bodies. Through the mid-1950s, the nuclear-weapon states tested vigorously, with little thought to environmental or security implications. By the late 1950s, public concern began to rise and serious proposals to ban nuclear tests emerged. The concern heightened after the United States conducted the 15-megaton Castle Bravo thermonuclear test at the Marshall Islands

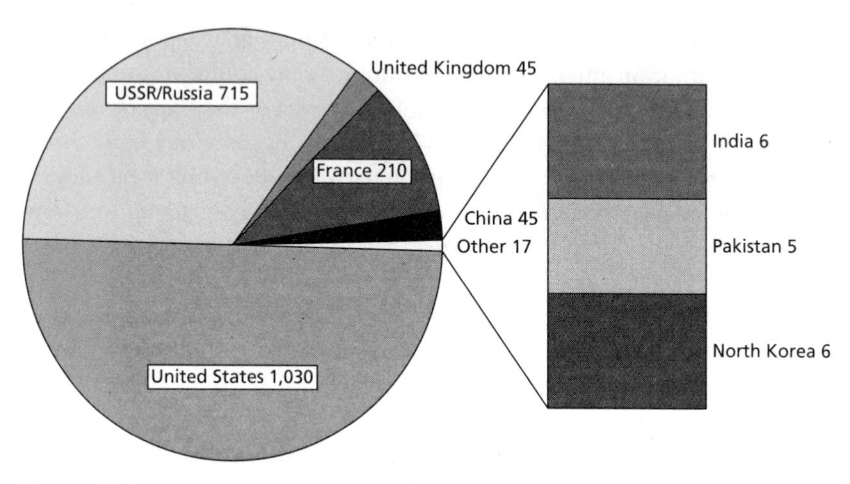

Figure 19.1
Nuclear tests conducted worldwide. USSR/Russia and the United States dominate with 85 percent of the world's tests. About 500 of the more than 2,000 tests were above ground. Here we count each individual explosion as a separate test, in contrast to treating salvos of simultaneous explosions—as happened with the Indian and Pakistani tests in 1998—as single tests. (Data source: Arms Control Association)

in 1954. A cloud of radioactive fallout rose as high as 40 kilometers (about 30 miles) and fallout contaminated crew on the Japanese fishing trawler *Lucky Dragon*. The Bravo test also affected indigenous people on the islands of Rongelap, Ailinginae, Rongerik, and Utirik as well as U.S. troops in the Marshall Islands. The fallout was so intense that inhabitants of Rongelap, who had never seen snow, believed it was snowing. The U.S. military gave no warning of the danger.

Today, Pacific Islanders suffer direct effects of climate change, but there's also concern that rising seas may result in exposure of radioactive materials from nuclear tests. A 1958 test on Runit Island in the Pacific atoll of Enewetak made a crater that was then used to store radioactive soil and debris from other nuclear tests. The U.S. Army eventually covered the crater with a concrete dome, trapping its radioactive contents (figure 19.2). As climate change raises sea level, there's been cracking of the dome and incursion of seawater. There's concern that radiation will leach into the surrounding ocean, although it should be sufficiently dilute as not to pose a hazard. Potential contamination of groundwater, however, remains a more serious concern.

In 1958, the Soviet Union declared a voluntary moratorium on testing, and the United States followed suit. A conference on test ban verification

Figure 19.2
Aerial view of the U.S. Army's dome on Runit Island. (U.S. Department of Defense)

produced agreement that all but underground tests could be detected with near certainty using then-existing technology, while underground tests would require intrusive, on-site inspections. Soviet refusal to allow such inspections, coupled with U.S. insistence on perfectly reliable verification techniques, led to a breakdown in the test moratorium. Then, on October 30, 1961, the Soviet Union detonated the largest-ever nuclear device, the 50-megaton "Tsar Bomba," at its Arctic test site. The bomb's shock wave reportedly shattered windows in Norway and Finland, some 600 miles distant.

Renewed atmospheric testing of high-yield weapons raised fears of global radioactive contamination, and the 1962 Cuban Missile Crisis again focused international attention on a possible test ban. The result was the **Partial Test Ban Treaty** (PTBT) of 1963, prohibiting all but underground nuclear tests. Although the Test Ban Treaty affected primarily the

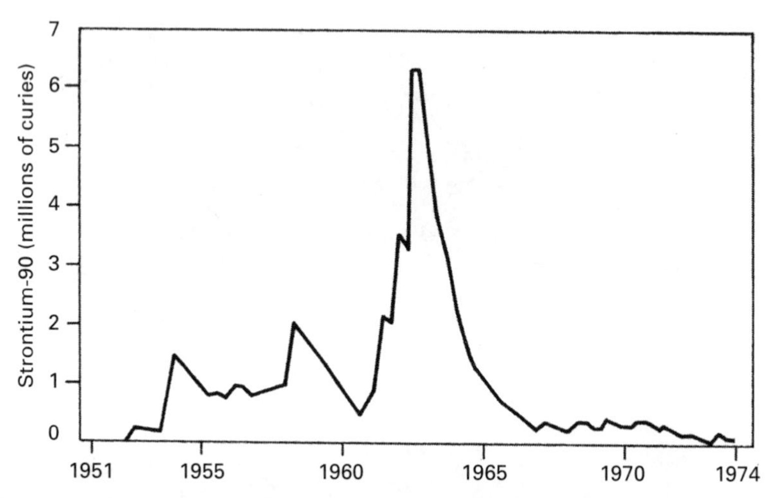

Figure 19.3
Atmospheric levels of strontium-90 rose dramatically with the nuclear tests of the 1950s and early 1960s, then fell abruptly when the 1963 Partial Test Ban Treaty went into effect. (From Samuel Glasstone and Philip J. Dolan, eds., *The Effects of Nuclear Weapons*, 3rd ed. [Washington, DC: U.S. Department of Defense and U.S. Department of Energy], p. 449.)

Soviet Union, the United States, and Great Britain, it has been signed by over 123 other nations. The Partial Test Ban Treaty was complemented in 1974 with the bilateral U.S.-Soviet **Threshold Test Ban Treaty** limiting underground tests to a maximum of 150 kilotons. Disagreement over verification left the Threshold Treaty unratified until 1990, although both sides continued to honor it.

Have the Partial and Threshold Test Ban Treaties been successful? The PTBT immediately forced U.S., Soviet, and British nuclear testing underground, resulting in a dramatic reduction in radioactive fallout (see figure 19.3). But nuclear testing continued unabated until 1996 when the **Comprehensive Nuclear Test Ban Treaty** (CTBT) opened for signature and a true moratorium took hold.

A Comprehensive Test Ban: Treaty and Verification

In 1954, Indian Prime Minister Jawaharlal Nehru proposed a ban on all nuclear testing, but serious negotiations on a comprehensive treaty didn't start until 1993, nearly 40 years later. A 1995 Non-Proliferation Treaty conference resolved to finalize the text of the Comprehensive Nuclear Test Ban Treaty by 1996. Negotiations at the **United Nations Conference on Disarmament** were contentious but met the deadline. The CTBT was opened for signature on September 24, 1996, and to date 184 countries have signed, with 168 of them having also ratified. Among those ratifying are three nuclear-weapon states: Russia, France, and the United Kingdom.

However, the treaty requires 44 so-called Annex 2 states—those who possessed nuclear reactors at the time of the treaty negotiations—to ratify before it goes into force. This still hasn't happened, and table 19.1 shows why: the United States and China, along with a handful of other nuclear-armed nations and others with nuclear ambitions, are holding back. However, the **Comprehensive Nuclear Test Ban Treaty Organization**[2] (CTBTO) is very much alive and working vigorously to promote the treaty's goals.

Detecting Nuclear Tests: The International Monitoring System

The key to a nuclear test ban—and to any future nuclear-weapons-free world—is for the international community to be sure it can detect clandestine nuclear explosions anywhere in, on, or above Earth. Although the Comprehensive Nuclear Test Ban Treaty isn't in force and may never be, the international community has developed a comprehensive sensor network known as the **International Monitoring System** (IMS) that makes

Table 19.1
Positions of Annex 2 states that have not ratified the CTBT

Hold-out Annex 2 states	Position on ratification
China	May need to test new weapons in the future or waiting for the United States to ratify.
United States	May need to test new weapons in the future. Treaty opponents argue that evasive testing remains possible.
India	May need to test new weapons and will likely not ratify before Pakistan and China do so.
Pakistan	Delay ratification until India ratifies. Concerned that parity with India will be lost.
Israel	Unlikely to ratify until the Middle East is more stable.
Egypt	Supports a Middle East nuclear-weapon-free zone but unlikely to ratify if Israel does not.
North Korea	May need to conduct more tests. Unlikely to ratify unless as a bargaining tool.
Iran	Unlikely to ratify unless Israel does so. Could use ratification as a bargaining tool in the future.

Source: Adapted from James Doyle, *Nuclear Safeguards, Security and Nonproliferation: Achieving Security with Technology and Policy* (Oxford, UK: Butterworth-Heinemann, 2011), 252.

Table 19.2
Technologies and their role (major, secondary, none) in detecting nuclear tests in various environments

Detection Technologies	Underground	Underwater	Atmosphere	Space
Seismic	Major	Major	Secondary	No role
Particulates	No role unless vented	Major	Major	No role
Radioxenon	Major	Major	Major	No role
Hydroacoustic	Secondary	Secondary	Secondary	No role
Infrasound	Secondary	Secondary	Major	No role
Bhangmeter	Secondary	Secondary	Major	Major
Satellite Imagery	Major	Major	Secondary	Secondary

Source: National Research Council, *The Comprehensive Nuclear Test Ban Treaty: Technical Issues for the United States* (Washington, DC: National Academies Press, 2012), Table 2–1; available at https://www.nap.edu/download/12849.

it extremely difficult for any nuclear test, no matter where it's conducted, to go undetected. The IMS network comprises four different sensor technologies, each with its own strengths for verifying whether a nuclear weapon is detonated underground, on Earth's surface, in the atmosphere, or in space (see table 19.2). Energy released in a nuclear explosion produces detectable signals, including low-frequency sound waves called *infrasonic waves*, seismic waves (as in earthquakes), hydroacoustic waves (underwater sound), visible light, and releases of radioactive isotopes.

Underground Tests: Seismic Detection

An underground nuclear explosion disturbs the medium around it and thus produces seismic waves. These travel outward like ripples on a pond, reaching seismic stations across the globe. The waves take complicated paths, with *body waves* traveling through Earth and *surface waves* along Earth's surface. The different waves have different periods and speeds. Seismic stations measure ground motion caused by passing waves and record its amplitude over time. Signals are classified according to the distance between source and sensor. *Teleseismic waves* exceed 1,000 miles (1,600 km) distance, while *regional waves* have shorter ranges. Regional waves carry a great deal of information but are harder to interpret because they're sensitive to characteristics of Earth's crust. In particular, regional waves help distinguish nuclear explosions from earthquakes. Teleseismic waves, in contrast, are less sensitive to the medium and can detect explosions at large distances. Algorithms process seismic information from many stations to pinpoint the source with an accuracy of about 10 miles.

An explosion's yield is determined from the seismic body-wave magnitude (similar to the Richter scale). However, the relevant formula varies for different locations depending on local rock type and explosion depth. The good news is that any nuclear explosion with a yield exceeding about half a kiloton (one-thirtieth of the Hiroshima bomb) will be detected at least 9 times out of 10 (figure 19.4).

How good does nuclear test verification need to be? A 2012 National Research Council report gives several reasons why and how a nation

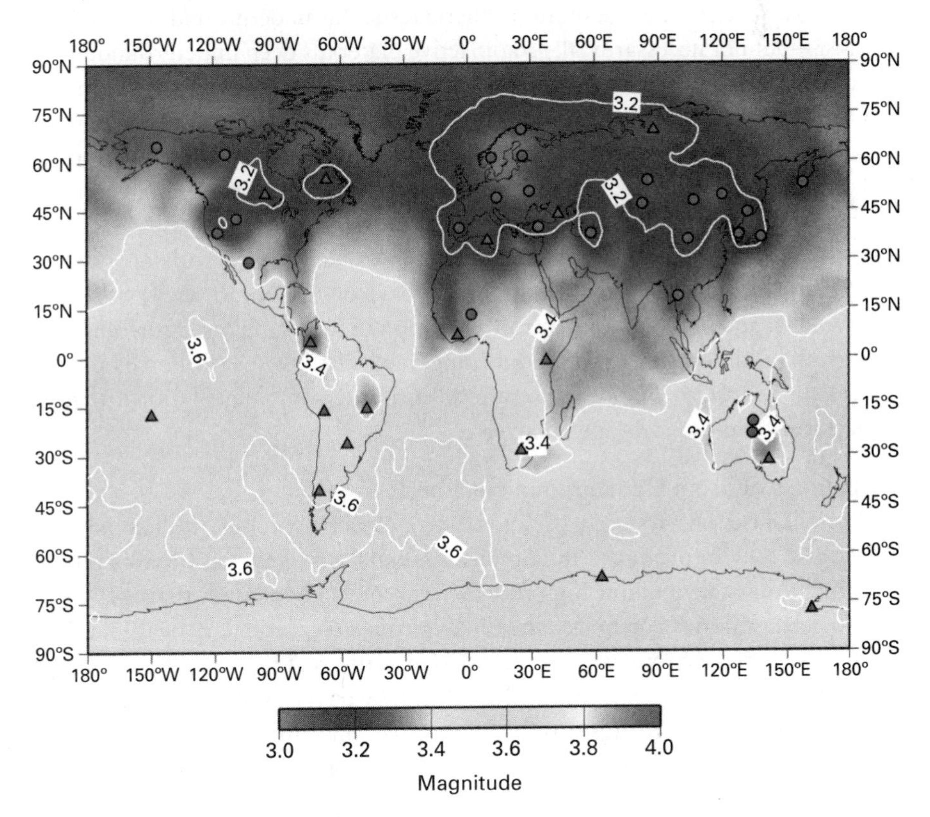

Figure 19.4
The IMS primary seismic network detection capability, assuming 43 stations. Circles mark individual stations. Labeled contours show the lowest-magnitude seismic event that would be identified, 9 out of 10 times, by three or more stations. (Courtesy of Tormod Kværna, NORSAR.)

might engage in nuclear tests.[3] These range from tests of non-nuclear explosive components to safety tests with yields under 0.1 ton TNT equivalent, through low-yield tests up to several kilotons (kt), and on to full-fledged nuclear explosions yielding more than 20 kt. The report concludes that very low yields, in the subkiloton range, may be concealable, but that full-fledged explosions aren't. Nevertheless, all these tests have military significance.

Detection of Radionuclides: Proof of Fission Explosion

The IMS seismic network can distinguish human-caused explosions from earthquakes and pinpoint where and when the event occurred. However, definite confirmation of a *nuclear* explosion, especially of low yield, requires detection of radioactive gases and particulates produced in the explosion. That was easy in the age of atmospheric tests. But underground tests trap most—but not necessarily all—radioactive materials deep underground. In addition to its seismic sensors, the Comprehensive Nuclear Test Ban Treaty Organization operates 80 sensors that process air samples to capture particulates, and half also extract radioactive noble gases such as xenon. Following a nuclear test, samples may contain radioisotopes that emit characteristic gamma rays, allowing identification of the isotopes.

The CTBTO has selected over 83 isotopes that cover a wide array of sources, including residue from components of the bomb itself, fission products, and products of neutron activation in the bomb or its surroundings.[4] In addition to CTBTO monitoring, the United States Air Force operates the *Constant Phoenix* aircraft, an airborne sampling platform that performs real-time isotopic analysis.

Radioxenon from Underground Nuclear Tests

The CTBTO has focused on radioactive xenon isotopes (**radioxenon**) because they're produced in copious amounts[5] and have half-lives short enough that they exhibit high radioactivity but long enough that they're still detectable days or weeks after an explosion. In particular, the CTBTO uses the ratios of four isotopes (Xe-131m, half-life 11.9 days; Xe-133m, 2.19 days; Xe-133, 5.25 days; Xe-135, 0.38 days) and a sophisticated algorithm that correlates an observed detection with a specific nuclear test to discriminate it from other sources.[6] Complications include 24-hour collection times for the detectors and the fact that radionuclides can leak out days or even months after an explosion. According to that 2012 National Resource Council study, the radioxenon network can detect a 1-kiloton nuclear explosion 9 times out of 10 within 14 days assuming that 10 percent of the radioactive gases escape. That escape may or may not happen, depending on the explosion depth and other factors.

An underground explosion produces a cavity whose size depends on explosive yield, type of rock, and depth. The shock wave from the explosion melts the rock in the immediate vicinity, opening a cavity that expands until there's no longer enough pressure to deform the rock (see figure 19.5). At this point, rock above the cavity may fall into it, creating a chimney that can stretch all the way to the surface, where it may form a subsidence crater. Reflection of the shock wave at the surface may also crack rock in the vicinity of the explosion. If the cavity doesn't collapse and remains impermeable, radioactive gases will be trapped and won't be detected. If the cavity walls crack, however, then gases can leak out and

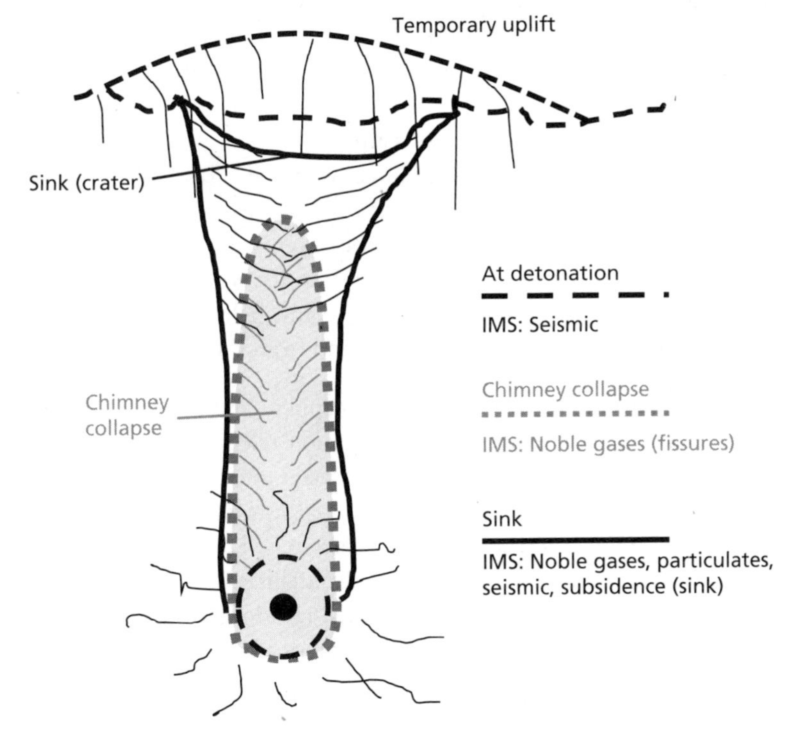

Figure 19.5
Illustration of underground testing phenomenology for three times following a high-yield explosion. After detonation the cavity expands and then contracts. Rock falls into the cavity, forming a chimney that can stretch all the way to the surface, resulting in a subsidence crater. This process can lead to venting of radioactive gases or even particulates. (Adapted from Ward Hawkins and Ken Wohletz, "Visual Inspection for CTBT Verification," figure 1, Los Alamos National Laboratory LA-13244-MS, 1997.)

be detected by radioxenon sensors. Additionally, a crater forming at the surface may be spotted by satellite imaging or radar.

Detecting Underwater Nuclear Explosions

IMS detects underwater nuclear tests using hydroacoustic sensors (hydrophones) in strategic locations around the planet. The publicly disclosed sensitivity of the network is below 1 ton TNT equivalent in most of the ocean. Hydrophones are deployed in the **SOFAR** (sound fixing and ranging) channel, a region deep in the ocean that allows efficient propagation of sound across vast distances. A 2008 experiment using a hydrophone off Chile detected the detonation of just 88 pounds of conventional explosive (40 kg) off the coast of Japan, some 10,000 miles (16,000 km) away.

The IMS hydroacoustic network consists of 11 observational stations. Since hydroacoustic sensors are so sensitive and sound travels so far, it's not necessary to have many of them. Each station deploys three hydrophones forming a triangle about 2 kilometers on a side, allowing speed and direction of the sound to be measured.

Underwater tests also produce radioactive noble gases and particulates, some of which are injected into the atmosphere. Since most of the ocean is international waters, air samples can be taken with sniffer aircraft and water samples with ships near the detonation site. Satellites can also provide evidence for underwater nuclear tests and help identify the testing country, as they track ships that travel to and from the explosion site.

Detecting Nuclear Explosions in the Atmosphere and in Space

As you saw in chapter 13, a nuclear explosion releases its vast energy as thermal energy, blast wave, radioactive materials, and electromagnetic radiation (visible light, X-rays, gamma rays). All of these forms of energy will be evident in atmospheric tests and most in space-based tests as well. So atmospheric and space tests are easily detected. Satellites monitor the globe to spot not only missile launches but also the tell-tale double flash of light from nuclear detonations. When a bomb explodes it produces a bright flash a few hundred microseconds after detonation. This light then dims but strengthens about one-tenth of a second later to another bright pulse. This unique double-hump in the light is detected using sensors known as **bhangmeters** (not, as its namer Frederick Reines, who won the 1995 Nobel Prize for discovery of the neutrino, reminds us, "bangmeter").

Explosions high in the atmosphere don't produce much seismic effect, and gamma rays and neutrons simply won't make it to the ground. So

Table 19.3
Capabilities of IMS techniques

Technology	Expected capability
Seismic	Better than 1 kiloton
Radioxenon	In 50% of cases greater than 1 kiloton
Hydroacoustic	Less than 1 ton for most oceans
Infrasound	Better than 1 kiloton across 90–95% of Earth's surface

Source: Based on the 2012 National Research Council study (see note 3).

seismic and direct radiation measurements aren't much help in detecting atmospheric tests.

One technique for identifying and locating atmospheric nuclear explosions is via infrasound signals that they generate. These very low frequency waves, far below what we can hear, propagate over long distances and are detected by sensitive microphones at Earth's surface. The system also "hears" non-nuclear sources of infrasound, including rocket launches, meteors, volcanoes, earthquakes, and storms. The IMS network now boasts 60 infrasound stations around the globe. With all 60 operational, they're expected to detect a 1-kiloton atmospheric explosion over 90 to 95 percent of Earth's surface.

Table 19.3 summarizes the capabilities of the four core techniques used in International Monitoring System.

The IMS has other uses than nuclear test detection. It provided valuable information on radioactive emissions from Fukushima and has been used to detect airplane crashes, volcanic eruptions, and low-frequency sounds from breaking icebergs and even whales. The 2013 asteroid explosion over the Russian city of Chelyabinsk was the equivalent of a 0.6-megaton nuclear blast. An asteroid-generated infrasound wave detected by IMS traveled around the globe twice!

Detection of North Korean Nuclear Tests

North Korea's recent nuclear testing provided a real-world trial for the International Monitoring System, whose seismic network detected all six tests. Even the weak 2006 explosion (0.6 kiloton, 1/25 of Hiroshima) was "seen" by 22 IMS seismic sensors. The 2009 test yielded about 3 kt and was detected by 61 sensors. The 2013 nuclear test, estimated at about 10 kt, was detected by 94 seismic sensors, two infrasound stations, and two radioxenon detectors.

However, only the 2006 and 2013 tests registered with the radionuclide network—with the 2006 test detected 18 days after the event at a station in Canada, more than 7,000 kilometers away! The stronger 2013 test was detected 55 days later in Japan and Russia. Surprisingly, the 2017 test, by far the strongest at around 250 kt and possibly thermonuclear, wasn't seen in radionuclide data. That's probably because intense heat glassified the test surroundings, sealing in even gaseous products. The test was, however, evident in satellite radar imagery showing first uplifting and then subsidence of the ground at the test site.[7]

The Future of the CTBT

The Comprehensive Nuclear Test Ban Treaty remains in limbo, but, as detection of the North Korean tests proved, the international community has a solid verification tool in the International Monitoring System. The United Nations Security Council recognized that fact in its 2016 Resolution 2310: "Even absent entry into force of the Treaty the monitoring and analytical elements of the verification regime ... contribute to regional stability as a significant confidence-building measure, and strengthen the nuclear non-proliferation and disarmament regime."

All signatories to the CTBT have access to IMS data, which is encrypted to make it essentially tamperproof. Therefore, data on a domestic incident can't be concealed, because (1) information will be released by another country, and (2) it will be suspicious if the country where the incident happened were to suddenly stop transmitting data. This happened when Russia stopped transmission from five radionuclide stations for several weeks after the August 2019 radiation-release incident that may have involved a nuclear-powered cruise missile (see chapter 14). The IMS transparency is akin to what open-source intelligence provides for societal verification. Today any country, regardless of income, can have access to near real-time state-of-the-art sensor data on its domestic and international environment as long as it has signed the CTBT (not necessarily ratified it).

Is Testing Necessary?

Testing is certainly helpful in developing and maintaining nuclear arsenals, but unfortunately it's not essential. Recall that the Hiroshima bomb's gun-type design was never tested, so confident were scientists that it would work. Today we have exquisite understanding of the physics of nuclear explosions, which we can model precisely with supercomputers enabled by the ongoing revolution in electronics and computing. Closely guarded "bomb codes" run on supercomputers at U.S. weapons labs and, presumably, in

Decoupled Testing

One argument against the CTBT is that states could cheat by testing nuclear weapons inside a hollow cavity. The idea is that the shock wave will be muffled by the space between the explosion site and the cavity walls, with energy going into increasing gas pressure in the cavity rather than melting and deforming rock and thus generating large seismic waves. The recent National Academies report is quite critical of such **decoupled testing**. Here we highlight the Academy's main observations, based largely on the only data point we have for decoupling: the 1966 Sterling test conducted in Mississippi (see figure 19.6).

Figure 19.6
A Baxterville, Mississippi, resident's homemade seismograph was an attempt to detect the nearby nuclear test. (Mississippi Department of Archives, Moncrief Photograph Collection)

(continued)

Decoupled Testing (*continued*)

The test was a mere 380-ton equivalent explosion detonated in a 34-meter-diameter (about 110 feet) cavity. The cavity itself was created six weeks earlier with a 5.3-kiloton nuclear explosion called Salmon.

There's an obvious chicken-or-egg problem here since a much larger nuclear test was necessary to produce the initial cavity in which the decoupled test took place. The Sterling test's decoupling was a success in that detection of the explosion gave yield estimates only 1/70 of the actual yield.

Motivation for the Sterling test was fear about potential Soviet cheating. Yet no country is known to have conducted a decoupled test to obfuscate on purpose. There have been efforts to supplement the data on the Sterling decoupled test with experiments using conventional explosives, as well as studies of Soviet tests that were partially decoupled, albeit less successfully than the Sterling test.

A concern sometimes raised by CTBT opponents goes as follows: "In a decoupled test, the seismic signal can be muffled by a factor of 70, especially in a salt cavity. Salt cavities also thwart air sampling by trapping radioactive debris. Iran has some of the world's largest salt cavities." But salt cavities can't sustain large, deep explosions without collapsing, and shallow explosions risk venting radioisotopes that are readily detected. So decoupled testing in salt cavities like Iran's presents serious challenges to the testing nation.

Also significant is that decoupling depends on the frequency of the seismic waves generated. That factor of 70 is at low frequencies, but decoupling drops quickly at higher frequencies. That gives a broadband seismic network like IMS good chance of detecting even decoupled explosions. That's especially true in regions where higher-frequency seismic waves propagate efficiently—which includes most of Russia, the Indian and Chinese test sites, and all of North Korea.

There's just not enough data on decoupling to convince a country that they wouldn't be caught conducting a decoupled test. Physicist David Hafemeister estimates the probability of a successful clandestine test at only about 15 percent, even for a small 1–2 kiloton explosion. With such low odds it's unlikely that a country would undertake a decoupled test for the purpose of concealing it.

other nuclear-armed nations and nuclear wannabees. And in the U.S., at least, we do real experiments with miniature nuclear detonations. That's the purpose of the National Ignition Facility, a laser-fusion experiment described in chapter 11 because of its role, now diminished, in exploring controlled nuclear fusion as an energy source. Together, computer simulations and experiments like NIF are used in **stockpile stewardship**—a program to ensure that U.S. nuclear weapons remain functional even as they age. Stockpile stewardship also involves disassembling and examining weapons to understand how they age, as well as materials-science studies of conventional high explosives, plutonium, and other weapons components. Computer simulations go even further, allowing weapons designers to confidently develop new weapons. So, no, testing isn't essential. But it would help, especially for a country lacking the United States' sophisticated simulation facilities.

Fissile Material Cut-Off Treaty

Fissile materials are the explosive fuel for nuclear weapons. You can't make a bomb without fissile material! The principal weapons materials are highly enriched uranium (HEU) and plutonium that's high in Pu-239 compared with other plutonium isotopes. The global stocks of these materials are staggering—about 1,340 tons of HEU and 520 tons of Pu-239. Given critical masses of 30 pounds for HEU and 5 pounds for Pu (see chapter 5), that's enough for some 300,000 fission weapons! These materials are distributed among various countries as shown in table 19.4. Approximately 99 percent of the HEU inventory is held by nuclear-weapon states, with Russia and the United States having the largest stocks. India, Pakistan, and North Korea are thought to have ongoing HEU production.

The key to preventing further proliferation of fissile materials is eliminating their production. That's the job of the **Fissile Material Cut-off Treaty,** a proposed agreement that would eliminate production of highly enriched uranium and plutonium. Negotiations on fissile materials have taken place

Table 19.4
Fissile material stocks as of 2017

Country	HEU, tons	Military Pu, tons	Civilian Pu, tons
Russia	679	128	59
United States	575	79.8	8
United Kingdom	21.2	3.2	110.3
France	30.6	6	65.4
China	14	2.9	0.04
Pakistan	3.4	0.28	-
India	4	7.07	0.4
Israel	0.3	0.9	-
North Korea	0	0.04	-
Others	15	-	49.3

Source: International Panel on Fissile Materials at http://fissilematerials.org.

at the United Nations Conference on Disarmament, a 65-member body that operates by consensus. All five official nuclear-weapon states except China have declared that they've ceased production of fissile materials for weapons, but there's been no independent verification. However, a fissile material treaty would impose fresh constraints on the nuclear-weapon states and the four non-NPT nuclear-armed countries (Israel, India, Pakistan, and North Korea). But India has declared that it won't sign a treaty unless a deadline is imposed on the nuclear-weapon states to fulfill their disarmament obligations under the Non-Proliferation Treaty. Moreover, Pakistan worries that a fissile material treaty would trap them in a disadvantageous situation vis-à-vis India's nuclear stockpile. Pakistan would therefore like the treaty to include existing stocks, rather than just restricting new fissile production—a stance that's shared by several other nations.

Where does a fissile material treaty stand now? In 2006, the George W. Bush administration submitted a proposal at the U.N. Conference on Disarmament that provided for a 15-year ban on production of HEU and plutonium. But it didn't include any verification measures and would have applied only to the five official nuclear-weapon states. President Obama stated in his famous Prague speech that "the United States will seek a new treaty that verifiably ends the production of fissile materials intended for use in state nuclear weapons." In March 2016, at the U.N. Conference on Disarmament, the United States put forward a proposal to set up a working group to negotiate a fissile material treaty. The Trump administration has stated that it would support such negotiation at meetings prior to reconsideration of the Non-Proliferation Treaty. Although discussions did take place in 2019, the United Nations Office of Disarmament Affairs concluded on a discouraging note, saying that although there's now the "necessary technical and substantive basis to move forward … What is lacking is political consensus."[8]

It's been more than 25 years since the Conference on Disarmament agreed to start negotiating a treaty on fissile materials, but there's been almost no substantive debate on such a treaty—whose adoption would be a vital step toward stopping nuclear weapons manufacturing worldwide. A fissile materials treaty would complement the Comprehensive Nuclear Test Ban Treaty, as argued by Mari Amano, the Japanese ambassador to the Conference on Disarmament. Once it enters into force, implied Amano, the test ban treaty would act as a "quality cap" on global nuclear weapons programs while the fissile material treaty would be a "quantity cap." After the CTBT, a treaty banning the production of fissile material is the next logical step toward a world free of nuclear weapons.

Treaty on the Prohibition of Nuclear Weapons

Widely supported efforts toward a legally binding ban on nuclear weapons altogether is the consequence of the disastrous humanitarian implications of nuclear weapons use as well as concern that nuclear danger is increasing. There's also widespread dissatisfaction that the nuclear-weapon states haven't met their disarmament obligations under the Non-Proliferation Treaty. These concerns led several nations, including Norway, Mexico, and Austria, to host three conferences on the humanitarian implications of nuclear weapons. Following the final conference in Vienna, 127 states endorsed the **Humanitarian Pledge**, calling on all NPT parties to renew their commitment to the disarmament provisions of the NPT's Article VI and to take interim steps to minimize the risk of nuclear use.

A more formal instrument is the **Treaty on the Prohibition of Nuclear Weapons**, negotiated by more than 130 states. The treaty is a good-faith effort to fulfill the NPT disarmament obligation. It opened for signature on September 20, 2017, and in 2020 it reached 50 ratifications—the threshold for it to become international law, binding on its signatories. The treaty then took effect in early 2021. It "prohibits nations from developing, testing, producing, manufacturing, transferring, possessing, stockpiling, using or threatening to use nuclear weapons, or allowing nuclear weapons to be stationed on their territory." It also prohibits them from "assisting, encouraging or inducing anyone to engage in any of these activities." The treaty strengthens commitments to the NPT and to the Comprehensive Nuclear Test Ban Treaty. Although it won't eliminate nuclear weapons, the Treaty on Prohibition of Nuclear Weapons may further delegitimize them and reinforce legal and political norms against their use. Unfortunately, neither the five nuclear-weapon states nor any NATO members have signed.

Summary

As we said in the preceding chapter's summary, horrifyingly destructive nuclear weapons have been with us for three-quarters of a century. We've put some international controls on these weapons and have succeeded in reducing the superpowers' nuclear arsenals while slowing but not stopping the spread of nuclear weapons. But could we eliminate nuclear weapons altogether? And if so, would that be a good thing? In this chapter, we've argued that, in the long run, the nonzero chance of civilization-annihilating nuclear war motivates serious efforts to ban nuclear weapons. Many of the world's nations agree, and negotiations to ban weapons testing, the production of fissile materials, and even the weapons themselves are under way. But it's going to take more political and moral will than we've seen so far, especially among the nuclear-armed nations, to bring these negotiations to fruition.

Glossary

bhangmeter A satellite-based light sensor that can detect the characteristic double flash of a nuclear explosion. The "h" belongs; it's not a "bangmeter"!

Comprehensive Nuclear Test Ban Treaty (CTBT) Opened for signature in 1996 at the U.N. General Assembly, the CTBT will prohibit all nuclear testing if and when it enters into force.

Comprehensive Nuclear Test Ban Treaty Organization (CTBTO) Actually, the Preparatory Commission of the CTBTO until such time as the Comprehensive Nuclear Test Ban Treaty is in force. The Preparatory Commission is an international organization tasked with building, certifying, and operating the infrastructure for detection and investigation of nuclear tests, preparing regulations, and leading activities that will facilitate the early entry into force of the CTBT.

decoupled testing A method of evasion where a nuclear test is conducted inside a hollow underground cavity so that the seismic magnitude is reduced.

Fissile Material Cut-off Treaty Treaty currently under discussion in the Conference on Disarmament to end the production of weapons-usable fissile material (highly enriched uranium and plutonium).

Humanitarian Pledge Nuclear weapons are the only weapons of mass destruction not yet explicitly prohibited under international law. The Humanitarian Pledge is intended to fill what many consider an unacceptable legal gap. It offers a platform for negotiations on a treaty banning nuclear weapons.

International Monitoring System (IMS) The IMS, the verification system for the Comprehensive Nuclear Test Ban Treaty, consists of 321 monitoring stations and 16 laboratories worldwide. These facilities monitor the planet for any sign of a nuclear explosion.

Partial Test Ban Treaty (PTBT) Treaty banning nuclear weapons tests in the atmosphere, in outer space. and underwater. While the treaty does not ban underground tests, it does prohibit underground explosions with "radioactive debris to be present outside the territorial limits of the State under whose jurisdiction or control" the explosion was conducted.

radioxenon Radioactive isotopes of the noble gas xenon, formed as fission products and useful in atmospheric sampling to verify a nuclear explosion.

SOFAR channel Acronym for Sound Fixing and Ranging channel, an undersea channel through which sound can travel long distances with little attenuation.

stockpile stewardship A U.S. program involving computer simulations as well as examinations of actual weapons and their materials to ensure that the existing nuclear arsenal remains viable.

Threshold Test Ban Treaty Bilateral treaty between the U.S. and Russia/USSR limiting underground tests to a maximum yield of 150 kilotons. In effect since 1974.

Treaty on the Prohibition of Nuclear Weapons A treaty that would ban nuclear weapons altogether. The treaty has been open for signatures since 2017 and already has at least 80 signatories and 34 ratifications.

United Nations Conference on Disarmament An international forum focused on multilateral disarmament efforts. Although it reports to the U.N. General Assembly and has a relationship with the United Nations, it adopts its own rules of procedure and agenda, giving it some degree of independence.

Notes

1. See Martha Smith-Norris, *Domination and Resistance: The United States and the Marshall Islands during the Cold War* (Honolulu: University of Hawaii Press, 2016). A look at the United States' relation to the Marshall Islanders in connection with both nuclear weapons tests and subsequent missile tests.

2. Actually, it's the *Preparatory Commission for the* Comprehensive Nuclear Test Ban Treaty Organization until the treaty is in force.

3. National Research Council, *The Comprehensive Nuclear Test Ban Treaty: Technical Issues for the United States* (Washington, DC: National Academies Press, 2012); see also Arms Control Association, "Technical Issues Related to the Comprehensive Nuclear Test Ban," https://www.armscontrol.org/act/2002-09/technical-issues-related-comprehensive-nuclear-test-ban.

4. K. M. Matthews, "The CTBT Verification Significance of Particulate Radionuclides Detected by the International Monitoring System," New Zealand Ministry of Health, National Radiation Laboratory, 2005, https://www.moh.govt.nz/notebook/nbbooks.nsf/0/856377d65427ad6fcc257346006f47e0/%24FILE/2005-1.pdf. See Table 5, page 29.

5. How copious? An example from Ted Bowyer: "If all the activity vented from a 1 kiloton nuclear detonation was diluted by Earth's entire atmosphere (10^{18} m^3) it would yield a concentration in the atmosphere of 0.5 Bq/m^3." The detection limit of radionuclide sampling is 50,000 times smaller! See https://elliott.gwu.edu/sites/g/files/zaxdzs2141/f/downloads/events/2.2%20Bowyer.pdf.

6. The CTBTO has found that different anthropogenic sources of radioactive xenon such as from civilian nuclear reactors can be discriminated by using the isotopic xenon ratios. See Martin B. Kalinowski et al., "Discrimination of Nuclear Explosions Against Civilian Sources Based on Atmospheric Xenon Isotopic Activity Ratios," *Pure and Applied Geophysics* 167, no. 4–5 (2010): 517–539.

7. For a description of this radar detection, and a recommendation that it be added to the arsenal of nuclear test detection techniques, see Teng Wang et al., "The Rise, Collapse, and Compaction of Mt. Mantap from the 3 September 2017 North Korean Nuclear Test," *Science* 631, no. 6398 (July 13, 2018): 166–170. Available at https://science.sciencemag.org/content/361/6398/166.

8. https://s3.amazonaws.com/unoda-web/wp-content/uploads/2019/05/Moving-forward-with-the-FMCT-Preparatory-Group-report-recommendations-in-advance-of-2020-NPT-Review-Conference1.pdf.

Further Reading

Carlson, John. *CTBT: Possible Measures to Bring the Provisions of the Treaty into Force and Strengthen the Norm against Nuclear Testing*. Vienna: Vienna Center for Disarmament and Non-Proliferation, April 25, 2019. https://vcdnp.org/ctbt-possible-measures-to-bring-the-provisions-of-the-treaty-into-force-strengthen-the-norm-against-nuclear-testing/. This report discusses possible options for dealing with the current impasse in getting the CTBT into force.

Dahlman, Ola, Jenifer Mackby, Svein Mykkeltveit, and Hein Haak. *Detect and Deter: Can Countries Verify the Nuclear Test Ban?* Dordrecht: Springer Science+Business Media, 2011. This is an excellent introduction to the capabilities of the International Monitoring System, as well as monitoring of nuclear explosions in general.

Feiveson, Harold A., Alexander Glaser, Zia Mian, and Frank N. Von Hippel. *Unmaking the Bomb: A Fissile Material Approach to Nuclear Disarmament and Nonproliferation*. Cambridge, MA: MIT Press, 2014. This book by scientists from Princeton University's Program on Science and Global Security is highly recommended for information on fissile materials, the Fissile Material Cut-off Treaty, and related topics.

Hu, Howard, Arjun Makhijani, and Katherine Yih, eds. *Nuclear Wastelands: A Global Guide to Nuclear Weapons Production and Its Health and Environmental Effects*. Cambridge, MA: MIT Press, 2000. Excellent resource on the quantitative health effects of nuclear testing.

International Campaign to Abolish Nuclear Weapons resource on the Treaty on the Prohibition of Nuclear Weapons. http://d3n8a8pro7vhmx.cloudfront.net/ican/legacy_url/1535/TPNW-English1.pdf?1582734783. ICAN is the organization that won the 2017 Nobel Peace Prize for their work on the treaty.

International Panel on Fissile Materials (IPFM). http://fissilematerials.org. More details on fissile materials, including data on worldwide stockpiles.

International Physicians for the Prevention of Nuclear War. *Radioactive Heaven and Earth: The Health and Environmental Effects of Nuclear Weapons Testing In, On, and Above the Earth*. Tacoma Park, MD: Institute for Energy and Environmental Research, 1991. PDF at https://ieer.org/wp/wp-content/uploads/1991/06/RadioactiveHeavenEarth1991.pdf. Excellent resource on the health effects of nuclear testing.

Krass, Allan S. *Verification: How Much Is Enough*? London: Taylor & Francis, 1985. Focuses on verification, monitoring, and trust. Although it's an old resource, it was written by a true authority and is still very relevant. Book is available free from SIPRI website: https://www.sipri.org/sites/default/files/files/books/SIPRI85Krass/SIPRI85Krass.pdf.

National Academy of Sciences, Committee on Technical Issues Related to Ratification of the Comprehensive Nuclear Test Ban Treaty. *Technical Issues Related to the Comprehensive Nuclear Test Ban Treaty*. Washington, DC: National Academy Press, 2002.

National Research Council. *The Comprehensive Nuclear Test Ban Treaty: Technical Issues for the United States*. Washington, DC: National Academies Press, 2012. This report and the one cited above from 2002 provide excellent descriptions of the capabilities of the International Monitoring System.

20

Nuclear Power, Nuclear Weapons, Nuclear Futures

Two major parts of this book have dealt with nuclear power and nuclear weapons. How valid is our separation of those topics? What will be the role of nuclear technology in the future? Can we continue to expand our use of nuclear technology without increasing the threat of nuclear annihilation?

These aren't easy questions, although experts of widely differing opinions claim to have answers. They range from assertions that nuclear power and nuclear weapons are unrelated to the notion that nuclear power plants are "bomb factories,"[1] and from belief that the age of nuclear power is essentially over to visions of a prosperous nuclear future. Some see the risk of nuclear war fading, while others are certain of a nuclear Armageddon.

Nuclear Power, Nuclear Weapons: The Connection

Even those who believe nuclear power to be safer and cleaner than many of its alternatives are often troubled by the possibility that the development of nuclear power may lead to the proliferation of nuclear weapons. All arguments in favor of nuclear power are vacuous if civilization is destroyed in a nuclear war made possible by the spread of nuclear power plants. *That* is one possibility we didn't weigh into chapter 8's comparison of coal and nuclear power. Should the potential for weapons proliferation and nuclear war change our assessment of nuclear power or, for that matter, of other "peaceful" nuclear technologies?

This book doesn't need a whole new part on the relation between nuclear power and nuclear weapons. Throughout the book we've stressed that there *is* a connection. Nuclear power and nuclear weapons use the same basic fissile materials. Both require mining and often enrichment of uranium. Reactors produce plutonium, and a bomb made from a few pounds of plutonium—probably even reactor-grade plutonium—can destroy a city.

You now know enough about nuclear technology to understand the several routes to nuclear weaponry, some of which might make use of nuclear power reactors. On the other hand, you've also seen some technical subtleties: the need for heavy water if one doesn't have enriched uranium; the challenge of Pu-240 contamination in fuel that stays a long time in a reactor; the technical difficulty of rapidly configuring an explosive critical mass. So what about the power-weapons connection? It's real, all right, but there's still room for controversy.

On the one hand, no nuclear-armed country built its first bomb with fissile material diverted from power reactors or from their fuel cycle. India came closest, but its reactor was a research device that made clandestine plutonium production easier than it is in a power plant. On the other hand, you've seen how the former Soviet Union used its Chernobyl-type RBMK reactors to simultaneously produce electricity for civil society and plutonium for nuclear weapons, and how Britain's early graphite-moderated reactors had a similar dual use. Even the United States, whose policy is to keep civilian nuclear power separate from the weapons establishment, has recently used civilian power plants to produce tritium for fusion-boosted fission weapons. And it's estimated that India has five or six tons of reactor-grade plutonium that's already been separated from spent reactor fuel and is held in military stockpiles. Indeed, any power reactor that's not under IAEA safeguards, anywhere in the world, must be considered a potential source of plutonium for nuclear weapons. Furthermore, the spread of nuclear reactors around the world, even if they are subject to safeguards, still raises concerns about potential weapons proliferation. A case in point is the United Arab Emirates' Barakah nuclear power station, where the first of four reactors went online in 2020. Although the UAE asserts that its venture into nuclear power is entirely for peaceful purposes, the reactor is the first on the Arabian Peninsula, and a nuclear expert concerned about the ecological fragility and geopolitical instability of the area has declared it "the wrong reactor, in the wrong place at the wrong time."[2]

The power-weapons connection has different implications on national and international scales. Even if the international spread of nuclear power plants raises the risk of nuclear weapons proliferation, should that affect the development of nuclear power within an already nuclear nation such as the United States? And there are degrees of risk. Breeder reactors and international plutonium trade clearly carry a greater proliferation risk than light-water reactors. So, does nuclear power itself increase the danger of nuclear weapons proliferation? Perhaps. But there are plenty of ways to acquire nuclear weapons that don't involve nuclear power plants.

The most significant role of nuclear power in weapons proliferation may be more subtle than direct production of weapons materials: nuclear plants require nuclear engineers and technicians, so they create a cadre of people with expertise in nuclear technology. Facilities for handling nuclear fuel and waste add further to a country's nuclear infrastructure. Things might stop there, but if the country showed any interest in nuclear weapons then that could push its nuclear establishment in a less benign direction.

There's another link between nuclear power and warfare: power plants could be tempting targets either for terrorists armed with conventional explosives or in a war between two nations that used only conventional weapons. Blasted by an accurate conventional explosive, a single nuclear plant could release over 30 times the radioactivity of the Chernobyl accident. Would a conventional attack on an opponent's nuclear power plants justify the use of nuclear weapons in response? In that sense, is the spread of nuclear power plants lowering the threshold for all-out nuclear war?

However, alternatives to nuclear power also carry catastrophic risks. What about major climatic upheaval due to continued use of fossil fuels, especially coal? Would it be better if we supplied nuclear power plants to developing countries to reduce their dependence on coal? Or does the risk from weapons proliferation outweigh even the rapidly intensifying impacts of climate change? What about a nuclear war that starts with a conflict over Middle Eastern oil supplies or resources made scarce by climate change? Do we weigh those *nuclear* dangers among the risks of *fossil* fuels? These are complex, international nuclear questions.

The Future of Nuclear Technology

Questions of nuclear technology and weapons proliferation hinge on more than just the power-weapons connection, although that remains a central concern. Even a seemingly benign process such as irradiation for food preservation, as you saw in chapter 4, may have a hidden weapons connection. And many of the other seemingly benevolent nuclear applications discussed in part 1 often make use of nuclear materials produced in research reactors like the one India used to make plutonium for its first nuclear explosive, or Iran's Teheran Research Reactor that produces, among other things, medical isotopes but also helps justify Iran's pursuit of uranium enrichment. Should the developed nuclear nations withhold nuclear technology from their less prosperous global neighbors because of the weapons connection? Denying technological benefits where they might be most needed hardly seems just. For the United States, Russia, Britain, France, and China, it's downright illegal: these nuclear-armed parties to the Non-Proliferation Treaty are treaty-bound to help non-nuclear-weapon states develop peaceful nuclear applications (see figure 20.1).

The general level of nuclear technology throughout the world continues to advance, and as it does, more and more nations will come within reach of nuclear weaponry. Controls on sensitive technology and nuclear materials can slow that trend, but it's still easy to imagine

Figure 20.1
This research reactor in the Philippines symbolizes the spread of nuclear technology in developing countries. This particular reactor aids research in biology, agriculture, and medicine; at the same time, it breeds nuclear materials and helps create nuclear expertise. (United Nations)

that soon almost any nation could build nuclear weapons if it wanted. Turning back the technological clock, or eliminating nuclear technology altogether, might prevent that state of affairs. Either of those options is unlikely. Instead, the world community must develop mechanisms that diminish not only the technological potential but also the political desire for nuclear weapons.

Nuclear technology can do great harm; no one doubts that all-out nuclear war would be a disaster of unprecedented and unimaginable proportions. Nuclear technology can also do great good; few would argue against the life-saving techniques of nuclear medicine. And nuclear technology is certainly capable of supplying some of the energy needed to run today's industrial civilization, although nuclear power remains mired in controversy. The potential of nuclear technologies—whether for good or for ill—rests ultimately in the nature of the atomic nucleus and the forces that bind it together. But the realization of that potential—good or bad—rests with us. Humankind is not going to forget the atomic nucleus but must learn to live safely with nuclear knowledge. That learning can't be left to the experts. Responsible citizens of Planet Earth need to make nuclear choices, and they need to make them after informed, nuanced, and critical analysis of complex and difficult nuclear questions.

Notes

1. Amory and Hunter Lovins, *Energy/War: Breaking the Nuclear Link* (San Francisco: Friends of the Earth, 1980), 19.

2. Paul Dorfman, University College London, quoted in Al Jazeera (https://www.aljazeera.com/amp/ajimpact/nuclear-gulf-experts-sound-alarm-uae-nuclear-reactors-200628194524692.html).

Further Reading

Aviation Week and Space Technology. Authoritative news on the aerospace industry, including the latest on nuclear weapons delivery systems. Current and back issues available at https://archive.aviationweek.com.

Bulletin of the Atomic Scientists. Established in 1945 by scientists who had worked on the Manhattan Project, this journal is, despite its technical-sounding title, actually a popular magazine dealing with political and technical issues of arms control. In recent years its scope has expanded to include climate change and other planetary threats. Its famous Doomsday Clock reflects the *Bulletin*'s assessment of the risk of nuclear war and other global catastrophes. Although its writing is authoritative, the *Bulletin* is decidedly with the antinuclear doves. Now published in digital format only, at https://thebulletin.org.

International Atomic Energy Agency. Vienna, Austria. The IAEA is a UN-affiliated international organization responsible for promoting the safe and peaceful use of nuclear technology. Visit their news page at https://www.iaea.org/news for updates on nuclear news, including new applications of nuclear technology.

Miller, Steven E., and Scott D. Sagan. "Nuclear Power without Nuclear Proliferation?" *Daedalus* (Fall 2009). One of many informative articles in this special issue with the theme "on the global nuclear future."

Nuclear News. Published by the American Nuclear Society and intended for nuclear professionals, *Nuclear News* features up-to-date articles on nuclear power, nuclear waste management, and related issues. The viewpoint is solidly with the nuclear industry, although nuclear accidents and other mishaps are reported. Available at http://www.ans.org.

Scientific American. This respected and authoritative monthly features frequent articles on nuclear technology and related issues, often by experts in their fields. The articles are objective and informative, although a range of nuclear opinions is clearly evident.

Union of Concerned Scientists. Founded in 1969 by MIT students and faculty, and supported by its extensive general membership, UCS brings evidence-based science to bear on issues at the intersection of science, technology, and public policy, including nuclear power and weapons issues. Solid and authoritative but with a decided bias against nuclear power. At https://www.ucsusa.org.

Weart, Spencer R. *The Rise of Nuclear Fear* (Cambridge, MA: Harvard University Press, 2012). A mix of history, anthropology, psychology, art, media culture, and science, Weart's book shows how factors well beyond science shape the public's perception of nuclear technology. Spencer Weart is a historian of science who served for many years as director of the Center for the History of Physics of the American Institute of Physics. His book, like the current *Nuclear Choices* that you're reading, is a thorough update of a book first published nearly 25 years earlier.

Index